Current Information Sources

in Mathematics

Current Information Sources

in Mathematics

An Annotated Guide to

Books and Periodicals, 1960-1972

ELIE M. DICK

Department of Mathematics
Louisiana State University in New Orleans

1973

Libraries Unlimited, Inc., Littleton, Colo.

Library of Congress Card Number 72-075143
International Standard Book Number 0-87287-047-2

LIBRARIES UNLIMITED, INC.
P.O. Box 263
Littleton, Colorado 80120

PREFACE

The purpose of this guide is to provide students, instructors, and research workers with an up-to-date bibliography of recent sources in mathematics, with an emphasis on monographic material. Few current comprehensive booklists are available. N. G. Parke's *Guide to the Literature of Mathematics and Physics Including Related Works on Engineering Science* (2nd rev. ed. Dover, 1958) covers mathematics and physics books only up to 1956. There are a number of more specialized bibliographies, prepared primarily for use in specific university courses; however, because they provide little information as to the contents of individual works, they serve as checklists rather than as evaluative tools. The large volume of mathematics literature published during the last ten years poses a serious problem to those who need quick access to the significant literature on a specific area. The aim of this bibliographic guide is to identify the major works in all branches of mathematics, and to describe their contents.

This annotated selective bibliography is confined to books which were published or reprinted in English or in English translation during the period from 1960 to mid-1972. Most of the books published before 1960 tend to be obsolete now, as a result of the continuous evolution in university curricula and the rapid increase in the amount of research done in mathematics. Moreover, many of these pre-1960 books can be located in Parke's work or in S. Goldman's *Guide to the Literature of Engineering, Mathematics, and the Physical Sciences* (Applied Physics Laboratory, Silver Spring, Maryland, 2nd ed., 1964, Report No. AD 608 053).

Selection was based on a book's potential value to research workers and students with emphasis on the latter. Objectivity was assured through the use of *Mathematical Reviews*, *Computing Reviews*, *SIAM Review*, and many other journals which review mathematics books. In addition, a number of professors and students at various colleges and universities were consulted and the libraries of several universities were checked.

The more than 1,600 entries of the book are divided into 37 chapters. The first 33 chapters are devoted to books in all branches of mathematics, with entries arranged alphabetically by author within each chapter. The remaining four chapters list periodicals (alphabetically by title), guides and directories (alphabetically by author), professional organizations and government agencies, and publishers.

Each entry includes complete bibliographic information—author, title, edition, place of publication, publisher, year of publication, number of pages, price (when available), an annotation based on the table of contents, and *Mathematical Reviews* number (when available). Occasionally a brief comment is added which reflects the opinions of professors and students surveyed, as well as the opinions of reviewers in various mathematical journals. Entries for periodicals include the journal title, year of first issue, frequency of publication, subscription rates, and name and address of the editor and the publisher.

A brief description of the chapters will clarify the subject classification:

Chapter 1, **General Elementary Mathematics**. Includes books designed to introduce college students to mathematics. Such books cover briefly many topics, such as algebra, trigonometry, logic, sets, functions, calculus, number theory, graphs, matrices, topology, geometry, computer programming, probability, statistics, etc.

Chapter 2, **Algebra and Trigonometry**. Topics discussed: the number systems, polynomials, exponents, sets, logic, graphs, the trigonometric functions, etc.

Chapter 3, **Calculus, Analytic Geometry, and Vector Analysis**. Topics discussed: limits, continuity, differentiation, integration, series, conic sections, vectors, vector functions, introduction to analysis, etc.

Chapter 4, **History, Biography and Reprints of Classics**. Topics discussed: history of mathematics, development of the mathematical disciplines, biographies of famous mathematicians, reprints of their works, etc.

Chapter 5, **Set Theory, Logic and Foundations**. Topics discussed: sets, relations, transfinite numbers, deductive systems, predicate and propositional calculus, intuitionism, modal logic, linguistics, algebraic logic, Boolean algebras, model theory, recursive analysis, algorithms and machines, abstract automation theory, etc.

Chapter 6, **Combinatorial and Graph Theory**. Topics discussed: combinatorial problems, combinatorial inequalities, block designs, difference sets, Latin squares, lattices, graphs, trees, chromatic theory, groups, etc.

Chapter 7, **Theory of Numbers**. Topics discussed: congruences, residues, special functions, sequences, Diophantine equations, p-adic theory, multi-linear forms, modular functions and groups, automorphic functions, geometric number theory, continued fractions, analytic number theory, algebraic number theory, probabilistic number theory, class field theory, finite fields, polynomials, etc.

Chapter 8, **General Abstract Algebra**. Topics discussed: sets, mappings, groups, rings, integral domains, fields, ideals, modules, vector spaces, matrices, homorphisms, isomorphisms, etc.

Chapter 9, **Linear and Multilinear Algebra and Matrix Theory**. Topics discussed: vector spaces, linear dependence, matrices, inverse of matrices, determinants, eigenvalues, eigenvectors, canonical forms, quadratic forms, transformations, tensor algebra, exterior algebra, Grassman algebras, etc.

Chapter 10, **Algebraic Structures and Ordered Algebraic Structures**. Topics discussed: lattices, Boolean algebras, Jordan algebras, Banach algebras, Lie algebras, Hopf algebras, rings, ideals, modules, domains, fields, etc. Books included in this chapter cover mainly one aspect of algebra, while those in Chapter 8 cover several topics. See also Chapter 11.

Chapter 11, **Group Theory and Generalizations**. Topics discussed: free groups, periodic groups, finite groups, abelian groups, lattices of groups, modular groups, representations of groups, semigroups, etc. Books included in this chapter are specifically on one aspect of group theory. See also Chapter 13.

Chapter 12, **Category Theory and Homological Algebra**. Topics discussed: abstract categories, functors, homological algebra, etc.

Chapter 13, **Topological Groups and Lie Theory**. Topics discussed: topological groups, compact groups, transformation groups, Lie groups, Lie algebras, etc.

Chapter 14, **Topology**. Topics discussed: topological spaces, convergence, separation axioms, function spaces, metric spaces, compactness, connectedness, mappings, continuity, algebraic topology, knots, complexes, homology, cohomology, K-theory, sheaves, dimension theory, homotopy, spectral sequences, generalized manifolds, etc.

Chapter 15, **Topology and Geometry of Manifolds**. Topics discussed: geometric topology, differentiable mappings, cobordism, fibre bundles, differentiable manifolds, Lie groups, transformation groups, etc.

Chapter 16, **Geometry**. Topics discussed: algebraic geometry, transformations, algebraic curves, and surfaces, abelian varieties, Euclidean and non-Euclidean geometry, projective geometry, convex structures, affine geometry, differential geometry, vector and tensor analysis, differentiable manifolds, Riemannian geometry, integral geometry, etc.

Chapter 17, **Real Analysis**. Topics discussed: limits, continuity, measure, differentiation, in. oration, classical Banach spaces, etc.

Chapter 18, **Measure and Integration**. Topics discussed: measures, measurable functions, Lebesgue integration, integration on locally compact spaces and groups, etc.

Chapter 19, **Complex Analysis**. Topics discussed: complex numbers, analytic functions, series, conformal mappings, complex integration, calculus of residues, analytic continuation, singularities, integral representations, analytic spaces, Stein spaces, capacity, harmonic measure, etc.

Chapter 20, **Potential Theory**. Topics discussed: harmonic functions, boundary-value problems, capacity, Poisson's equation, etc.

Chapter 21, **Special Functions**. Topics discussed: Bessel functions, hypergeometric functions, Lamé functions, Heun functions, Mathieu functions, the gamma function, etc.

Chapter 22, **Functional Analysis**. Topics discussed: topological linear spaces, normed spaces, Banach spaces, inner product spaces, Hilbert spaces, function spaces, distributions, Banach algebras, C*-algebras, W*-algebras, operators, spectral theory, perturbation theory, semigroups of operators, etc.

Chapter 23, **Differential and Integral Equations**. Topics discussed: methods for solving differential equations, existence and uniqueness theorems, intital-value problems, boundary-value problems, asymptotic expansions, stability, dynamical systems, classification of differential equations, differential operators, equations in function spaces, Fredholm integral equations, Volterra integral equations, integro-differential equations, etc.

Chapter 24, **Fourier Analysis and Harmonic Analysis**. Topics discussed: trigonometric polynomials and series, orthogonal series, almost periodic functions, Fourier transform, convolutions, positive definite functions, measure algebras, ideals, etc.

Chapter 25, **Integral Transforms and Operational Calculus**. Topics discussed: Laplace transform, Fourier transform, Mellin transform, Hankel transform, K-transform, Weierstrass transform, Radon transform, convolutions, generalized functions, distributions, etc.

Chapter 26, **Calculus of Variations**. Topics discussed: the Euler and Weierstrass conditions, Hamilton-Jacobi theory, Lagrange's problem, Pontryagin minimum principle, etc.

Chapter 27, **Numerical Analysis**. Topics discussed: solutions of equations, numerical differentiation, numerical integration, interpolation, quadrature, approximation, difference equations, difference operators, iterative methods, inversion of matrices, error analysis, etc.

Chapter 28, **Computer Science**. Topics discussed: computer programming, computer languages, artificial intelligence, assemblers, loaders, automata theory, data processing, etc.

Chapter 29, **Probability and Statistics**. Topics discussed: distributions, limit theorems, laws of large numbers, stochastic processes, Markov processes, random walks, branching processes, renewal theory, queueing, estimation, hypothesis testing, ranking, decision theory, Bayesian inference, multi-variate analysis, variance analysis, sampling theory, biometrics, etc.

Chapter 33, **Miscellaneous**. This chapter includes books which cover various fields of mathematics and cannot be classified under any of the other chapters. In addition, it includes books on applied mathematics, such as elasticity, fluid mechanics, thermodynamics, quantum mechanics, astronomy, wave propagation, economics, operations research, programming, biology, systems, control, information theory, circuits, automata theory, etc. The reader is referred to the Subject Index for other topics.

Chapter 30 **(Bibliographies)**, Chapter 31 **(Dictionaries)**, Chapter 32 **(Tables)**, Chapter 34 **(Periodicals)**, Chapter 35 **(Directories and Guides)**, and Chapter 36 **(Professional Organizations and Government Agencies)** are self-explanatory. Chapter 37 **(A Selected List of Publishers)** lists the addresses of the most active mathematics publishers.

The Author Index includes names of authors, joint authors, and editors of collective works. The Subject Index provides access to books through their major and secondary topics. In both the author and the subject indexes, references are to item number rather than to page number.

It is not possible to acknowledge individually all those who have provided help as this book was being prepared. I would like, however, to express my appreciation to the editorial staff of Libraries Unlimited, Inc., for their assistance in the project. I am also very grateful to Professors Joel Cohen, Stanley Gudder, David Hector, James LaVita, Jean-Paul Marchand, Michael Martin, Peter Warren, and Burton Wendroff, all of the University of Denver, for their valuable suggestions and criticism.

November 15, 1972   Elie Dick

TABLE OF CONTENTS

CHAPTER 1

GENERAL ELEMENTARY MATHEMATICS

1. Allendoerfer, Carl B., and Cletus O. Oakley. **Fundamentals of Freshman Mathematics**. 3rd ed. New York: McGraw-Hill, 1972. 636p. $10.95.
Contents: Mathematical method. The number system. Polynomials. Algebraic fractions. Exponents and radicals. Equations. Vectors and matrices. Inequalities. Functions and relations. Exponential and logarithmic functions. Trigonometric functions. Analytic geometry. Differentiation and integration. Hyperbolic functions. Tables.

2. Allendoerfer, Carl B., and Cletus O. Oakley. **Principles of Mathematics**. 3rd ed. New York: McGraw-Hill, 1969. 540p. $9.95.
Contents: Logic and sets. Number fields. The integers. Vectors and matrices. Groups. Linear equations and inequalities. Equations and inequalities of higher degree. Functions. Exponential and logarithmic functions. Trigonometric functions. Analytic geometry. Limits. The calculus. Probability. Boolean algebra.

3. Ayres, William L., Cleota G. Fry, and Harold F. S. Jonah. **General College Mathematics**. 3rd ed. New York: McGraw-Hill, 1970. 327p. $9.95.
Contents: Mathematics—what and why. Ratio, proportion, and variation. The concept of function. Linear equations and relations. Translation into equations. Quadratic equations and relations. Geometry of triangles. Definition of the trigonometric functions. General right triangles. General triangles. General angles. Graphs of the trigonometric functions. Exponents and logarithms. Finance. Growth. Probability. Introduction to statistics. Frequencies and distributions. The normal curve and sampling. The basis of our number system. Linear programming. The theory of numbers. Topology. Logic—the art of reasoning.

4. Banks, J. Houston. **Elements of Mathematics**. 3rd ed. Boston: Allyn & Bacon, 1969. 470p.
Contents: Sets. Logic. Numeration systems. The finite cardinal numbers. The algebra of rational numbers. Representation of functions. Real and complex numbers. Other algebras. Systems of measurements. Measurement computation. Statistical measures. Geometric systems. Mathematical functions.

5. Barkan, Herbert, and Carl Konove. **Introduction to Mathematics with Applications**. New York: Van Nostrand-Reinhold, 1965. 384p. $8.95.
Contents: Measurement, dimensions, and significant figures. Fundamental ideas in plane analytic geometry. Simultaneous linear equations and simultaneous linear inequalities. Functions and graphs. Conic sections. Other methods for describing sets of points. Transcendental functions. Graphic solutions of equations. Some concepts of calculus. Curve fitting. Three dimensional coordinate geometry. Vectors and matrices. Algebra of sets. Probability. The nature of

1

mathematics. Powers and roots. Common logarithms. Circular functions. Exponential functions.

6. Barr, Donald R., and Floyd E. Willmore. **College and University Mathematics: A Functional Approach.** Boston: Allyn & Bacon, 1968. 745p.
Contents: Number systems. Graphing techniques. Relations and functions. Counting methods. Probability models. The circular functions. Linearity. Polynomial functions.

7. Berlinghoff, William P. **Mathematics: The Art of Reason.** Lexington, Mass., D. C. Heath, 1968. 280p.
Contents: On the nature of mathematics. The foundations of mathematics. Sets, relations, and functions. Algebraic structures. The mathematics of uncertainty. Kinds of infinity. A very brief history of mathematics.

8. Britton, Jack R., R. Ben Kriegh, and Leon W. Rutland. **University Mathematics.** San Francisco: W. H. Freeman, 1965. Vol. I, 662p.; Vol. II, 650p. $9.50ea.
Contents: VOLUME I. The complex number system. Algebraic expressions, equations, and inequalities. Geometry and the real numbers. Relations and functions. Linear and quadratic functions. Trigonometric functions. Limits and continuity. The derivative and the inverse derivative. Theorems on derivatives. The definite integral. Exponential and logarithmic functions. The calculus of trigonometric and hyperbolic functions. Formal integration and applications. VOLUME II. Coordinate geometry. Vectors and three-dimensional geometry. Matrices. Linear transformations. Surfaces and curves in three dimensions. Vector functions and applications. Partial differentiation. Multiple integrals. Infinite series. Power series and expansion of functions. Differential equations. Linear differential equations. The Laplace transform. Elementary probability theory.

9. Brown, J. Davis, and Frank J. Palas. **Basic Mathematical Concepts.** Lexington, Mass., D. C. Heath, 1970. 405p.
Contents: Logic. Set theory. The real number system. Graphs of relations and functions. Analytic geometry. Limits, continuity, and differentiation. The definite integral. Methods of counting. Probability. Some basic statistics.

10. Crowdis, David G., and Brandon W. Wheeler. **Introduction to Mathematical Ideas.** New York: McGraw-Hill, 1969. 304p. $8.50.
Contents: Numbers, numerals, and symbols. Computation. Problem solving with computers. Computers and programming. Symbolic logic. Argument and axiomatics. Sets. Boolean algebra. The laws of chance. The interpretation and use of the laws of chance.

11. Eaves, Edgar D. **Introductory Mathematical Analysis.** 3rd ed. Boston: Allyn & Bacon, 1969. 556p.
Contents: Logic and sets. Number systems. Elementary algebraic processes. Polynomials, equations, and inequalities. Relations, functions, and graphs.

Average rates, instantaneous rates, and limits. Differentiation. Integration. Simultaneous, linear equations and inequalities, linear programming. Trigonometric functions. Simple and compound interest. Sequences, series, and annuities. Exponential and logarithmic functions. Probability and statistics.

12. Hart, William L. **Algebra, Elementary Functions, and Probability.** Lexington, Mass.: D. C. Heath, 1965. 547p.
Contents: Sets and linear inequalities in one variable. Functions: linear statements in two variables. Quadratic functions and equations in one variable. Sequences and series. Quadratic equations in two variables. Complex numbers. Theory of equations. Exponential and logarithmic functions. Mathematical induction. General systems of linear equations. Combinations and permutations. Probability. Applications of probability in statistics. Introduction to linear programming. Analytic trigonometry. Vectors in space or in a plane. Lines and planes in space.

13. Hatton, M. D. **Elementary Mathematics for Scientists and Engineers.** Elmsford, N.Y.: Pergamon, 1965. 412p.
Contents: Differentiation. Infinite series. Indefinite integration. Curve sketching and conics. Applications of integration. Polar coordinates. Curvature. Taylor's series and approximations. Complex numbers. Differential equations and applications.

14. Hight, Donald W. **A Concept of Limits.** Englewood Cliffs, N.J.: Prentice-Hall, 1966. 138p.
Contents: Sequences and their limits. Functions and their limits. Generalization and application of the limit concept.

15. Johnson, Wendell G., and Luke N. Zaccaro. **Modern Introductory Mathematics.** New York: McGraw-Hill, 1966. 576p. $8.95.
Contents: An informal introduction to sets. The logical nature of mathematics. The real numbers. Mathematical induction. Topics of algebra. Complex numbers. Algebraic functions. Exponential and logarithmic functions. Trigonometry. Theory of equations. Vectors, matrices, determinants. Systems of equations. Elementary combinational analysis and the binomial theorem. Probability. Sequences and series.

16. Kattsoff, Louis O., and Albert J. Simone. **Foundations of Contemporary Mathematics: With Applications in the Social and Management Sciences.** New York: McGraw-Hill, 1967. 576p. $9.50.
Contents: Statements. Sets. Counting-permutations and combinations. Probability. Relations and functions. Inequalities and introduction to linear programming. Linear equations, matrices and determinants. Some nonlinear relations: quadratic, exponential and logarithmic graphs. Sequences, limits and summation. Limits of functions of one variable. Derivatives. Maxima and minima: theory and application. The definite integral. Integral functions. Matrices, vector spaces, linear dependence, and bases. Rank and the general solution to linear equations. Simplex method of linear programming: theory and appli-

3

cations. Statistical-decision theory. Convex sets and Bayes decision rules.

17. Kelley, John L., and Donald R. Richert. **Mathematics for Elementary Teachers**. San Francisco: Holden-Day, 1970. 300p.
Contents: Sets, numbers, numerals. The counting numbers. Clock arithmetic. Primes. The integers. The rational numbers. The real numbers. Functions. Operations. Open sentences. Length. Area. Volume and weight.

18. Kemeny, John G., A. Scheleifer, Jr., J. L. Snell, and G. L. Thompson. **Finite Mathematics with Business Applications**. 2nd ed. Englewood Cliffs, N.J.: Prentice-Hall, 1972. 544p.
Contents: Statements and sets. Counting problems. Probability theory. Vectors and matrices. Linear programming. Decision theory and analysis. Mathematics of finance. Markov decision processes. The theory of games.

19. Leavitt, Teddy C. J. **Limits and Continuity**. New York: McGraw-Hill, 1967. 177p. $3.50; $2.50pa.
Contents: An intuitive approach to continuity. Limits of a sequence. Continuity. Limits. Theorems on continuity and limits. Preview of the calculus.

20. Leonhardy, A. **Introductory College Mathematics**. 2nd ed. New York: John Wiley, 1963. 487p. $8.95.
Contents: What is mathematics? The algebra of numbers. Numbers in exponential form. Measurement and computation. The comparison of quantities. Functions, relations, and their graphs. Variation. The rate of change of a function. Exponential and logarithmic functions. Periodic functions. Simple statistical methods. Probability.

21. Levi, Howard. **Polynomials, Power Series, and Calculus**. New York: Van Nostrand-Reinhold, 1968. 168p. $5.75.
Contents: Background material. Polynomials. Functions having polynomial approximations. Applications of the derivative and antiderivative. Infinite sequences and infinite series. Functions defined by power series. Series expansions of the elementary functions.

22. Maddox, Thomas K., and Lawrence H. Davis. **Elementary Functions**. Englewood Cliffs, N.J.: Prentice-Hall, 1969. 270p.
Contents: Number lines and coordinate planes. The idea of a function and simple examples. Linear definitions. A family of absolute value functions. Quadratic functions. Using functions to construct other functions. Exponential functions. Periodic functions.

23. McFarland, Dora, and Eunice M. Lewis. **Introduction to Modern Mathematics**. Lexington, Mass.: D. C. Heath, 1968. 406p.
Contents: An introduction to sets and numbers. Historical background for numeration. Systems of numeration. Sets of points. The system of natural numbers. The system of whole numbers. Arithmetic of the whole numbers. The

system of integers. Metric geometry. The system of rational numbers. The system of real numbers. Mathematical systems. Relations and functions.

24. Monjallon, A. **Introduction to Modern Mathematics**. 3rd ed. New York: John Wiley, 1967. 226p. $3.00.
Contents: Sets. Discussion of sets. Operations on sets. Relations. Functions. The language of mathematics. Introduction to axiomatics. The commutative group.

25. Moore, Charles G., and Charles E. Little. **Basic Concepts of Mathematics**. New York: McGraw-Hill, 1967. 462p. $8.50.
Contents: Deduction and symbolic logic. Set concepts. Number systems. Equations and applications. Measurement and approximate numbers. Statistics. Probability. Exponents and place value notation. Logarithms. Basic concepts of geometry. Circles. Trigonometric functions. Applications of trigonometric functions.

26. Polya, G. **Mathematical Discovery on Understanding, Learning and Teaching Problem Solving**. New York: John Wiley, 1962, 1965. Vol. 1, 216p.; Vol. 2, 191p. $7.50ea.
Contents: VOLUME 1. The pattern of two loci. The Cartesian pattern. Recursion. Superposition. Problems: widening the scope. VOLUME 2. Geometric representation of the progress of the solution. Plans and programs. Problems within problems. The coming of the idea. The working of the mind. The discipline of the mind. Rules of discovery. On learning, teaching and learning teaching. Guessing and scientific method. RECOMMENDED.

27. Pownall, Malcolm W. **A Prelude to the Calculus**. New York: McGraw-Hill, 1967. 315p. $6.95.
Contents: Structure of the real number system. Relations. Functions. Limits. Two geometric problems.

28. Rees, Paul K. **Principles of Mathematics**. Englewood Cliffs, N.J.: Prentice-Hall, 1965. 383p.
Contents: Arithmetic. Formulas, charts, functions, graphs, introduction to sets. Linear equations and systems of linear equations. Products and factors. Algebraic fractions, fractional equations. The trigonometric functions. Right triangles. Introduction to polar coordinates. More exponents and some radicals. Logarithms. Laws of sines and cosines. Quadratic equations. Inequalities. Compound interest. Geometric progression, simple annuities. Ratio proportion, variation. Introductory concepts of analytic geometry. Statistics. Permutations, combinations, probability. A glimpse at the calculus.

29. Ribenboim, P. **Functions, Limits, and Continuity**. New York: John Wiley, 1964. 140p. $6.95.
Contents: Sets, correspondences. Integers and fractions. Construction of real numbers. Bounded sets, accumulation points. Sequences of real numbers. Functions. Limits of functions. Continuous functions. Uniform continuity. Other methods for defining real numbers. Cardinal numbers of sets.

30. Richardson, William H. **Finite Mathematics**. New York: Harper & Row, 1968. 191p. $8.95.
Contents: Mathematical logic. Set theory. Counting and the binomial theorem. Probability. What is a proof? Mathematical induction.

31. Rose, I. H. **Algebra: An Introduction to Finite Mathematics**. New York: John Wiley, 1963. 489p. $8.50.
Contents: Sets and their numbers. Functions, operations and relations. Reason and irrationality. The real number system. Complex numbers and the unary operation. Exponential and logarithmic functions. Polynomials and fractional expressions. Equations and inequalities. Progressions and the mathematics of finance. Statistics and probability. Matrices and determinants. Logic and Boolean algebra. Groups.

32. Sachs, Jerome M., Ruth B. Rasmusen, and William J. Purcell. **Basic College Mathematics**. 2nd ed. Boston, Allyn & Bacon, 1965. 391p.
Contents: Mathematics, sets and logical systems. Introduction to arithmetic. Number basis and arithmetic applications. Modular arithmetic and groups. Introduction to algebra. Two variables, graphing and functions. Exponents and algebraic techniques. Geometry. Statistics.

33. Sanders, Paul, and Arnold D. McEntire. **Elementary Mathematics: A Logical Approach**. 2nd ed. Scranton, Penn.: Intext Educational Publishers, 1969. 351p. $7.95.
Contents: Preliminary notions. Logic and proof. Numbers and numerals. Real numbers. Special properties of the real numbers. Abstract systems. Algebraic processes. Equations and the binomial theorem. Analytic geometry. Linear equations. Vectors and matrices. Linear programming. Probability. Beginning statistics.

34. Schaaf, W. L. **Basic Concepts of Elementary Mathematics**. 3rd ed. New York: John Wiley, 1969. 500p. $8.95.
Contents: Modern mathematics. Elementary logic. Euclidean geometry. Numeration systems. The natural numbers and the integers. Real and complex numbers. Equations, inequalities, and formulas. Relations and functions. Measurement. Mensuration. Another look at geometry. Arrangements, selections and probability.

35. Wade, Thomas L., and Howard E. Taylor. **Fundamental Mathematics**. 3rd ed. New York: McGraw-Hill, 1967. 518p. $8.50.
Contents: Sets and natural numbers. The arithmetic and algebra of integers. The arithmetic and algebra of rational numbers. First-degree equations and inequalities. Equations with two variables. Further extensions of the number system. Polynomials. Relations, functions, graphs, ratios, proportion, variation. Basic trigonometry, logarithms. Compound interest and annuities. Permutations, combinations, and probability. Mathematical induction and the binomial theorem. Some topics in statistics. The binomial distribution and the normal distribution.

36. Willerding, M. F. **Elementary Mathematics: Its Structure and Concepts.** 2nd ed. New York: John Wiley, 1970. 474p. $8.95.
Contents: Sets. Logic. The system of whole numbers. Systems of numeration. The algorithms for operations with whole numbers. Number sentences. Topics from number theory. Topics from geometry. Fractions. Non-negative rational numbers. The integers. The rational numbers. The real numbers. Mathematical systems. Measurement of geometric figures. Probability.

37. Wren, F. Lynwood. **Basic Mathematical Concepts.** New York: McGraw-Hill, 1965. 398p. $7.95.
Contents: The nature of number. Systems of numeration. The natural number system. The domain of integers. The field of rational numbers. The number field of elementary mathematics. Modular arithmetic. The concepts of position, shape, and size. The concept of measurement. The concepts of relation and function.

38. Young, F. H. **The Nature of Mathematics.** New York: John Wiley, 1968. 407p. $8.95.
Contents: The natural numbers. From the concrete to the abstract and back again. Congruence modulo m. Cartesian co-ordinate system. Relations and functions. Limits. Ordered n-tuples. Early history of mechanical aids to computation.

CHAPTER 2

ALGEBRA AND TRIGONOMETRY

39. Barnett, Raymond A. **Elementary Algebra: Structure and Use.** New York: McGraw-Hill, 1968. 402p. $7.95.
Contents: Natural numbers. Integers. Rational and real numbers. First degree polynomial equations in one and two variables. First degree polynomial inequalities in one and two variables. Polynomials. Rational expressions. Exponents and radicals. Second degree polynomial equations in one variable. Relations and functions. The postulational method.

40. Blanton, Floyd Lamar, and James Earl Perry. **Modern College Algebra.** New York: McGraw-Hill, 1967. 269p. $7.50.
Contents: Logic and set terminology. The system of complex numbers. Relations and functions. Equations and inequalities. Special topics.

41. Bruce, William J., and E. Phibbs. **Algebra and Trigonometry.** New York: Appleton-Century-Crofts, 1971. 448p. $8.95.
Contents: Set theory. The number system. Functions as operations on sets. Identities and equations. Trigonometric functions. Functions of sums and related functions. Operations in linear algebra. Graphs of trigonometric functions. Sequences and series. Exponential and logarithmic functions. Solutions of triangles. Permutations. Combinations and the binomial theorem. Polynomial algebra. Inverse trigonometric functions. Polar coordinates and complex numbers. Tables.

42. Drooyan, I., and W. Wooton. **Elementary Algebra for College Students.** 2nd ed. New York: John Wiley, 1968. 302p. $7.95.
Contents: Natural numbers and their representation. The integers—signed numbers. First-degree equations. Products and factors. First-degree equations in two variables. Quadratic equations. Radical expressions. Solution of quadratic equations by other methods. Number systems.

43. Dubisch, R., and V. E. Howes. **Intermediate Algebra.** 2nd ed. New York: John Wiley, 1969. 351p. $7.95.
Contents: Fundamental concepts. Polynomials. Rational algebraic expressions. Word problems. First-degree equations. Equations of higher degree. Functions, graphs, and inequalities. Exponents and logarithms. Families of curves. Sequences, series, and the binomial theorem. Complex numbers.

44. Dupree, Daniel, and Frank Harmon. **College Algebra and Trigonometry.** Englewood Cliffs, N.J.: Prentice-Hall, 1968. 288p.
Contents: Logic, sets, and the real number system. Functions, inverse functions and graphs of functions. Equations, inequalities, and identities. Appli-

cations of the algebraic and trigonometric functions. Determinants and matrices. The exponential and hyperbolic trigonometric functions.

45. Eulenberg, M. D., and T. S. Sunko. **Introductory Algebra: A College Approach.** 2nd ed. New York: John Wiley, 1968. 317p. $7.95.
Contents: The meaning of algebra. The set of integers. Algebraic expressions. Simple equations. Factoring and the solution of quadratic equations. Fractions and the solution of fractional equations. Functional relationships. Measurement. Logarithms. Variation. Introduction to the mathematics of sets.

46. Fuller, Gordon C. **College Algebra.** 3rd ed. New York: Van Nostrand-Reinhold, 1969. 320p.
Contents: Sets and operations. The real numbers. Operations on algebraic expressions. Factoring and operations on fractions. Functions and relations. Linear equations. Exponents and radicals. Quadratic equations. Systems involving quadratic equations. Inequalities. Progressions. Permutations, combinations and probability. The binomial theorem and mathematical induction. Theory of equations. Complex numbers. Matrices and determinants. Logarithms. Partial fractions.

47. Fuller, Gordon. **Plane Trigonometry with Tables.** 4th rev. ed. New York: McGraw-Hill, 1972. 271p. $8.95.
Contents: The trigonometric functions. Graphs. Identities. Equations. Inverses. Logarithms. Complex numbers. Tables. ARBA 73.

48. Hart, William L. **College Algebra: A Contemporary Approach.** 5th ed. Lexington, Mass.: D. C. Heath, 1966. 422p.
Contents: Basic review. Sets and linear inequalities in one variable. Functions: linear statements in two variables. Quadratic functions and equations in one variable. Sequences and series. Quadratic equations in two variables. Complex numbers. Theory of equations. Exponential and logarithmic functions. Mathematical induction. General systems of linear equations. Combinations and permutations. Probability. Introduction to linear programming.

49. Heineman, E. Richard. **Plane Trigonometry with Tables.** 3rd ed. New York: McGraw-Hill, 1964. 324p. $6.95.
Contents: The trigonometric functions. Trigonometric functions of an acute angle. Trigonometric identities. Related angles. Radian measure. Graphs of the trigonometric functions. Functions of two angles. Trigonometric equations. Graphical methods. Logarithms. Right triangles. Oblique triangles. Inverse trigonometric functions. Complex numbers.

50. Hillman, Abraham P., and Gerald L. Alexanderson. **Functional Trigonometry.** 2nd ed. Boston: Allyn & Bacon, 1966. 370p., 75p. of Tables.
Contents. The geometrical background. Vectors, points and numbers. Complex numbers. The trigonometric ratios. Functions and graphs. Solving triangles. Table I: Logarithms of numbers. Table II: Five-place logarithms of the trigonometric ratios. Table III: Natural trigonometric functions to four places.

51. Kelley, John L. **Algebra: A Modern Introduction**. New York: Van Nostrand-Reinhold, 1965. 335p. $7.50.
Contents: Axioms for numbers. Sets and numbers. Vectors and lines. The complex plane. Inner products. Vector geometry. Linear algebra.

52. Leonhardy, A. **College Algebra**. 2nd ed. New York: John Wiley, 1968. 468p. $8.95.
Contents: The set of integers. The set of rational numbers. The set of real numbers. The set of complex numbers. Mathematical induction. Equations and inequalities. Functions and relations and their graphs. Systems of equations. Matrices and determinants. Theory of equations. Exponential and logarithmic functions. Circular and trigonometric functions. Permutations, combinations, and the binomial theorem. The theory of probability.

53. Mancill, Julian D., and Mario O. Gonzalez. **Basic College Algebra**. Rev. ed. Boston: Allyn & Bacon, 1968. 470p.
Contents: Sets and set operations. The real number system. Algebraic expressions. Fundamental operations on polynomials. Rational fractions. Exponents and radicals. Relations and functions. Complex numbers. Algebraic equations. Matrices, determinants, and systems of linear equations. Properties of polynomials in one variable. Solution of real numerical polynomial equations—systems of equations. Permutations, combinations, and probabilities. Inequalities. Sequences and series. Numerical approximations. Exponential and logarithmic functions.

54. Mueller, Francis J. **Elements of Algebra**. 2nd ed. Englewood Cliffs, N.J.: Prentice-Hall, 1969. 356p.
Contents: From arithmetic to algebra. Elementary operations. First degree equations and inequalities. Problem solving by equation. Special products and factoring. Fractions. Exponents, radicals and complex numbers. Quadratic equations and inequalities. Graphing and systems of equations and inequalities. Logarithm.

55. Niles, N. O. **Plane Trigonometry**. 2nd ed. New York: John Wiley, 1968. 282p. $6.95.
Contents: Fundamental concepts. Trigonometric functions (angles). The right triangle. Trigonometric functions (real numbers). Fundamental identities. Variations and graphs of the trigonometric functions. Trigonometric functions (composite angles). Logarithms. Oblique triangles. Inverse trigonometric functions. Trigonometric equations. Vectors and complex numbers.

56. Peterson, Thurman S. **Intermediate Algebra for College Students**. 3rd ed. New York: Harper & Row, 1967. 383p. $8.95.
Contents: Signed numbers. Polynomials. Equations and stated problems. Factoring. Fractions. Exponents, roots, and radicals. Graphical methods. Systems of linear equations. Quadratic equation. Systems involving quadratic equations. Ratio, variation, and binomial theorem. Logarithms. Progressions.

10

57. Rees, Paul K., and Fred W. Sparks. **Algebra and Trigonometry.**
2nd ed. New York: McGraw-Hill, 1969. 544p. $9.50.
Contents: Sets and set notation. The number system and fundamental operations. Fractions. Linear and fractional equations. Exponents and radicals.
Functions and graphs. Angular measure and the trigonometric functions. Quadratic equations. Systems of equations. Elementary theory of matrices and
determinants. Complex numbers. Higher-degree equations. Logarithms. Right
triangles. Oblique triangles. The progressions. Mathematical induction. The
binomial theorem.

58. Rees, Paul K., and Fred W. Sparks. **College Algebra.** 6th ed. New
York: McGraw-Hill, 1972. 564p. $9.95.
Contents: The number systems. The four fundamental operations. Products
and factoring. Fractions. Exponents and radicals. Equations and inequali-
ties. Functions, relations and graphs. Systems of equations and inequalities.
Matrices and determinants. Complex numbers. Logarithms. Progressions.
Mathematical induction. The binomial theorem. Permutations and combina-
tions. Probability.

59. Robinson, Thomas J. **Algebra and Trigonometry.** New York: Harper
& Row, 1970. 407p. $8.95.
Contents: Logic, sets, relations, and functions. The real number system.
Factoring, radicals, and exponents. Graphs of functions and relations, and
linear and quadratic functions. Logarithms and exponents. Trigonometry.
Permutations, combinations, and probability. Progression, sequences, and
series. Inverses of functions and trigonometric equations. Complex numbers.
Systems of equations and inequalities. Matrices and determinants. Theory
of equations. Tables.

60. Robison, J. Vincent. **Modern Algebra and Trigonometry.** New York:
McGraw-Hill, 1966. 401p. $7.95.
Contents: Sets and operations. Real numbers. First-degree equations. Functions and graphs. Trigonometric functions. Factoring polynomials. Operations
with fractions. Exponents. Exponential functions. Logarithms. Logarithmic
functions. Solution of triangles. Vectors. Trigonometric functions of sums and
differences. Complex numbers. Polynomial functions. Inverse trigonometric
functions. Systems of equations. Matrices. Sequences. Permutations. Combinations. Probability.

61. Russell, Donald S., and M. Collins. **Elementary Algebra.** 4th ed.
Boston: Allyn & Bacon, 1972. 384p.
Contents: The number system. Addition and subtraction of signed numbers.
Multiplication and division. First-degree equations. Inequalities. Special products
and factorization. Fractions. Changing the subject of a formula. Relations,
functions, and graphs. Systems of equations. Exponents and radicals. Quadratic equations. Ratio, proportion, and variation.

62. Willerding, M. F., and S. Hoffman, Jr. **Modern Intermediate Algebra.** New York: John Wiley, 1969. 308p. $7.95.

Contents: Real numbers and their properties. Exponents and radicals. Polynomials. Rational expressions. First degree equations and inequalities. Relations and functions. Complex numbers. Quadratic functions. Quadratic equations and inequalities. Systems of equations. Sequences and series. Exponential and logarithmic functions.

CHAPTER 3

CALCULUS, ANALYTIC GEOMETRY, AND VECTOR ANALYSIS

63. Apostol, T. M. **Mathematical Analysis: A Modern Approach to Advanced Calculus.** 4th ed. Reading, Mass.: Addison-Wesley, 1964. 559p. $12.75.
Contents: The real and complex number systems. Set theory. Point set topology. Limits and continuity. Differentiation. Functions of bounded variation. Rectifiable curves and connected sets. The Riemann-Stieltjes integration. Multiple integrals. Line integrals. Vector analysis. Infinite series and infinite products. Sequences of functions. Improper Riemann-Stieltjes integrals. Fourier series and Fourier integrals. Cauchy's theorem and the residue calculus.

64. Ayre, H. Glenn, Rothwell Stephens, and Gordon D. Mock. **Analytic Geometry: Two and Three Dimensions.** 2nd ed. New York: Van Nostrand-Reinhold, 1967. 352p. $7.95.
Contents: Coordinate systems. Lines and related topics. The plane. Vectors. The circle, sphere, cylinder, and cone. The ellipse and the ellipsoid. The parabola and paraboloid. The hyperbola and the hyperboloid. The general equation of second degree. Graphs.

65. Barnett, R. A., and J. N. Fujii. **Vectors.** New York: John Wiley, 1963. 132p. $4.50.
Contents: Vectors and scalars. Vector addition and multiplication by a scalar. Collinear and coplanar vectors. The scalar product and base vectors. The scalar (dot) product. Base vectors and rectangular coordinate systems. Direction cosines and direction numbers. The vector product, triple products, and identities. The vector (cross) product. The vector product in component form. Triple products. Vector identities. Vector equations, curves and surfaces. Vector functions.

66. Bedford, F. W., and T. D. Dwivedi. **Vector Calculus.** New York: McGraw-Hill, 1970. 756p. $13.50.
Contents: Vectors in the plane. Vectors in three space. The algebra of vectors. The outer product. Planes and lines. Triple products. Surfaces. The algebra of matrices. Systems of linear equations. Linear transformations. Vector differential calculus for curves. Scalar functions of several variables. Vector differential calculus for surfaces. Multiple integrals. Vector integral calculus.

67. Bell, Stoughton, J. R. Blum, J. Vernon, Lewis Rosenblatt, and Judah Rosenblatt. **Modern University Calculus.** San Francisco: Holden-Day, 1966. 974p. $11.95.

Contents: The number system. Error analysis. Functions. Continuity and limits. Integrals. Derivatives and differentials. Maxima and minima and numerical analysis. Fundamental theorem of calculus and rules of integration. The logarithmic and exponential functions. The mean value theorem. Parametric representation of curves and arc lengths. Polynomial approximations. Vectors and projections. Differential calculus of several variables. Integral calculus in several variables.

68. Bers, L. **Calculus**. New York: Holt, Rinehart, and Winston, 1969. 932p.
Contents: Numbers. Coordinates. Functions. Derivatives. Integrals. Transcendental functions. Techniques of integration. Series. Vectors. Quadrics. Vector-valued functions. Partial derivatives. Multiple integrals.

69. Blakeley, W. R. **Calculus for Engineering Technology**. New York: John Wiley, 1968. 441p. $10.50.
Contents: Functions. Basic analytic geometry. Rate of change and the derivative. Maximum and minimum. The differential and small changes. Products, quotients, and functions of functions. Integration. Transcendental functions. Trigonometric functions. Exponential and logarithmic functions. Partial derivatives. Power series. First-order differential equations. Second-order differential equations. Laplace transforms. Fourier analysis.

70. Bourne, D. E., and P. C. Kendall. **Vector Analysis**. Boston: Allyn & Bacon, 1968. 219p.
Contents: Rectangular Cartesian coordinates and rotation of axes. Scalar and vector algebra. Vector functions of a real variable. Differential geometry of curves. Scalar and vector fields. Line, surface and volume integrals. Integral theorems.

71. Britton, Jack R., R. Ben Kriegh, and Leon W. Rutland. **Calculus and Analytic Geometry**. San Francisco: W. H. Freeman, 1966. 1069p. $12.50.
Contents: Functions and graphs. Linear and quadratic functions. Limits and continuity. The derivative and the inverse derivative. Theorems on derivatives. The definite integral. Exponential and logarithmic functions. The calculus of trigonometric and hyperbolic functions. Vectors and complex-valued functions. Formal integration. Coordinate geometry. Vectors and three-dimensional geometry. Matrices. Linear transformations. Surfaces and curves. Vector functions and applications. Partial differentiation. Multiple integrals. Infinite series. Power series and expansion of functions. Differential equations. Linear differential equations.

72. Buck, R. Creighton. **Advanced Calculus**. 2nd ed. New York: Mc-Graw-Hill, 1965. 527p. $10.50.
Contents: Sets and functions. Continuity. Integration. Convergence. Differentiation. Applications to geometry and analysis. Differential geometry and vector calculus. Special applications.

73. Burdette, A. C. **An Introduction to Analytic Geometry and Calculus.**
New York: Academic Press, 1967. 352p. $9.50.
Contents: The coordinate system—fundamental relations. The straight line.
Non-linear equations and graphs. Functions and limits. The derivative. The
integral. The trigonometric functions. The exponential and logarithm functions. Methods of integration. Parametric equations—polar coordinates. Functions of several variables. Infinite series.

74. Courant, R., and F. John. **Introduction to Calculus and Analysis.**
New York: John Wiley, 1965. 661p. $13.50 (vol. 1).
Contents: The fundamental ideas of the integral and differential calculus.
The techniques of calculus. Applications in physics and geometry. Taylor's
expansion. Numerical methods. Infinite sums and products. Trigonometric
series. Differential equations for the simplest types of vibration.
Volume 2 is in preparation.

75. Cronin-Scanlon, Jane. **Advanced Calculus: A Start in Analysis.** Rev.
ed. Lexington, Mass.: D. C. Heath, 1969. 320p.
Contents: Real numbers and Euclidean n-space. Continuous and differentiable
functions. Integration theory. The Riemann integral. The Lebesgue integral.
Infinite series of functions. Power series. Fourier series. Curves. Surfaces.

76. Crowell, R. H., and W. E. Slesnick. **Calculus with Analytic Geometry.**
New York: W. W. Norton, 1968. 727p. $9.95.
Contents: Functions, limits, and derivatives. Applications of the derivatives.
Conic sections. Integration. Logarithms and exponential functions. Trigo-
nometric functions. Techniques of integration. The definite integral. Infinite
series. Geometry in the plane. Differential equations. RECOMMENDED.

77. Davis, Harry F. **Introduction to Vector Analysis.** 2nd ed. Boston:
Allyn & Bacon, 1967. 349p.
Contents: Vector algebra. Vector functions of a single variable. Scalar and
vector fields. Line and surface integrals. Differential forms, matrices, and
tensors.

78. Davis, Thomas A. **Analytic Geometry.** New York: McGraw-Hill,
1967. 350p. $6.50; $4.50pa.
Contents: One-dimensional coordinate systems. Inequalities and intervals.
Absolute values and inequalitites, Directed distance. Two-dimensional
coordinate system. Distance in the plane. Midpoint. Inclination
and slope. Parallel and perpendicular lines. The angle from one line to another.
Graphs and equations. Intercepts, symmetry, and asymptotes. The straight
line. Conics. The circle. The parabola. The ellipse. The hyperbola. Translation
of axes. Eccentricity.

79. Devinatz, Allen. **Advanced Calculus.** New York: Holt, Rinehart, &
Winston, 1968. 488p. $10.95.
Contents: The real number system. Sets. Countability. Functions. Limits.

Limits superior and inferior. The Heine-Borel theorem. Continuity. Uniform continuity. Infinite series. Differentiation. Riemann integration. The Riemann-Stieltjes integrals. Vector spaces. Topology in n-dimensional spaces. Linear transformations. Determinants. Inverse and implicit function theorems. Differentiation and integration in n-dimensional spaces.

80. Fadell, Albert G. **Vector Calculus and Differential Equations.** New York: Van Nostrand-Reinhold, 1968. 576p. $11.95.
Contents: Vectors. Vector functions. Differential calculus of R^n. Multiple integrals. Infinite series. Line integrals. Improper integrals. Integrals with parameter. Differential equations. The Laplace transform. Fourier series.

81. Fisher, Robert C., and Allen D. Ziebur. **Calculus and Analytic Geometry.** 2nd ed. Englewood Cliffs, N.J.: Prentice-Hall, 1965. 768p.
Contents: Limits. The derivative. Applications of the derivative. The conics. The integral. Exponential, logarithmic and inverse functions.

82. Flanders, H., R. R. Korfhage, and J. J. Price. **Calculus.** New York: Academic Press, 1970. 969p. $13.95.
Contents: The derivative. Curve sketching. Maxima and minima. Direction fields. Velocity and acceleration. Circular functions. The exponential function. Integration. Applications of integration. Numerical integration. Space geometry. Volume. Partial derivatives. Approximation methods. Techniques of differentiation. Applications of derivatives. Inverse functions. Logarithms. Trigonometric functions. Techniques of integration. Interpolation and numerical integration. First order differential equations. Second order linear equations. Vectors. Applications of vectors. Several variables. Double integrals. Taylor approximations. Power series. Improper integrals. Approximate solutions of differential equations. Complex numbers. Higher partial derivatives. Vector operations. Multiple integrals. Applications of multiple integrals. Derivatives.

83. Friedman, A. **Advanced Calculus.** New York: Holt, Rinehart, and Winston, 1971. 423p.
Contents: Numbers and sequences. Continuous functions. Differentiable functions. Integration. Sequences and series of functions. Space of several variables and continuous functions on it. Differentiation. Multiple integrals. Line and surface integrals.

84. Fulks, W. **Advanced Calculus: An Introduction to Analysis.** 2nd ed. New York: John Wiley, 1969. 597p. $11.95.
Contents: The number system. Functions, sequences and limits. Continuity and differentiability. Integration. The elementary transcendental functions. Limits and continuity. Properties of differentiable functions. Vectors and curves. Functions of several variables, limits and continuity. Differentiable functions. Transformations and implicit functions, extreme values. Multiple integrals. Line and surface integrals. Infinite series. Sequences and series of functions. Uniform convergence. The Taylor series. Improper integrals. Inte-

gral representation of functions. Gamma and beta functions. Laplace's method and Stirling's formula. Fourier series. RECOMMENDED.

85. Goffman, Casper. **Calculus of Several Variables.** New York: Harper & Row, 1965. 182p. $9.95.
Contents: Euclidean space. Orthonormal basis. Open sets. Closed sets. Completeness. Continuous mappings. Definition of differential. Taylor's theorem for one variable. Taylor's theorem for n variables. Absolute maxima and minima. Implicit function theorem. Differentiable manifolds. Differentiable functions and mappings. Gradient, divergence, and curl.

86. Greenspan, Donald. **Introduction to Calculus.** New York: Harper & Row, 1968. 439p. $11.95.
Contents: Functions and differences. Limits and derivatives. Maxima and minima. Sums, integrals, and the fundamental theorem. Techniques of integration. Interval arithmetic and difference equations. Differential equations.

87. Hildebrand, Francis B. **Advanced Calculus for Applications.** Englewood Cliffs, N.J.: Prentice-Hall, 1962. 646p.
Contents: Ordinary differential equations. The Laplace transform. Numerical methods for solving ordinary differential equations. Series solutions of differential equations. Special functions. Boundary-value problems and characteristic-function representations. Vector analysis. Topics in higher-dimensional calculus. Partial differential equations. Solutions of partial differential equations of mathematical physics. Functions of a complex variable.

88. Hoffman, Stephen, Jr. **Advanced Calculus.** Englewood Cliffs, N.J., Prentice-Hall, 1970. 352p.
Contents: Curves and surfaces. Line and surface integrals. The gradient, divergence and curl. Stokes' theorem. Integration. Limits and convergence. Differentiation. Improper integrals. Representation of functions.

89. Hummel, James. **Introduction to Vector Functions.** Reading, Mass.: Addison-Wesley, 1967. 372p. $9.75.
Contents: Vector spaces. Matrices. Linear transformations. Orthogonality. Diagonalization. Groups. Matrix groups. Scalar-valued functions of vectors. Vector-valued functions of vectors. The implicit function theorem. Line integrals. Surface integrals. Multiple integrals. Vector analysis.

90. Hunt, Burrowes. **Calculus and Linear Algebra.** San Francisco: W. H. Freeman, 1967. 401p. $9.50.
Contents: The natural numbers. The real numbers. Sequences and real functions. Elements of analytic geometry. Linear algebra in R^2 and R^3. Curves and continuity. Integration problems. Integration over intervals. Differentiation.

91. Johnson, Richard E., and Fred L. Kiokemeister. **Calculus with Analytic Geometry.** 4th ed. Boston: Allyn & Bacon, 1969. 956p.

Contents: Analytic geometry. Functions. Limits. Derivatives. Integrals. Exponential and logarithmic functions. Trigonometric functions. Formal integration. Improper integrals and Taylor's formula. Infinite series. Polar coordinates. Differential calculus of functions of several variables. Multiple integration.

92. Jones, D. S., and D. W. Jordan. **Introductory Analysis.** New York: John Wiley, 1969, 1970. Vol. 1, 482p., $15.00. Vol. 2, 394p., $13.95. Contents: VOLUME 1. Conics. Functions and graphs. Limits and continuity. The derivative. Maxima and minima. The mean-value theorem. Sequences. Infinite series. Functions defined by series. Elementary integration methods. The definite integral. VOLUME 2. Differential equations of the first order. Systematic integration and applications. "Improper" integrals: an extension of the possibilities in integration. Vectors and curves. Complex numbers. Linear differential equations of the second order with constant coefficients. Solid geometry and vectors. Functions of two variables.

93. Kaplan, W., and D. J. Lewis. **Calculus and Linear Algebra. Vol. 1: Vectors in the Plane and One-Variable Calculus.** New York: John Wiley, 1970. 690p. $9.95. Contents: Geometry and trigonometry. Two-dimensional vector geometry. Limits. Differential calculus. Integral calculus. The concepts of integration. The indefinite integral. The definite integral. The elementary transcendental functions. Applications. Volume 2 (**Vector Spaces, Many-Variable Calculus and Differential Equations**) in preparation.

94. Loomis, Lynn H., and S. Sternberg. **Advanced Calculus.** Reading, Mass.: Addison-Wesley, 1968. 580p. $13.95. Contents: Vector spaces. Differential calculus. Compactness. Completeness. Inner product spaces. Differential equations. Multilinear functionals. Integration. Differentiable manifolds. Exterior calculus. Potential theory. Classical mechanics.

95. Marder, L. **Vector Analysis.** New York: American Elsevier, 1970. 177p. $7.50. Contents: Vector algebra. Differentiation and integration of vectors. Scalar and vector fields. Integration. The gradient vector. Electrostatic applications. Divergence. Gauss's theorem. Green's integral theorems. Curl. Stokes' theorem. Scalar and vector potentials. Orthogonal curvilinear coordinate systems.

96. McQuisten, R. B. **Scalar and Vector Fields: A Physical Interpretation.** New York: John Wiley, 1965. 314p. $6.95; $4.95pa. Contents: The concept of scalar and vector fields. Vector algebra. Vector calculus. Vector field representation and coordinate transformations. The gradient of a scalar field. The divergence. The curl of a vector field. Differential field operators. The integral properties of fields– Stokes', Gauss's, and Green's

theorems. Field operators in orthogonal curvilinear coordinate systems. Field potentials. Time-dependent fields.

97. Meyer, Herman. **Introduction to Modern Calculus.** New York: McGraw-Hill, 1969. 512p. $10.50.
Contents: Real numbers and real-valued functions. The theory of limits. The derivative. The definite integral. Composition of functions. Transcendental functions. Improper integrals and infinite series. Differential equations. Techniques of differentiation and integration. Applications.

98. Middlemiss, Ross R., John L. Marks, and James R. Smart. **Analytic Geometry.** 3rd ed. New York: McGraw-Hill, 1968. 434p. $8.50.
Contents: Rectangular coordinates. Angles and directions. Relations, graphs, and loci. The line. Transformation of coordinates. The circles and the ellipse. The parabola and the hyperbola. Polynomials. Algebraic curves of higher degree. The trigonometric curves. The exponential and logarithmic curves. Curve fitting. Analytic geometry of three dimensions.

99. Murdoch, D. C. **Analytic Geometry with an Introduction to Vectors and Matrices.** New York: John Wiley, 1966. 294p. $7.95.
Contents: Coordinates, functions, and graphs. Straight lines and circles in the plane. Vectors. Planes, lines, and spheres. Conics and quadrics. Vector spaces and transformations of coordinates. Matrices, determinants, and linear equations. Orthogonal transformations and rotations. Polar coordinates and complex numbers.

100. Niven, Ivan. **Calculus: An Introductory Approach.** 2nd ed. New York: Van Nostrand-Reinhold, 1966. 224p. $6.00.
Contents: Limits. Integration. Differentiation. The fundamental theorem. The trigonometric functions. The logarithmic and exponential functions.

101. Pignani, Tullio J., and Paul W. Haggard. **Modern Analytic Geometry.** Lexington, Mass.: D. C. Heath, 1970. 305p.
Contents: The straight line. The conic sections. Matrices and determinants. Planar vectors. Polar coordinates. Parametric equations. Locus sketching. Space geometry. Space vectors and lines.

102. Protter, M. H., and C. B. Morrey, Jr. **Modern Mathematical Analysis.** Reading, Mass.: Addison-Wesley, 1965. 790p. $10.95.
Contents: Solid analytic geometry. Vectors in three dimensions. Infinite series. Partial differentiation. Multiple integration. Linear algebra. Vector spaces. Eigenvalue problems. Fourier series. Implicit function theorems. Transformations. Functions defined by integrals. Vector field theory. Green's and Stokes' theorems. Differential equations.

103. Rees, Paul K., and Fred W. Sparks. **Calculus with Analytic Geometry.** New York: McGraw-Hill, 1969. 576p. $10.95.
Contents: Lines, circles, rational functions. Limits. The derivative of a

function. Differentiation of algebraic functions. The differential. The indefinite integral. The definite integral. The conics. The trigonometric functions. The exponential and logarithmic functions. Rolle's theorem, mean value theorem, indeterminate forms. Polar coordinates and parametric equations. Derivatives from parametric equations, curvature, vectors. Partial derivatives. Multiple integrals. Infinite series. Differential equations.

104. Rossi, Hugo. **Advanced Calculus: Problems and Applications to Science and Engineering**. New York: W. A. Benjamin, 1970. 732p. $17.50.
Contents: Linear functions. Ordinary differential equations. Particle motion. Series expansion in the complex domain. Fourier series and their application in partial differential equations. Integration. Stokes' theorem. Geodesics and Dirichlet's principle.

105. Schachter, Harold. **Calculus and Analytic Geometry**. New York: McGraw-Hill, 1972. 307p. $8.95.
Contents: Analytic geometry. Equations of curves. Functions. Derivatives. Applications. Antidifferentiation. The definite integral. Transcendental functions. Conic sections. Polar coordinates.

106. Schwartz, Abraham. **Calculus and Analytic Geometry**. 2nd rev. ed. New York: Holt, Rinehart, and Winston, 1967. 1008p.
Contents: The derivative. Integration. Plane analytic geometry. The trigonometric functions. The logarithmic and exponential functions. On the evaluation of integrals. On the underlying theory. More applications. Polar coordinates. Solid analytic geometry. Partial differentiation. Multiple integrals. Infinite series. Differential equations.

107. Seeley, Robert T. **Calculus of One Variable**. 2nd ed. Glenview, Ill.: Scott, Foresman and Company, 1972. 729p.
Contents: Derivatives. Applications. Theory of maxima. Integration. Vectors and the laws of motion. Complex numbers. Approximation. Infinite sequences and series. Numbers. Proof of the basic propositions of calculus.

108. Spain, Barry. **Vector Analysis**. New York: Van Nostrand-Reinhold, 1965. 114p. $6.25.
Contents: Scalar and vector products. Derivatives of vectors. Taylor's theorem for a vector function. Line integrals. Surface integrals. Volume integrals. Directional derivatives. Gradient. Divergence. Gauss's theorem. Curl. Stokes' theorem. Green's theorems.

109. Stein, Sherman K. **Calculus: In the First Three Dimensions**. New York: McGraw-Hill, 1967. 613p. $9.95.
Contents: The definite integral. The derivative. Limits and continuous functions. The law of the mean. The fundamental theorem of calculus. Maximum and minimum of a function. Taylor's series. Partial derivatives. The derivative of a vector function. Green's theorem.

110. Tallack, J. C. **Introduction to Elementary Vector Analysis.** New York: Cambridge University Press, 1966. 138p. $9.50.
Contents: Introduction to vectors through displacements. Addition and subtraction of vectors. Multiplication and division of a vector by a number. Position vectors. Centroid. Projection and components of a vector. Differentiation and integration. The scalar product.

111. Thomas, George. **Calculus and Analytic Geometry.** 4th ed. Reading, Mass.: Addison-Wesley, 1968.
Contents: Slope of a curve. Derivative. Limits. Continuity. Mean value theorem. Integration. Fundamental theorems of calculus. Transcendental functions. Hyperbolic functions. Polar coordinates. Vectors. Parametric equations determinants. Elementary linear algebra. Vector functions and their derivatives. Partial differentiation. Multiple integration. Vector analysis. Green's theorem. Stokes' theorem. Divergence theorem. Power series. Taylor series. Fourier series. Tests for convergence. Complex numbers. Elementary differential equations.

112. Tierney, John A. **Calculus and Analytic Geometry.** 2nd ed. Boston: Allyn & Bacon, 1972. 641p. $13.95.
Contents: Plane analytic geometry. The differential calculus. The integral calculus. Graph of an equation. Trigonometric and inverse trigonometric functions. The logarithmic and exponential functions. Techniques of integration. Parametric equations. Polar coordinates. Hyperbolic functions. Multiple integrals. Infinite series. Differential equations.

113. Trench, William F., and Bernard Kolman. **Multivariable Calculus with Linear Algebra and Series.** New York: Academic Press, 1972. 758p.
Contents: Linear equations and matrices. Vector spaces and linear transformations. Vectors and analytic geometry. Differential calculus of real-valued and vector-valued functions. Integration. Series.

114. Widder, David. **Advanced Calculus.** 2nd ed. Englewood Cliffs, N.J.: Prentice-Hall, 1961. 520p.
Contents: Partial differentiation. Vectors. Differential geometry. Stieltjes integral. Multiple integrals. Line and surface integrals. Limits and indeterminate forms. Infinite series. Convergence of improper integrals. The gamma function. Evaluation of definite integrals. Fourier series. The Laplace transform.

115. Williamson, Richard E., Richard H. Crowell, and H. F. Trotter. **Calculus of Vector Functions.** 3rd ed. Englewood Cliffs, N.J.: Prentice-Hall, 1972. 576p.
Contents: Linear algebra. Vector calculus. Real-valued functions. Integration. Vector field theory. RECOMMENDED.

116. Youse, Bevan K. **Calculus with Analytic Geometry.** Scranton, Penn.: Intext Educational Publishers, 1968. 271p. $8.50.
Contents: The real number system. Functions and their graphs. Derivative and the limit concept. The mean value theorem. Riemann integral. Transcendental functions. Sequences and series. Functions of several variables.

CHAPTER 4

HISTORY, BIOGRAPHY,
AND REPRINTS OF CLASSICS

117. Baron, Margaret E. **The Origins of the Infinitesimal Calculus.** Elmsford,
N.Y.: Pergamon, 1969. 304p. $13.00.
Contents: The ancient Greeks. The works of Newton, Leibniz, and others.
Emphasis on the sixteenth and seventeenth centuries. M.R. 41 No. 1471.

118. Beckman, P. **A History of** Π **(Pi).** Boulder, Colo.: Golem Press, 1970.
190p.
Contents: The discovery of Π and its measurement from the early Greeks to
the present.

119. Bowne, G. D. **The Philosophy of Logic, 1880-1908.** The Hague:
Mouton, 1966. 157p.
Contents: A survey of the arguments used by the mathematical philosophers
from 1880 to 1908.

120. Boyer, C. B. **A History of Mathematics.** New York: John Wiley, 1968.
717p. $12.95.
Contents: Primitive origins. Egypt. Mesopotamia. Ionia and the Pythagoreans.
The heroic age. The age of Plato and Aristotle. Euclid of Alexandria. Archimedes
of Syracuse. Apollonius of Perga. Greek trigonometry and mensuration. Revival
and decline of Greek mathematics. China and India. The Arabic hegemony.
Europe in the Middle Ages. The Renaissance. Prelude to modern mathematics.
The time of Fermat and Descartes. A transitional period. Newton and Leibniz.
The Bernoulli era. The age of Euler. Mathematicians of the French Revolution.
The time of Gauss and Cauchy. The heroic age in geometry. The arithmetization
of analysis. The rise of abstract algebra. Aspects of the twentieth century. M.R.
38 No. 3105.

121. Brewster, David. **Memoirs of the Life, Writing, and Discoveries of Sir
Isaac Newton.** New York: Johnson Reprint Corp., 1965. Vol. I, 478p.; Vol. II,
564p. $35.00set.
Contents: This is a reprint of the 1855 edition, with an introduction by Richard
S. Wesfall.

122. Crowe, Michael J. **A History of Vector Analysis: The Evolution of the
Idea of a Vectorial System.** Notre Dame, Ind.: University of Notre Dame Press,
1967. 270p. $12.95.
Contents: The works of Argand, Bessel, Gauss, Hamilton, Grassman, Maxwell,
Gibbs, and Heaviside.

123. De Moivre, A. **The Doctrine of Chances: A Method of Calculating the Probabilities of Events in Play.** New York: Chelsea, 1967. 368p. $7.50.
Contents: This is a reprint of the 1756 edition, with a bibliography by H. M. Walker.

124. De Morgan, Augustus. **"On the Syllogism" and Other Logical Writings.** Edited by Peter Heath. New Haven, Conn.: Yale University Press, 1966. 355p. $10.00.
Contents: Some of the works of De Morgan. M.R.37 No. 6147.

125. Dickson, Leonard E. **History of the Theory of Numbers.** New York: Chelsea, 1966. 3v. $22.50set.
Contents: Vol. I: Divisibility and primarility. Vol. II: Diophantine analysis. Vol. III: Quadratic and higher forms. M.R.39 No. 6807 a, b, c.

126. Grattan-Guinness, Ivor. **The Development of the Foundations of Mathematical Analysis from Euler to Riemann.** Cambridge, Mass.: M.I.T. Press, 1970. 186p.
Contents: Analysis in the eighteenth century. The theory of limits. Convergence. Fourier analysis.

127. Halberstam, H., and R. E. Ingram, editors. **The Mathematical Papers of Sir William Rowan Hamilton. Vol. III: Algebra.** New York: Cambridge University Press, 1967. 672p. $37.50.
Contents: All the works of Hamilton in algebra: complex numbers as ordered pairs, quaternions, algebraic equations, and calculus. M.R.35 No. 4081.

128. Hall, Tord. **Carl Friedrich Gauss: A Biography.** Trans. from the Swedish by A. Froderberg. Cambridge, Mass.: M.I.T. Press, 1970. 176p. $7.95.
Contents: Areas Gauss worked in and his main results.

129. Hebroni, P. **Works on Mathematics.** Jerusalem: The Public Committee for publishing the works of P. Hebroni, 1968. English text: 191p. Hebrew text, 79p.
Contents: The major works of Hebroni, with a biography by J. J. Rivlin.

130. Herivel, John. **The Background to Newton's Principia: A Study of Newton's Dynamical Researches in the Years 1664-1684.** Oxford: Clarendon Press, 1965. 337p. $12.75.
Contents: A discussion of Newton's works on dynamics with reproduction of these works. Translations of his works written in Latin.

131. Lanczos, C. **Space Through the Ages.** New York: Academic Press, 1970. 320p. $11.50.
Contents: Early beginnings. Greek geometry. The evolution of metrical geometry. Tensor algebra. Tensor analysis. The geometry of Gauss and Riemann. Einstein's theory of gravitation. Abstract spaces. Projective geometry.

132. Lebesgue, Henri. **Measure and the Integral**. Trans. by Scripta Technica and edited by Kenneth O. May. San Francisco: Holden-Day, 1966. 216p.
Contents: Biographical sketch of Lebesgue. Measure of magnitudes. Collections. Whole numbers. Lengths. Numbers. Plane areas. Volumes. Length of curves. Areas of surfaces. Measurable magnitudes. Integration and differentiation. Development of the integral.

133. Manheim, Jerome H. **The Genesis of Point Set Topology**. New York: Macmillan, 1964. 166p. $3.50.
Contents: A historical study of the works of Cantor on point sets and Fourier series. M.R.37 No. 2561.

134. Meschkowski, Herbert. **Ways of Thought of Great Mathematicians**. Trans. by John Dyer-Bennett. San Francisco: Holden-Day, 1964. 126p.
Contents: The Pythagoreans. Archimedes. Nicholas of Cusa. Blaise Pascal. Gottfried Wilhelm Leibniz. Carl Friedrich Gauss. George Boole. Weierstrass and his school. Georg Cantor.

135. Meschkowski, Herbert. **Evolution of Mathematical Thought**. Transl. by J. H. Gayl. San Francisco: Holden-Day, 1965. 126p.
Contents: The problem. Foundations of Greek mathematics. The road to non-Euclidean geometry. The problem of infinity. Cantor's foundation of the theory of sets. Antimonies and paradoxes. Intuitionism. Geometry and experience. Problems of mathematical logic. Formalism. Decision problems. Operative mathematics. The philosophical harvest from research in the foundations of mathematics.

136. Poincare, H. **New Methods of Celestial Mechanics**. Trans. from the French. Washington, D.C.: National Aeronautics and Space Administration, 1967. Vol. I: 332p., $3.00. Vol. II: 413p., $3.00. Vol. III: 401p., $3.00.
Contents: VOLUME I. Periodic solutions. The non-existence of integral invariants. Asymptotic solutions. VOLUME II. Methods of Newcomb, Gylden, Lindstet, and Bohlin. VOLUME III. Integral invariants. Periodic solutions of the second type. Doubly asymptotic solutions.
The three volumes originally appeared in 1892, 1893, and 1899, respectively.

137. Pullan, J. M. **The History of the Abacus**. New York: Praeger, 1968. 127p. $4.95.
Contents: Introduction. The origins of arithmetic. Calculus and abacus. Roman numerals to Arabic figures. Counter-casting to pen-reckoning. The abacus method. Jettons. The abacus in archeology. The abacus in education. Conclusion. M.R.39 No. 2573.

138. Rashevesky, N. **Looking at History through Mathematics**. Cambridge, Mass.: M.I.T. Press, 1968. 199p. $10.00.
Contents: An attempt to show how mathematical reasoning could be used to explain history.

139. Reid, Constance. **Hilbert.** New York: Springer-Verlag, 1970. 290p. $8.80.
Contents: Detailed biography of Hilbert.

140. The Royal Society. **Biographical Memoirs of Fellows of the Royal Society.** London: The Royal Society, 1969. 266p. $6.50.
Contents: Memoirs of L. E. J. Brower, L. D. Landau, and others.

141. Scott, J. F. **A History of Mathematics: From Antiquity to the Beginning of the Nineteenth Century.** 2nd ed. New York: Barnes & Noble, 1969. 266p. $12.50.
Contents: Mesopotamian mathematics. Greek mathematics. Mathematics in the Orient. The invention of trigonometry. Napier's logarithms. The work of Gauss. M.R.41 No. 1473.

142. Scriba, Christoph J. **The Concept of Number: A Chapter in the History of Mathematics, with Applications of Interest to Teachers.** Mannheim: Bibliographisches Institut, 1968. 216p.
Contents: The number system. Evolution of the concept of numbers. Operations with numbers in ancient Greece, Rome, and Europe. Higher arithmetic. Algebra. Number theory. Development of the concept of the number system since the nineteenth century.

143. Srinivasiengar, C. N. **The History of Ancient Indian Mathematics.** Calcutta: World Press Private, 1967. 157p. $5.00.
Contents: The decimal system of numeration. The sulua sutras. The mathematics of the Jainas. The Bakhshali manuscript. Arya Bhata. Brahmagupta. Mahaviracharya. Bhaskara. Kuttaka. Varga-Prakriti or the equation of the multiplied square. Miscellaneous equations. Indian mathematics after Bhaskara.

144. Sullivan, Mark W. **Apuleian Logic: The Nature, Sources, and Influence of Apuleius's Peri Hermeneias.** Amsterdam: North-Holland Pub. Co., 1967. 265p. $12.00.
Contents: The work of Apuleius of Madaura (125-171 A.D.) on elementary syllogistic logic, with analysis.

145. Szabo, M. E. **The Collected Papers of Gerhard Gentzen: Studies in Logic and the Foundations of Mathematics.** Amsterdam: North-Holland Pub. Co., 1969. 338p. $20.20.
Contents: The complete works of Gerhard Gentzen translated into English, with an index and a biography of Gentzen. M.R.41 No. 6660.

146. Taylor, E. G. R. **The Mathematical Practitioners of Hanoverian England 1714-1840.** New York: Cambridge University Press, 1966. 518p. $14.50.
Contents: The transitional years, 1714-1730. A break-through, 1730-1740. The mid-century, 1740-1760. The new reign, 1760-1780. The great instrument-makers. The surveyors and their contemporaries, 1780-1800. Early years of the new century, 1800-1820. The practitioners disappear, 1820-1840.

147. Truesdell, C. **Essays in the History of Mechanics.** New York: Springer-Verlag, 1968. 384p. $19.50.
Contents: The mechanics of Leonardo da Vinci. Newton's mechanics. Kinetic theories of gases. Rational mechanics from Helmholtz to Hincin. M.R.39 No. 3946.

148. Venn, J. **The Logic of Chance.** 4th ed. Bronx, N.Y.: Chelsea, 1962. 508p. $6.50; $2.95pa.
Contents: Foundations of the science of probability. Applications.

149. Von Neumann, John. **The Collected Works of John von Neumann.**
Edited by A. H. Taub. Elmsford, N.Y.: Pergamon, 1963. 6v.
Volume 1: **Logic, Theory of Sets, and Quantum Mechanics.** 664p.
Volume 2: **Operators, Ergodic Theory, and Almost Periodic Functions in a Group.** 578p.
Volume 3: **Rings of Operators.** 564p.
Volume 4: **Continuous Geometry and Other Topics.** 526p.
Volume 5: **Design of Computers, Theory of Automata and Numerical Analysis.** 784p.
Volume 6: **Theory of Games, Astrophysics, Hydrodynamics and Meteorology.** 548p.

150. Whitehead, J. H. C. **The Mathematical Works of J. H. C. Whitehead.**
Edited by I. M. James. Elmsford, N.Y.: Pergamon, 1962. 4v.
Volume 1: **Introduction: Differential Geometry.** 394p.
Volume 2: **Complexes and Manifolds.** 448p.
Volume 3: **Homotopy Theory.** 464p.
Volume 4: **Algebraic and Classical Topology.** 360p.

151. Whiteside, D. T., editor. **The Mathematical Works of Isaac Newton.**
New York: Johnson Reprint Corp., 1967. 2v. $17.50 (Vol. 1).
Contents: Newton's lectures on algebra and analytic geometry, cubics, and finite differences. Vol. 1: M.R. 32 No. 5479.

152. Wilder, R. L. **Evolution of Mathematical Concepts: An Elementary Study.**
New York: John Wiley, 1968. 224p. $8.00.
Contents: Early evolution of numbers. Evolution of geometry. The real numbers. Conquest of the infinite. The processes of evolution. Evolutionary aspects of modern mathematics.

CHAPTER 5

SET THEORY, LOGIC AND FOUNDATIONS

153. Abian, Alexander. **The Theory of Sets and Transfinite Arithmetic.**
Philadelphia: W. B. Saunders, 1965. 405p.
Contents: Sets. Truth tables. Quantifiers. Axioms of set theory. Axiom of
Choice. Algebra of sets. Ordered sets. Zorn's lemma. The real numbers.
Equipollence of sets. Similarity. Ordinal and cardinal numbers and their
arithmetic.

154. Bourbaki, Nicolas. **Elements of Mathematics: Theory of Sets.** Trans.
from the French. Reading, Mass.: Addison-Wesley, 1968. 414p.
Contents: Description of formal mathematics. Theory of sets. Ordered sets.
Cardinals. Integers. Structures. Summary of results.

155. Christian, Robert R. **Introduction to Logic and Sets.** 2nd ed. New
York: Blaisdell, 1965. 116p.
Contents: The concept of equality. Propositions and truth values. Truth
tables. Equivalence of propositions. Algebra of logic. Sets. De Morgan's rules.
Quantifiers. Implications. Contains problems with answers and solutions.

156. Cohen, Paul J. **Set Theory and the Continuum Hypothesis.** New
York: W. A. Benjamin, 1966. 154p. $8.00; $3.95pa.
Contents: A survey of logic. The Zermelo-Fraenkel set theory. Godel's
concept of constructibility. The axiom of constructibility. The generalized
continuum hypothesis and the axiom of choice.

157. Curry, Haskell B. **Foundations of Mathematical Logic.** New York:
McGraw-Hill, 1963. 420p. $13.00.
Contents: Introduction. Formal systems. Epitheory. Relational logical algebra.
The theory of implication. Negation. Quantification. Modality.

158. Delong, Howard. **A Profile of Mathematical Logic.** Reading, Mass.:
Addison-Wesley, 1970. 304p. $10.95.
Contents: Historic background. Propositional calculus. The predicate calculus.
Set theory. Post's completeness proof. Godel's completeness proof. Godel's
incompleteness proof. M.R.41 No. 3230.

159. Eilenberg, Samuel, and C. C. Elgot. **Recursiveness.** New York: Academic Press, 1970. 89p. $6.50.
Contents: Preface. Preliminaries. Primitive recursive functions. Recursive
isomorphisms. Recursive relations. Recursive functions. Recursively enumerable
sets.

160. Fitting, Melvin C. **Intuitionistic Logic, Model Theory and Forcing Studies in Logic and the Foundations of Mathematics.** Amsterdam: North-Holland Pub. Co., 1969. 191p. $9.75.
Contents: Elementary semantics. Proof theory of first order intuitionistic logic. Intuitionistic models of Zermelo-Fraenkel set theory. M.R.41 No. 6666.

161. Fraenkel, Abraham A. **Set Theory and Logic.** Reading, Mass.: Addison-Wesley, 1966. 102p.
Contents: Equivalence and cardinal numbers. Countable and uncountable sets. Transfinite cardinals. The Axiom of Choice. Operations with cardinals ordering. Well ordering. No previous mathematical knowledge is needed.

162. Ginzburg, Abraham. **Algebraic Theory of Automata.** New York: Academic Press, 1968. 165p. $9.00.
Contents: Sets. Relations. Functions. Groups. Semigroups. Finite automata. Semi-automata. Congruences. Recognizer. Initial and final states. The minimality theorem. Regular expressions. Covering. Decomposition theory and the Krohn-Rhodes decomposition. M.R.39 No. 4009.

163. Halmos, Paul R. **Naive Set Theory.** New York: Van Nostrand-Reinhold, 1960. 104p. $3.95.
Contents: The Axiom of Extension. The Axiom of Specification. Unordered pairs. Unions and intersections. Complements and powers. Ordered pairs. Relations. Functions. Families. Inverses and composites. Numbers. The Peano axioms. Arithmetic. Order. The Axiom of Choice. Zorn's lemma. Well ordering. Transfinite recursion. Ordinal numbers. Sets of ordinal numbers. Ordinal arithmetic. The Schröder-Bernstein theorem. Countable sets. Cardinal arithmetic. Cardinal numbers.

164. Hao, Wang. **A Survey of Mathematical Logic.** Peking: Science Press, 1964. 651p.
Contents: The axiomatic method. Calculating machines. Formal number theory. Predicative and impredicative theory.

165. Hatcher, Williams. **Foundations of Mathematics.** Philadelphia: W. B. Saunders, 1968. 327p. $12.75.
Contents: Historical account. The first-order predicate logic. Frege's set theory. Type theory. Zermelo-Fraenkel set theory. Hilbert's program. Godel's incompleteness theorems. Quine's set theory and category theory. M.R.38 No. 5610.

166. Hayden, Seymour, and J. F. Kennison. **Zermelo-Fraenkel Set Theory.** Columbus, Ohio: Charles E. Merrill, 1968. 164p.
Contents: Zermelo-Fraenkel assumptions for the construction of sets. Relations and functions. Linear and well ordering groups. Subgroups. Semi groups. Rings. Integral domains. Fields. Ordinal and cardinal numbers. The Axiom of Choice. The Axiom of Infinity. Peano's axioms. The integers. The real and complex numbers. Transfinite arithmetic. M.R.38 No. 2025.

167. Kenelly, John W. **Informal Logic.** Boston: Allyn & Bacon, 1967. 134p.
Contents: Statement variables and operators. Formulas and truth tables.
Algebra of statements. Applications. The function calculus of one variable. The
function calculus of many variables.

168. Kleene, Stephen C. **Mathematical Logic.** New York: John Wiley, 1967.
398p. $12.95.
Contents: Elementary mathematical logic. The propositional calculus. The
predicate calculus. The predicate calculus with equality. Mathematical logic and
the foundations of mathematics. The foundations of mathematics. Computability
and decidability.

169. Kleene, Stephen C., and R. E. Vesley. **The Foundations of Intuitionistic
Mathematics, Especially in Relation to Recursive Functions.** Amsterdam: North-
Holland Pub. Co., 1965. 206p.
Contents: Notions of realizability and the intuitionistic continuum.

170. Kneebone, G. T. **Mathematical Logic and the Foundations of Mathematics: An Introductory Survey.** Princeton, N.J.: D. Van Nostrand, 1963. 435p.
Contents: Traditional logic. Symbolic logic. Foundations of mathematics.
Godel's theorems. Intuitionism. Recursive arithmetic. Set theory. The philosophy
of mathematics.

171. Kopperman, Ralph D. **Model Theory and Its Applications.** Boston: Allyn
& Bacon, 1971. 256p.
Contents: Language and truth. Lowenheim-Skolem theorem. Ultraproducts.
Relationships with logic. Application to algebra and analysis.

172. Korfhage, Robert R. **Logic and Algorithms, With Applications to the
Computer and Information Sciences.** New York: John Wiley, 1966. 194p.
$7.95.
Contents: Sets. Relations and mappings. Boolean algebras. The propositional
calculus. A view of binary vectors. Algorithms and computing machines. The
first-order predicate calculus. Formal languages. A brief history.

173. Kreisel, G., and J. J. Krivine. **Elements of Mathematical Logic (Model
Theory).** Amsterdam: North-Holland Pub. Co., 1967. 220p.
Contents: Predicate calculus. Quantifiers. Definability. The axiomatic method.
Foundations of mathematics. M.R.36 No. 2463.

174. Kruse, Arthur H. **Localization and Iteration of Axiomatic Set Theory.**
Detroit, Mich.: Wayne State University Press, 1969. 226p.
Contents: Predicates. Semiformalization. Formalizations. Ordering relations.
Lower and upper closures. Vertical closures. M.R.39 No. 5351.

175. Leblanc, Hughes. **Techniques of Deductive Inference.** Englewood Cliffs,
N.J.: Prentice-Hall, 1966. 216p.
Contents: The logic of connectives. The sentential calculus. Rules of inference.
Validity. Satisfiability. Implication. The logic of quantifiers.

176. Leisenring, A. C. **Mathematical Logic and Hilbert's ϵ-Symbol.** New York: Gordon & Breach, 1970. 142p.
Contents: Syntax and semantics. Formal systems. The ϵ-theorems. Formal theories. The cut elimination theorem.

177. Lemmon, E. J. **Introduction to Axiomatic Set Theory.** New York: Dover, 1969. 131p.
Contents: The general theory of classes. Sets. Relations. Functions. Equipollence. Ordinals. Cardinals. The Axiom of Choice.

178. Lightstone, A. H. **The Axiomatic Method: An Introduction to Mathematical Logic.** Englewood Cliffs, N.J.: Prentice-Hall, 1964. 246p.
Contents: Symbolic logic. Set theory. The axiomatic method and abstract algebra. Mathematical logic—the propositional calculus. The predicate calculus. The completeness of the predicate calculus.

179. Lightstone, A. H. **Symbolic Logic and the Real Number System: An Introduction to the Foundations of Number Systems.** New York: Harper & Row, 1965. 225p. $8.95.
Contents: Symbolic logic. The logical connectives. Truth tables. Tautologies and valid arguments. Quantifiers. Sets. Russell's paradox. Groups and semigroups. Rings. Integral domains. Fields, ordered fields. The natural numbers. The integers. The rational numbers. The real number system. The complex number system.

180. Lukasiewicz, J. **Elements of Mathematical Logic.** 2nd ed. Elmsford, N.Y.: Pergamon, 1964. 134p.
Contents: Rules of inference. The sentential calculus. The consistency and the independence of sentential calculus. Many-valued logic. Quantifiers. Aristotle's syllogistic.

181. Margaris, Angelo. **First Order Mathematical Logic.** Waltham, Mass.: Blaisdell, 1967. 211p. $6.75.
Contents: Set theory. Axiomatic theories. The rules of inference. Predicates. Quantifiers. Connectives. Predicate calculus. The completeness theorem. First-order theories. Model theory. Godel's theorem.

182. Mendelson, Elliott. **Introduction to Mathematical Logic.** New York: Van Nostrand-Reinhold, 1964. 300p. $7.95.
Contents: The propositional calculus. Quantification theory. Formal number theory. Axiomatic set theory. Effective computability. Appendix: A consistency proof for formal number theory.

183. Monk, J. Donald. **Introduction to Set Theory.** New York: McGraw-Hill, 1969. 193p. $10.95.
Contents: The axioms. Boolean algebra of classes. Algebra of relations. Functions. Infinite Boolean operations. Direct products, power classes. Equivalence relations. Ordering. Ordinals: basic definitions. Transfinite induction. The

natural numbers. Sequences and normal functions. Recursion. Ordinal arithmetic. Special topics. The Axiom of Choice. Equivalents of the Axiom of Choice. Application of the Axiom of Choice. Cardinals: basic definitions. Finite and infinite sets. Cardinal addition. Cardinal multiplication: cardinal exponentiation. Regular and singular cardinals. RECOMMENDED.

184. Morse, Anthony P. **A Theory of Sets**. New York: Academic Press, 1965. 130p. $7.95.
Contents: Language and inference. Logic. Set theory. Some aspects of equality. Classification. The role of replacement. Singletons. Ordered pairs. Substitution. Unicity. Relations. Functions. Ordinals. Definition by induction. Choice. Maximality. Well ordering. Natural numbers. Sequences. Reiteration. Set functions and fixed points. Equinumerosity. Cardinals. Cardinality. Cardinal arithmetic. Direct extensions. Families of sets. Tuples. The construction of definitions.

185. Mostowski, A. **Constructible Sets with Applications: Studies in Logic and the Foundations of Mathematics**. Amsterdam: North-Holland Pub. Co., 1969. 269p. $7.70.
Contents: The class of sets constructible from a sequence, consistency of the Axiom of Choice. M.R.41 No. 52.

186. Nidditch, P. H. **Elementary Logic of Science and Mathematics**. Glencoe, Ill.: The Free Press of Glencoe, 1960. 371p.
Contents: Propositional calculus. Functional calculus. Boolean algebra. Probability. Experimentation. Deduction and hypothesis in the inductive sciences.

187. Novikov, P. S. **Elements of Mathematical Logic**. Trans. from Russian by Leo Boron. Reading, Mass.: Addison-Wesley, 1964. 296p.
Contents: Propositional algebra. Propositional calculus. Predicate logic. Predicate calculus. Proof theory.

188. Peter, R. **Recursive Functions**. 3rd rev. ed. New York: Academic Press, 1967. 300p. $13.50.
Contents: Recursive functions and relations. The simultaneous recursion. Recursion with respect to several variables. Reductions. Nested recursion. Multiple recursions. Transfinite recursions. The normal form for multiple recursions. The explicit form for general recursive functions. Calculable functions. Extension of the concept of recursivity. M.R.36 No. 2496.

189. Quine, Willard Van Orman. **Set Theory and Its Logic**. Rev. ed. Cambridge, Mass.: Belknap Press of Harvard University, 1969. 361p.
Contents: Logic. Real classes. Classes of classes. Natural numbers. Iteration and arithmetic. Ordinal and cardinal numbers. Transfinite recursion. The Axiom of Choice. Russell's theory of types. Axioms of infinity. The Von Neumann-Bernays system.

190. Rieger, Ladislav. **Algebraic Methods of Mathematical Logic.** Trans. from Czechoslovakian by Michael Basch. New York: Academic Press, 1967. 210p. $10.00.
Contents: The language of mathematics and its symbolization. Recursive construction of the relation of consequence. Expressive possibilities of the present symbolization. Intuitive and mathematical notions of an idealized axiomatic mathematical theory. The algebraic theory of elementary predicate logic. Foundations of the algebraic theory of logical syntax. Algebraic laws of semantics of first-order predicate logic.

191. Robbin, Joel W. **Mathematical Logic: A First Course.** New Amsterdam: W. A. Benjamin, 1969. 212p. $12.50.
Contents: First-order logic. Recursive arithmetic. Incompleteness theorems. Second-order logic. M.R.40 No. 4078.

192. Robison, Gerson B. **An Introduction to Mathematical Logic.** Englewood Cliffs, N.J.: Prentice-Hall, 1969. 212p.
Contents: Mathematics and reality. Mathematical systems. Truth tables: the sentential calculus. The deeper structure of statements. The demonstration. Two major theorems: the lemma theorem, the general deduction theorem. Supplementary inference rules. Universal theorems. Terms and definitions. The Boolean approach.

193. Rogers, Hartley, Jr. **Theory of Recursive Functions and Effective Computability.** New York: McGraw-Hill, 1967. 528p. $14.75.
Contents: Recursive functions. Unsolvable problems. Purposes. Recursive invariance. Recursive and recursively enumerable sets. Reducibilities. One-one reducibility. Many-one reducibility. Creative sets. Truth-table reducibilities. Simple sets. Turing reducibility. Hyper-simple sets. Post's problem. Incomplete sets. The recursion theorem. Recursively enumerable sets as lattice. Degrees of unsolvability. M.R.37 No. 61.

194. Rosser, J. Barkley. **Simplified Independence Proofs: Boolean Valued Models of Set Theory.** New York: Academic Press, 1969. 217p. $10.00.
Contents: Boolean algebra. The basic model. The independence of V = L. Analogies with forcing. The independence of A x C. The independence of the continuum hypothesis. The generalized GCH—the bounded case. The generalized GCH—the unbounded case. M.R.40 No. 2536.

195. Rotman, B., and G. T. Kneebone. **The Theory of Sets and Transfinite Numbers.** London: Old Bourne, 1966. 144p.
Contents: Sets. Operations on sets. Relations and functions. Ordering relations. Surjections. Injections. Bijections. Isomorphisms. Theory of ordinal and cardinal numbers. The Axiom of Choice. Zorn's lemma.

196. Rubin, Jean E. **Set Theory for the Mathematician.** San Francisco: Holden-Day, 1967. 398p.
Contents: Class algebra. Functions and relations. Natural numbers. Finite and

infinite classes. The rational and real numbers. Ordering relations. Ordinal numbers. Ordinal number theory. Propositions equivalent to the Axiom of Choice. Cardinal numbers. Cardinal numbers and the Axiom of Choice. The generalized continuum hypothesis. Additional axioms.

197. Sheng, C. L. **Threshold Logic**. New York: Academic Press, 1969. 206p. $12.00.
Contents: Switching or Boolean functions. Definition of threshold function and basic concepts of linear separation. Characterization of threshold functions. Unateness, monotonicity, and assumability. Admissible transformations, partial ordering, and weights. Testing and realization of threshold functions. The successive higher ordering method for testing and realization. Compound synthesis. Sensitivity of weights and threshold. Applications of threshold logic element. Threshold logic elements for probability transformation.

198. Shoenfield, Joseph R. **Mathematical Logic**. Reading, Mass.: Addison-Wesley, 1967. 344p. $12.75.
Contents: First-order model theory. Recursion theory. Set theory. Higman's theorem. Proof theory. M.R.37 No. 1224.

199. Slisenko, A. O., editor. **Studies in Constructive Mathematics and Mathematical Logic, Part I**. Trans. from Russian. New York: Plenum, 1969. 88p. $12.50.
Contents: Method of establishing deducibility in classical predicate calculus, G. V. Davydov. On the correction of unprovable formulas, G. V. Davydov. Lebesgue integral in constructive analysis, O. Demuth. Sufficient conditions of incompleteness for the formalization of parts of arithmetic, N. K. Kossovsky. Normal forms for deduction in predicate calculus with equality and functional symbols, V. A. Lifshits. Some reduction classes and undecidable theories, V. A. Lifshits. Deductive validity and reduction classes, V. A. Lifshits. Decision problem for some constructive theory of equalities, V. A. Lifshits. On constructive groups, V. A. Lifshits. Invertible sequential variant of constructive predicate calculus, S. Yu. Maslov. Choice of terms in quantifier rules of constructive predicate calculus, G. E. Mints. Analog of Hebrand's theorem for prenex formulas of constructive predicate calculus, G. E. Mints. Imbedding operations associated with Kripke's "semantics," G. E. Mints. On imbedding operations, G. E. Mints and V. P. Orevkov. Unsolvability, in modal predicate calculus, of a class of formulas containing only one monadic predicate variable, V. P. Orevkov. Sequential variant of constructive logic calculus for normal formulas of constructive predicate calculus, R. A. Plyushkevichus. On sequential variants of applied predicate calculus, M. G. Rogova. On maximal regulators of continuity for constructive functions, A. O. Slisenko. On representability of algorithmic decidable predicates by Rabin machines, R. I. Freidzon.

200. Slupecki, J., and L. Borkowski. **Elements of Mathematical Logic and Set Theory**. Elmsford, N.Y.: Pergamon, 1967. 362p.
Contents: The sentential calculus. The functional calculus. General set theory. Ordered sets. Semantic categories. The theory of logical types. Certain set-

theoretical concepts as defined in logic. Axiomatic set theory. Set theory and arithmetic. The philosophical aspects of the concept of a set.

201. Snyder, D. Paul. **Modal Logic and Its Applications.** New York: Van Nostrand-Reinhold, 1971. 335p. $12.50.
Contents: The scope of modal logic. The basic proof techniques. The monadic modalities. Quantified monadic modal logic. Species and genera of monadic modality. The dyadic modalities. Modality.

202. Stoll, Robert R. **Set Theory and Logic.** San Francisco: W. H. Freeman, 1963. 474p. $9.50.
Contents: Sets and relations. The natural number sequence and its generalizations. The extension of the natural numbers to the real numbers. Logic. Informal axiomatic mathematics. Boolean algebras. Informal axiomatic set theory. Several algebraic theories. First-order theories.

203. Suppes, Patrick. **Axiomatic Set Theory.** New York: Van Nostrand-Reinhold, 1960. 265p. $7.75.
Contents: General developments. Relations and functions. Equipollence, finite sets, and cardinal numbers. Finite ordinals and denumerable sets. Rational numbers and real numbers. Transfinite induction and ordinal arithmetic. The Axiom of Choice. RECOMMENDED.

204. Van Heijenoort, Jean, editor. **Frege and Gödel: Two Fundamental Texts in Mathematical Logic.** Cambridge, Mass.: Harvard University Press, 1970. 117p. $6.00.
Contents: G. Frege, Begriffsschrift, a formula language, modeled upon that of arithmetic, for pure thought. K. Gödel, Some mathematical results on completeness and consistency. K. Gödel, On formally undecidable propositions of *Principia Mathematica* and related systems. K. Gödel, On completeness and consistency.
Originally appeared as **From Frege to Gödel Source Book.**

205. Wilder, R. L. **Introduction to the Foundations of Mathematics.** 2nd ed. New York: John Wiley, 1965. 327p. $10.95.
Contents: The axiomatic method. Analysis of the axiomatic method. Theory of sets. Infinite sets. Well-ordered sets. Ordinal numbers. The linear continuum and the real number system. Groups and their significance for the foundations. The early developments. The Frege-Russe thesis: mathematics an extension of logic. Intuitionism. Formal systems. Mathematical logic. The cultural setting of mathematics. RECOMMENDED.

206. Zehna, Peter W., and Robert L. Johnson. **Elements of Set Theory.** 2nd ed. Boston: Allyn & Bacon, 1972. 194p. $8.95.
Contents: Elementary logic. Sets and their properties. Sets and functions. Finite and infinite sets. Cardinal numbers.

CHAPTER 6

COMBINATORIAL AND GRAPH THEORY

207. Beckenbach, E. F., editor. **Applied Combinatorial Mathematics.**
New York: John Wiley, 1964. 608p. $16.95.
Contents: The machine tools of combinatorics, Derrick H. Lehmer. Techniques
for simplifying logical networks, Montgomery Phister, Jr. Generating functions,
John Riordan. Lattice statistics, Elliott W. Montroll. Polya's theory of counting, Nicholas G. De Bruijn. Combinatorial problems in graphical enumeration, Frank Harary. Dynamic programming and Markovian decision processes,
with particular application to baseball and chess, Richard Bellman. Graph
theory and automatic control, Robert Kalaba. Optimum multivariable control, Edwin L. Peterson. Stopping-rule problems, Leo Breiman. Combinatorial
algebra of matrix games and linear programs, Albert W. Tucker. Network flow
problems, Edwin F. Beckenbach. Block design, Marshall Hall, Jr. Introduction to information theory, Jacob Wolfowitz. Sperner's lemma and some extensions, Charles B. Tompkins. Crystallography, Kenneth N. Trueblood.
Combinatorial principles in genetics, George Gamow. Appendices, Herman
Weyl.

208. Berge, C. **Principles of Combinatorics.** New York: Academic Press,
1971. 186p. $10.00.
Contents: The elementary counting functions. Partition problems. Inversion
formulas and their applications. Permutation groups. Polya's theorem.

209. Busacker, R. G., and T. L. Saaty. **Finite Graphs and Networks: An
Introduction with Applications.** New York: McGraw-Hill, 1965. 294p. $11.50.
Contents: Directed and undirected graphs. Partitions and distances in graphs.
Planar and non-planar graphs. Coloring theorems. Matrix representations. Net-
work flows. Applications to economics, operations research, games, engineer-
ing, and others. M.R.35 No. 79.

210. Eisen, Martin. **Elementary Combinatorial Analysis.** New York:
Gordon & Breach, 1969. 246p.
Contents: Permutations and combinations. The multinomial theorem. Generating functions. The principle of inclusion and exclusion. Application of
combinatorial analysis to probability theory. Möbius function and Polya's
theorem.

211. Hall, Marshall, Jr. **Combinatorial Theory.** Waltham, Mass.: Blaisdell,
1967. 310p. $9.50.
Contents: Permutations. Combinations. The principle of inclusion and exclusion. Möbius inversion. Generating functions. Recursions. Partitions. The
theorems of P. Hall, D. Konig, Ramsey and Dilworth. Convex spaces. Linear
programming. Graphs. Block designs. Difference sets. Singer's theorem. The

multiplier theorem. Finite geometries. Orthogonal Latin squares. Orthogonal arrays. The Euler conjecture. Hadamard's matrices. The Hanani theorems. Copositive and completely positive quadratic forms. Incidence matrices. Incidence equations. M.R.37 No. 80.

212. Harary, Frank. **Graph Theory**. Reading, Mass.: Addison-Wesley, 1969. 274p. $12.50.
Contents: History of graph theory. Ulam's conjecture. Blocks and cutpoints. Cycles and cocycles. Matroids. Vertex and edge connectivity. Menger's theorem and Dirac's generalization. Eulerian and Hamiltonian graphs. Line graphs. 1- and 2- factors. Covering sets. Independent sets. Matchings in a graph. The 4-color problem. Automorphism group of a graph. Directed graphs. M.R.41 No. 1566.

213. Harary, Frank, editor. **Graph Theory and Theoretical Physics**. New York: Academic Press, 1967. 358p. $14.50.
Contents: Graphical enumeration problems, F. Harary. Graph theory and crystal physics, P. W. Kasteleyn. Graph theory applied to electrical networks, P. R. Bryant. The decomposition of complete graphs into planar subgraphs, L. W. Beineke. Some classes of perfect graphs, C. Berge. Graphs and matrices, A. L. Dulmage and N. S. Mendelsohn. Estimation methods for Mayer's graphical expansions, J. Groeneveld. Enumerating labelled trees, J. W. Moon. Some applications of a theorem of de Bruijn, R. C. Read. Generating functionals and graphs, G. Stell. Topics in graph theory, T. Tutte. The headwood gay colouring conjecture, J. W. T. Youngs.

214. Harary, Frank, editor. **Proof Techniques in Graph Theory**. New York: Academic Press, 1969. 330p. $14.50.
Contents: The four color conjecture and other graphical diseases, Frank Harary. Several proofs of the number of labeled 2-dimensional trees, Lowell W. Beincke and John W. Moon. On the chromatic number of permutation graphs, Gary Chartrand and Joseph B. Frechen. The expanding unicurse, Blanche Descartes. Problems and results in graph chromatic theory, P. Erdös. Forbidden subgraphs, Dennis Geller. A proof technique in graph theory, Dennis Geller and Stephen Hedetniemi. Independence and covering numbers of line graphs and total graphs, R. P. Gupta. The decline and fall of Zarankiewicz's theorem, Richard K. Guy. On the intersection number of a graph, Frank Harary. On endomorphisms of graphs and their homomorphic images, Z. Hedrlin. Counterexamples in the theory of well-quasi-ordered sets, T. A. Jenkyns and C. St.J. A. Nash-Williams. On the existence of certain minimal regular n-systems with given girth, W. Kuich and N. Sauer. Reconstruction of unicyclic graphs, Bennet Manvel. The group of a graph whose adjacency matrix has all distinct eigenvalues, Abbe Mowshowitz. Extremal nonseparable graphs of diameter 2, U. S. R. Murty. A class of strongly regular graphs, E. A. Nordhaus. The exponentiation group as the automorphism group of a graph, Edgar M. Palmer. Remarks on the Heawood conjecture, Gerhard Ringel and J. W. T. Youngs. Indifference graphs, Fred S. Roberts. Enumeration of Euler graphs, Robert W. Robinson. A graph-theoretical model for

periodic discrete structures, James Turner. Even and odd 4-colorings, W. T. Tutte. A theorem on Tait colorings with an application to the generalized Petersen graphs, Mark E. Watkins. The Möbius function in combinatorial analysis and chromatic graph theory, Herbert S. Wilf. Key-word indexed bibliography of graph theory, James Turner. M.R.40 No. 4162.

215. Harris, Bernard, editor. **Graph Theory and Its Applications.** New York: Academic Press, 1970. 262p. $5.00.
Contents: Graph theory as a structural model in the social sciences, Frank Harary. The mystery of the Heawood conjecture, J. W. T. Youngs. Graph theory algorithms, Ronald C. Read. On eigenvalues and colorings of graphs, Alan J. Hoffman. Blocking polyhedra, D. R. Fulkerson. Connectivity in matroids, W. T. Tutte. The use of circuit codes in analog-to-digital conversion, Victor Klee. Some mapping problems for tournaments, J. W. Moon. On some connections between graph theory and experimental designs and some recent existence results, D. K. Ray Chaudhuri. On the foundations of combinational theory, Gian-Carlo Rota and Ronald Mullin. A composition theorem for the enumeration of certain subsets of the symmetric semigroup, Bernard Harris and Lowell Schoenfeld. Symbols for the Harris-Schoenfeld paper.

216. Liu, C. L. **Introduction to Combinatorial Mathematics.** New York: McGraw-Hill, 1968. 393p. $13.50.
Contents: Permutations and combinations. Generating functions. Recurrence. Inclusion and exclusion principles. Polya's theory. Trees. Circuits. Planar graphs. Dual graphs. Domination. Chromatic numbers. Transport networks. Linear and dynamic programming. Block designs.

217. Mirsky, L. **Transversal Theory.** New York: Academic Press, 1971. 255p. Contents: Sets. Functions. Zorn's lemma. Topological spaces. Graphs. Duality. Hall's theorem. Rado's selection principle. Independent transversals. Combinatorial properties of matrices. The Nash-William formula.

218. Moon, John W. **Topics on Tournaments.** New York: Holt, Rinehart, and Winston, 1968. 104p. $5.95.
Contents: Mainly elementary results on tournaments. M.R.41 No. 1574.

219. Ore, Oystein. **The Four-Color Problem.** New York: Academic Press, 1967. 259p. $12.00.
Contents: Planar graphs. Bridges and circuits. Dual graphs. Euler's formula and its consequences. Large circuits. Colorations. Color functions. Formulations of the four-color problem. Cubic graphs. Hadwiger's conjecture. Critical graphs. Planar 5-chromatic graphs. Three colors. Edge coloration.

220. Pereus, J. K. **Combinatorial Methods.** New York: University of New York Press, 1969. 212p. $4.75.
Contents: Counting and enumeration on a set. Set generating functions. Numerical generating functions. Counting with restrictions. Partitions. Compositions. Decompositions. Distribution of labelled and unlabelled objects.

The Polya theorem. Counting and enumeration on a regular lattice. Random walk on lattices. One dimensional lattices. Two dimensional lattices. Counting patterns on two dimensional lattices. The Ising model.

221. Riordan, John. **Combinatorial Identities.** New York: John Wiley, 1968. 256p. $15.00.
Contents: Recurrence. Inverse relations. Abel's generalization of the generating functions. Partition polynomials. Differential and difference operators. M.R.38 No. 53.

222. Ryser, H. J. **Combinatorial Mathematics.** New York: John Wiley, 1963. 154p. $6.00.
Contents: Fundamentals of combinatorial mathematics. The principle of inclusion and exclusion. Recurrence relations. A theorem of Ramsey. Systems of district representatives. Matrices of zeros and ones. Orthogonal Latin squares. Combinatorial designs. Perfect difference sets.

223. Tutte, W. T., editor. **Recent Progress in Combinatorics: Proceedings of the Third Conference on Combinatorics held at the University of Waterloo, Waterloo, Ontario, Canada.** New York: Academic Press, 1969. 347p. $16.00.
Contents: *Instructional Courses and Related Papers.* A survey of coding theory, E. R. Berlekamp. The Greek alphabet of "graph theory," Frank Harary. Covering theorems in groups (or: How to win at football pools), R. G. Stanton. Some new results for the covering problem, J. G. Kalbfleisch and P. H. Weiland. *Invited Papers.* The rank of a family of sets and some applications to graph theory, Claude Berge. Enumeration of tree-shaped molecules, N. G. de Bruijn. A combinatorial approach to the bell polynomials and their generalizations, Roberto Frucht. The number of subspaces of a vector space, Jay Goldman and Gian-Carlo Rota. Graphs, complexes, and polytopes, Branko Grünbaum. On the structure of n-connected graphs, R. Halin. On unions and intersections of cones, A. J. Hoffman. Optimal distance configurations, Leroy M. Kelly. Combinatorial designs as models of universal algebras, N. S. Mendelsohn. Well-balanced orientations of finite graphs and unobstrusive odd-vertex pairings, C. St.J. A. Nash-Williams. The partition calculus, R. Rado. Teaching graph theory to a computer, Ronald C. Read. Finite graphs (investigations and generalizations concerning the construction of finite graphs having given chromatic numbers and no triangles), Horst Sachs. Strongly regular graphs, J. J. Seidel. Projective geometry and the 4-color problem, W. T. Tutte. *Contributed Papers.* Remark on a combinatorial theorem of Erdös and Rado, H. L. Abbott and B. Gardner. The four leading coefficients of the chromatic polynomials $Q_n(u)$ and $R_n(x)$ and the Birkhoff-Lewis conjecture, Ruth Bari. Simple proofs for the Ramanujan congruences $p(5m + 4) \equiv 0 \pmod 5$ and $p(7m + 5) \equiv 0 \pmod 7$, J. M. Gandhi. Connected extremal edge graphs having symmetric automorphism group, Allan Gewirtz and Louis V. Quintas. Bounds on the chromatic and achromatic numbers of complementary graphs, Ram Prakash Gupta. A problem of Zarankjewicz, Richard K. Guy and Stefan Znám. Infinite graphs and matroids, D. A. Higgs. A survey of thickness, Arthur M. Hobbs. Line-symmetric tournaments, Michel Jean. On the Ramsey Number N(4,4:3), J. G. Kalbfleisch. Sylvester matroids, U. S. R. Murty. Cyclic coloration of plane graphs, Oystein Ore and Michael D.

Plummer. Pseudosymmetry, circuit-symmetry, and path-symmetry of a digraph, J. M. S. Simoes Pereira. On the boxicity and cubicity of a graph, Fred S. Roberts. On the number of mutually disjoint triples in Steiner systems and related maximal packing and minimal covering systems, J. Schonheim. On a problem of Erdös, David P. Sumner. Some classes of hypoconnected vertex-transitive graphs, Mark E. Watkins. Maximum families of disjoint directed cut sets, D. H. Younger. *Abstracts. Unsolved Problems.* M.R.40 No. 4128.

224. Vajda, S. **The Mathematics of Experimental Design: Incomplete Block Designs and Latin Squares.** New York: Hafner, 1967. 110p. $4.75.
Contents: Finite groups. Galois fields. Finite projective geometries. Euclidean spaces. Difference sets. Balanced incomplete block designs. Latin squares. Orthogonal arrays. Partially balanced incomplete block designs. M.R.37 No. 3928.

225. Vajda, S. **Patterns and Configurations in Finite Spaces.** New York: Hafner, 1967. 120p. $4.95.
Contents: Finite fields. Finite planes. Finite spaces. Configurations.

226. Vilenkin, N. Ya. **Combinatorics.** Trans. from Russian by A. Shenitzer and Sara Shenitzer. New York: Academic Press, 1971. 310p. $12.00.
Contents: General rules of combinatorics. Samples, permutations, and combinations. Combinatorial problems with restrictions. Distributions and partitions. Combinatorics on a chessboard. Recurrence relations. Combinatorics and series. Problems in combinatorics.

CHAPTER 7

THEORY OF NUMBERS

227. Archibald, R. G. **An Introduction to the Theory of Numbers.** Columbus, Ohio: C. E. Merrill, 1970. 305p.
Contents: Introduction. Divisibility. Congruences. Some significant functions in the theory of numbers. Primitive roots and indices. Quadratic congruences. Elementary considerations on the distribution of primes and composites. Continued fractions. Certain Diophantine equations and sums of squares.

228. Artin, E., and J. Tate. **Class Field Theory**. New York: W. A. Benjamin, 1968. 259p. $9.50; $3.95pa.
Contents: The first fundamental inequality. Second fundamental inequality. Reciprocity law. The existence theorem. Connected component of ideal classes. Grundwald's theorem. Higher ramification theory. Explicit reciprocity laws. Group extensions. Abstract class field theory. Weil groups.

229. Atkin, A. O. L., and B. J. Birch, editors. **Computers in Number Theory.** New York: Academic Press, 1971. 434p. $23.00.
Contents: The economics of number theoretic computation. Linear relations connecting the imaginary parts of the zeros of the zeta function. An explanation of some exotic continued fractions found by Brillhart. Some numerical computations relating to automorphic functions. Automorphic integrals with pre-assigned period polynomials and the Eichler cohomology. The number of conjugacy classes of certain finite matrix groups. Hilbert's theorem 94. Reducibility of polynomials. Sums of squares in the function field $R(x.y)$. The location of four squares in an arithmetic progression, with some applications. Diophantine equations involving generalized triangular and tetrahedral numbers. On the representation of an integer as the sum of four integer cubes. Arithmetic properties of linear recurrences. Some relationships satisfied by additive and multiplicative recurrent congruential sequences, with implications for pseudo-random number generation. Investigation of T-numbers and E-sequences. Use of computers in cyclotomy. Calculation of the first factor of the cyclotomic class number. A numerical study of units in composite real quartic and octic fields. Class numbers and units of complex quartic fields. The Diophantine equation $x^3 - y^2 = k$. A non-trivial solution of the Diophantine equation $-9(x^2 + 7y^2)^2 - 7(v^2 + 7v^2)^2 = 2$. On the Fermat quotient. The inhomogeneous minima of some totally real cubic fields. The products of three and of four linear forms. Computations relating to cubic fields. The enumeration of perfect forms. Two problems in Diophantine approximation. The improbable behaviour of Ulam's summation sequence. Non-repetitive sequences. On sorting by comparisons. Spectra of determinant values in (0,1) matrices. On the non-existence of certain perfect codes. A class of theorems in additive number theory which lend themselves to computer proof. Differences bases: three problems in additive number theory. A natural generalisation of Steiner triple systems. Doubly periodic

arrays. Counting polyominoes. Combinatorial analysis with values in a semi-group. The use of computers in search of identities of the Rogers-Ramanujan type. Computers in the theory of partitions. Binary partitions. Multiplanar partitions. Some problems in number theory. Some unsolved problems. Languages.

230. Bachman, George. **Introduction to p-Adic Numbers and Valuation Theory**. New York: Academic Press, 1964. 173p. $6.50.
Contents: Valuations of rank one. Complete fields and the field of p-Adic numbers. Valuation rings, places, and valuations. Normed linear spaces. Extensions of valuations.

231. Borevich, Z. I., and I. R. Shafarevich. **Number Theory**. Trans. from the Russian. New York: Academic Press, 1966. 435p. $15.50.
Contents: Congruences. Representation of numbers by decomposable forms. The theory of divisibility. Local methods. Analytic methods. Algebraic supplement. Tables. RECOMMENDED.

232. Cohn, Paul M. **Lectures on Algebraic Numbers and Algebraic Functions**. Kingston, Ontario: Queen's University, 1969. 174p. $3.00.
Contents: Absolute value. Dedekind domains. The unit theorem. Function fields. Riemann-Roch theorem. Elliptic function fields. Abel's theorem. M.R.39 No. 5572.

233. Davenport, Harold. **Multiplicative Number Theory**. Chicago: Markham, 1967. 189p.
Contents: Dirichlet's theorem on primes. Dirichlet's class number formula. The distribution of zeros of L-functions. The prime number theorem for arithmetic progressions. Siegel's theorem. Linnik and Renyi large sieve method. RECOMMENDED. M.R.36 No. 117.

234. Dickson, L. E., H. H. Mitchell, H. S. Vandiver, and G. E. Wahlin. **Algebraic Numbers**. New York: Chelsea, 1967. 211p. $4.95.
Contents: Algebraic numbers. Cyclotomy. Hensel's p-adic numbers. Fields of functions.

235. Dudley, Underwood. **Elementary Number Theory**. San Francisco: W. H. Freeman, 1969. 262p. $8.50.
Contents: Integers. Unique factorization. Linear Diophantine equations. Congruences. Fermat's and Wilson's theorems. The divisors of an integer. Perfect numbers. Euler's theorem and function. Primitive roots and indices. Quadratic congruences. Quadratic reciprocity. Numbers in other bases. Duodecimals. Decimals. Pythagorean triangles. Infinite descent and Fermat's conjecture. Sums of two squares. Sums of four squares. $x^2 - Ny^2 = 1$. Formulas for primes. Bounds for $\Pi(x)$. Tables: Factor table for integers less than 10,000.

236. Dynkin, E. B., and V. A. Uspenskii. **Problems in the Theory of Numbers**. Lexington, Mass.: D. C. Heath, 1963. 117p.

Contents: Introduction to the theory of algebraic numbers. Relationship to Fibonacci sequences and Pascal's triangle.

237. Eames, W. P. **The Elementary Theory of Numbers, Polynomials and Rational Functions**. New York: American Elsevier, 1968. 143p. $6.50.
Contents: Real and complex numbers. The Peano axioms. Groups. Rings. Fields. Homomorphisms. Polynomials. Partial functions. Symmetric functions. M.R.37 No. 6137.

238. Eichler, Martin. **Introduction to the Theory of Algebraic Numbers and Functions**. Trans. from the German by George Striker. New York: Academic Press, 1966. 324p. $14.50.
Contents: Linear algebra. Appendix: The theta function. Ideals and divisors. Appendix: topics from the theory of algebraic number fields. Algebraic functions and differentials. Algebraic functions over the complex number field. Correspondences between fields of algebraic functions.

239. Eynden, Charles Vanden. **Number Theory: An Introduction to Proof**. Scranton, Penn.: Intext Educational Publishers, 1970. 158p. $7.50.
Contents: Divisibility. Numerical functions. The arithmetic of congruence classes. Solving congruences. The distribution of powers into congruence classes. The theory of primitive roots. Quadratic reciprocity.

240. Gelfond, A. O. **The Solution of Equations in Integers**. Trans. from the Russian and edited by J. B. Roberts. San Francisco: W. H. Freeman, 1961. 63p. $1.00pa.
Contents: Equations in one unknown. Equations of the first degree in two unknowns. Examples of equations of second degree in three unknowns. Finding all solutions of equations of the form $x^2 - Ay^2 = 1$. The general case of equations of second degree with two unknowns. Equations in two unknowns of degree higher than the second. Algebraic equations of degree higher than the second with three unknowns and an exponential equation.

241. Gelfond, A. O., and U. V. Linnik. **Elementary Methods in Analytic Number Theory**. Trans. from the Russian by Amiel Feinstein. Edited by L. J. Mordell. New York: McGraw-Hill, 1967. 242p. $9.50.
Contents: Additive properties of numbers. Solution of Waring's problem and the problem of Hilbert-Kamke. The problem of the distribution of prime numbers. Derivation of the distribution law of prime Gaussian numbers. The sieve of Eratosthenes. The method of Atle Selberg. On the distribution of the fractional parts of numerical sequences. Calculation of the number of integer points within contours. On the distribution of power residues. Elementary proof Hasse's theorem. Elementary proof of Siegel's theorem. The transcendence of certain number classes.

242. Gioia, Anthony A. **The Theory of Numbers: An Introduction**. Chicago: Markham, 1970. 185p. $8.50.
Contents: Number theoretic functions. Congruences. Residues. Sums of squares.

Continued fractions. Farey sequences. Pell's equation. The prime number theorem. Geometry of numbers. HIGHLY RECOMMENDED. M.R.41 No. 3366.

243. Greenberg, Marvin J. **Lectures on Forms in Many Variables**. New York: W. A. Benjamin, 1969. 167p. $12.50; $3.95pa.
Contents: Tsen's theorem. Chevalley-Waring's theorem. Wedderburn's theorem. Artin's conjecture. Finite fields. Function fields. Complete discrete valuation rings. Witt vectors. p-Adic fields. RECOMMENDED.

244. Halberstam, H., and K. F. Roth. **Sequences**. Vol. I. Oxford: Clarendon Press, 1966. 291p. $10.00.
Contents: Density theorems for sums of sequences. Representation of positive integers as sums of summands of given sequences. Representations functions by probability methods. The sieve method. The methods of Viggo Brun and Atle Selberg. The large sieve. Primitive sequences. Sets of multiples. M.R.35 No. 1565.

245. Khinchin, A. Ya. **Continued Fractions**. Chicago: University of Chicago Press, 1964. 95p. $1.95.
Contents: Properties of the apparatus. The representation of numbers by continued fractions. The measure theory of continued fractions.

246. Knopp, Marvin I. **Modular Functions in Analytic Number Theory**. Chicago: Markham, 1970. 150p. $11.50.
Contents: The modular group and certain subgroups. Modular functions and forms. The modular forms $\eta(\tau)$ and $\theta(\tau)$. The multiplier systems v_η and v_θ. Sums of squares. The order of magnitude of p(n). The Ramanujan congruences for p(n). Proof of Ramanujan's conjectures for powers of 5 and 7.

247. LeVeque, William J., editor. **Studies in Number Theory. MAA Studies in Mathematics, Volume VI**. Englewood Cliffs, N.J.: Prentice-Hall, 1969. 212p.
Contents: Brief survey of Diophantine equations, W. J. LeVeque. Diophantine equations: p-Adic methods, D. S. Lewis. Diophantine problems, Julia Robinson. Computer technology applied to the theory of numbers, D. H. Lehmer. Asymptotic distribution of Beurling's generalized prime numbers, P. T. Bateman and H. G. Diamond.

248. Long, Calvin T. **Elementary Introduction to Number Theory**. Lexington, Mass.: D. C. Heath, 1965. 150p.
Contents: Divisibility properties of integers. Prime numbers. Congruences. Conditional congruences. Multiplicative number—theoretic functions. Table of prime numbers less than ten thousand.

249. Maass, Hans. **Lectures on Modular Functions of One Complex Variable**. Tata Institute of Fundamental Research Lectures on Mathematics, No. 29. Bombay: Tata Institute of Fundamental Research, 1964. 269p. $2.00.
Contents: Horocyclic groups. Automorphic forms and cusp forms. The modular group and its subgroups. Elliptic modular functions. Modular forms of real

dimension. Theory of non-analytic modular forms. RECOMMENDED AS REFERENCE. M.R.36 No. 1392.

250. Malyshev, A. V., editor. **Studies in Number Theory.** Trans. from Russian. New York: Plenum, 1968. 66p. $12.50.
Contents: On the weighted number of integer points on a quadric, A. V. Malyshev. Distribution of integer points on multidimensional hyperboloids and cones, B. Z. Moroz. On an explicit form of the Kummer-Takagi reciprocity law, D. K. Faddeev. On the "Algorithmus der Erhohung" of B. N. Delaunay, D. K. Faddeev. On a paper of A. Baker, D. K. Faddeev. On Fourier coefficients of modular forms, A. V. Malyshev.

251. Mann, H. B. **Addition Theorems: The Addition Theorems of Group Theory and Number Theory.** New York: John Wiley, 1965. 114p. $8.75.
Contents: The fundamental inequalities for the addition of sets of group elements. Applications. The addition theorems of number theory. Asymptotic density and the theorems of Ostmann and Kneser. The theorems of Erdös and Kasch. Difference sets in Abelian groups. Necessary conditions for the existence of difference sets. Difference sets from powers in a finite field. Decomposition theorems.

252. Mordell, L. J. **Diophantine Equations.** New York: Academic Press, 1969. 312p. $13.50.
Contents: Equations proved impossible by congruence considerations. Equations involving sums of squares. Quartic equations with only trivial solutions. Some linear equations. Properties of congruences. Homogeneous equations of the second degree. Pell's equation. Rational solutions derived from given ones. Rational points on some cubic curves. Rational points on cubic surfaces. Rational and integer points on quartic surfaces. Integer solutions of some cubic equations in three variables. Simple algebraic considerations. Applications of algebraic number theory. Finite basis theorem for the rational points on a cubic curve $f(x,y,z)$ of Genus One. Rational points on curves of Genus $g = 0$ or 1 and $g > 1$. Representation of numbers by homogeneous forms in two variables. Representation of numbers by special binary quadratic and quaternary quadratic forms. Representation of numbers by homogeneous forms in several variables. Representation of numbers by polynomials. Thue's theorem on the integer solutions of $f(x,y) = m$. Local methods or p-adic applications. Binary cubic forms. Binary quartic forms. The equation $y^2 = x^3 + k$. The equation $y^2 = ax^3 + cx + d$. Some equations of degree > 3. Fermat's last theorem. M.R.40 No. 2600.

253. Narasimhan, Raghavan, S. Raghavan, S. S. Rangachari, and L. Sunder. **Algebraic Number Theory.** Bombay: Tata Institute of Fundamental Research, 1966. 98p. $1.25.
Contents: Set theory. Algebra. Elementary number theory. General theory of algebraic number fields. The unique factorization theorem for ideals. Dirichlet's theorem on units. Quadratic fields. Dirichlet's theorem on the infinitude of prime numbers in an arithmetic progress. HIGHLY RECOMMENDED. M.R.41 No. 1682.

254. Niven, I. **Diophantine Approximations.** New York: John Wiley, 1963. 68p. $5.50.
Contents: The approximation of irrationals by rationals. The product of linear forms. The multiples of an irrational number. The approximation of complex numbers. The product of complex linear forms.

255. Niven, I., and H. S. Zuckerman. **An Introduction to the Theory of Numbers.** 2nd ed. New York: John Wiley, 1966. 280p. $9.95.
Contents: Divisibility. Congruences. Quadratic reciprocity. Some functions of number theory. Some Diophantine equations. Farey fractions. Simple continued fractions. Elementary remarks on the distribution of primes. Algebraic numbers. The partition function. Density of sequences of integers.

256. Pettofrezzo, Anthony J., and Donald R. Byrkit. **Elements of Number Theory.** Englewood Cliffs, N.J.: Prentice-Hall, 1970. 208p.
Contents: Preliminary considerations. Divisibility properties of integers. The theory of congruences. Continued fractions.

257. Postnikov, A. G. **Ergodic Problems in the Theory of Congruences and of Diophantine Approximations.** Proceedings of the Steklov Institute of Mathematics, No. 82. Trans. from the Russian by B. Volkmann. Providence, R.I.: American Mathematical Society, 1967. 128p. $5.80.
Contents: Distribution of residues modulo a prime. A. Weil estimate. Metric theory of dynamical systems. Uniform distribution.

258. Rankin, Robert A. **The Modular Group and Its Subgroups.** Madras: The Ramanujan Institute, 1969. 85p. $2.00.
Contents: The modular group and subgroup. Normal and commutator subgroups. Level of a subgroup. Congruence groups. Lattice subgroups of free subgroups of the modular group. Subgroups of small index. RECOMMENDED.

259. Shockley, J. E. **Introduction to Number Theory.** New York: Holt, Rinehart, and Winston, 1967. 247p.
Contents: Basic concepts. Linear Diophantine equations. Introduction to the theory of congruences. The Euler-Fermat theorem. Decimal expansion of rational numbers. Perfect numbers. Arithmetic functions. Elementary results on the distribution of prime numbers. Quadratic residues. The two- and four-square problems. Some nonlinear Diophantine problems. Continued fractions and Pell's equation. M.R.35 No. 1535.

260. Siegel, Carl Ludwig. **Lectures on Advanced Analytic Number Theory.** Bombay: Tata Institute of Fundamental Research, 1965. 331p. $2.00.
Contents: Kronecker limit formulas and applications to algebraic number theory. The elliptic theta function. The Epstein theta function. Modular functions and algebraic number theory. The Hilbert modular group. Euler summation formula. The Riemann theta function. The Poisson summation formula. The modular group. HIGHLY RECOMMENDED.

261. Sierpinski, W. **250 Problems in Elementary Number Theory.** New York: American Elsevier, 1970. 125p.
Contents: Divisibility. Prime numbers. Composite numbers. Relatively prime numbers. Diophantine equations.

262. Stark, Harold M. **An Introduction to Number Theory.** Chicago: Markham, 1970. 347p. $8.50.
Contents: Elements of number theory. The Euclidean algorithm. Unique factorization. Congruences. Magic squares. Diophantine equations. Rational and irrational numbers. Continued fractions. Quadratic fields.

263. Stewart, B. M. **Theory of Numbers.** 2nd ed. New York: Macmillan, 1964. 383p. $8.50.
Contents: Mathematical induction. Linear congruences. Linear Diophantine equations. M.R.14 No. 353.

264. Storer, Thomas. **Cyclotomy and Difference Sets.** Chicago: Markham, 1967. 134p. $5.00.
Contents: Historical account of cyclotomy and the construction of difference sets. Finite fields. Galois domains. A list of unsolved problems and conjectures.

265. Turan, P., editor. **Number Theory and Analysis: A Collection of Papers in Honor of Edmund Landau (1877-1938).** New York: Plenum, 1969. 355p. $19.50.
Contents: On the large sieve method, E. Bombieri and H. Davenport. Uber binäre additive probleme gemischter art, B. M. Bredihin, J. V. Linnik, and N. G. Tschudakoff. How to extend a calculus, J. G. van der Corput. On the representation of positive integers as sums of three cubes of positive rational numbers, H. Davenport and E. Landau. Analytische klassenzahlformeln, M. Deuring. Uber folgen ganzer zahlen, P. Erdös, A. Sárközi, and E. Szemerédi. On the average length of a class of finite continued fractions, H. Heilbronn. Interpolation analytischer funktionen auf dem einheitskreis, E. Hlawka. On the high-indices theorem for Borel summability, A. E. Ingham. Bemerkungen zu landauschen methoden in der gitterpunkthlehre, V. Jarnik. Uber einige fragen der vergleichenden primzahl-theorie, S. Knapowski and P. Turán. On local theorems for additive number-theoretic functions, J. Kubilius. The "Pits effect" for the integral function $f(z) = \Sigma \exp [- \vartheta^{-1}(n \log n\text{-}n) + \pi i \alpha n^2] z^n$ $\alpha = \frac{1}{2}(\sqrt{5}\text{-}1)$, J. E. Littlewood. On numbers which can be expressed as a sum of powers, L. J. Mordell. On some Diophantine equations $Y^2 = X^3 + K$ with no rational solutions (II), L. J. Mordell. Uber das vorbzeichen des restgliedes im primzahlsatz, G. Polya. A measure for the differential-transcendence of the zeta-function of Riemann, J. Popken. Comments on Euler's "De mirabilibus proprietatibus numerorum pentagonalium," H. Rademacher. On the distribution of numbers prime to n, A. Rényi. Spline interpolation and the higher derivatives, I. J. Schoenberg. Zu den beweisen des vorbereitungssatzes von weierstrass, C. L. Siegel. Uber gitterpunkte in mehrdimensionalen kugeln IV, Arnold Walfisz and Anna Walfisz. Publications of Edmund Landau, I. J. Schoenberg.

266. Vorobyov, N. N. **The Fibonacci Numbers**. Lexington, Mass.: D. C. Heath, 1963. 47p.
Contents: Properties of Fibonacci numbers and their relationship to continued fractions.

267. Weil, Andre. **Basic Number Theory**. New York: Springer-Verlag, 1967. 294p. $12.00.
Contents: PART I. Elementary theory. Locally compact fields. Lattices and duality over local fields. Places of A-fields. Adeles. Algebraic number fields. The theorem of Riemann-Roch. Zeta functions of A-fields. Traces and norms. PART II. Class field theory. Simple algebras. Simple algebras over local fields. Simple algebras over A-fields. Local class field theory. Global class field theory. M.R.38 No. 3244.

268. Weiss, Edwin. **Algebraic Number Theory**. New York: McGraw-Hill, 1963. 256p. $11.50.
Contents: Elementary valuation theory. Extension of valuations. Local fields. Ordinary arithmetic fields. Global fields. Quadratic fields. Cyclotomic fields.

CHAPTER 8

GENERAL ABSTRACT ALGEBRA

269. Albert, A. A., editor. **Studies in Modern Algebra.** Englewood Cliffs, N.J.:
Prentice-Hall, 1963. 190p.
Contents: Introduction, A. A. Albert. Some recent advances in algebra, Saunders
MacLane. Some additional advances in algebra, Saunders MacLane. What is a
loop?, R. H. Bruck. The four and eight square problems and division algebras,
Charles W. Curtis. A characterization of the Cayley numbers, Erwin Kleinfeld.
Jordan algebras, Lowell J. Paige.

270. American Mathematical Society. **Fifteen Papers on Algebra.** American
Mathematical Society Translations: Series II, Volume 50. Providence, R.I.:
American Mathematical Society, 1966. 316p. $16.50.
Contents: A continual analogue of some theorems on Toeplitz matrices,
N. I. Ahiezer. Indecomposable representations of finite groups over the ring of
p-adic integers, S. D. Berman and P. M. Gudivok. On the equivalence of two
ideals in an algebraic field of order n, K. K. Billevic. On the structure of the
cohomology rings of a Sylow subgroup of the symmetric group, I. V. Bogacenko.
On 2-extensions of a local field, S. P. Demuskin. Subalgebras of nonassociative
free sums of algebras with arbitrary amalgamated subalgebra, C. E. Dididze.
Natural mappings of functors in the category of topological spaces, D. B. Fuks.
Algebras freely generated by finite amalgams, V. E. Govorov. Direct decompositions with indecomposable summands in algebraic categories, A. H. Livsic.
Rational points of algebraic curves over function fields, Ju. I. Manin. Torsion
of the special Lie groups, A. L. Oniscik. Inclusion relations among transitive
compact transformation groups, A. L. Oniscik. Some arithmetic properties of
the modular characters of p-solvable groups, A. V. Rukolaine. On the transcendence and algebraic independence of values of E-functions related by an artibrary
number of algebraic equations over the field of rational functions, A. B. Sid-
lovskii.

271. American Mathematical Society. **Seven Papers on Algebra.** American
Mathematical Society Translations: Series II, Volume 69. Providence, R.I.:
American Mathematical Society, 1968. 256p. $13.50.
Contents: Algebraic theory of linear inequalities, S. N. Cernikov. On subfields
of hyperelliptic fields, I, A. I. Lapin. Representations of the cyclic groups of
prime order p over residue classes mod p, V. S. Drobotenko, E. S. Drobotenko,
Z. P. Zilinskaja, and E. Ja. Pogoriljak. Generalized nilpotent algebras and their
adjoint groups, A. I. Mal'cev. Galois theory of transcendental extensions and
uniformization, I. I. Pjateckii-Sapiro and I. R. Safarevic. Theory of algebraic
linear groups and periodic groups, V. P. Platonov. Projections and isomorphisms
of nilpotent groups, L. E. Sadovskii.

272. Ames, Dennis B. **An Introduction to Abstract Algebra.** Scranton, Penn.: Intext Educational Publishers, 1969. 368p. $10.00.
Contents: Groups. Vector spaces and linear transformations. Structure of groups. Rings. Factorization and ideals. Modules. Algebras. Field theory and Galois theory. Homological algebra. Elementary structure theory of rings. RECOMMENDED. M.R.39 No. 1249.

273. Birkhoff, Garrett, and Thomas C. Bartee. **Modern Applied Algebra.** New York: McGraw-Hill, 1970. 416p. $11.50.
Contents: Sets and functions. Binary relations and graphs. Finite state machines. Programming languages. Boolean algebra. Optimization and computer design. Monoids and groups. Binary group codes. Lattices. Rings and ideals. Polynomial rings and polynomial codes. Finite fields. Recurrent sequences. Computability.

274. Dean, R. A. **Elements of Abstract Algebra.** New York: John Wiley, 1966. 324p. $9.95.
Contents: Groups. Rings. The integers. Fields. Euclidean domains. Polynomials. Vector spaces. Field extensions and finite fields. Algebraic extensions. Finite groups. Galois theory.

275. Dubisch, R. **Introduction to Abstract Algebra.** New York: John Wiley, 1965. 193p. $8.95.
Contents: Sets. The natural numbers. Equivalent pairs of natural numbers. Equivalence classes and the integers. Integral domains. The rational numbers. Groups and fields. The real numbers. Rings, ideals, and homomorphisms. Complex numbers and quaternions. Vector spaces. Polynomials.

276. Goldhaber, Jacob K., and Gertrude Ehrlich. **Algebra.** New York: Macmillan, 1970. 418p. $11.95.
Contents: Groups. Rings. Integral domains. Modules. Finite dimensional vector spaces. Fields. Fields with valuations. Noetherian rings. Dedekind rings. M.R.41 No. 1459.

277. Herstein, I. N. **Topics in Algebra.** Waltham, Mass.: Blaisdell, 1964. 342p. $9.50.
Contents: Group theory. Ring theory. Vector spaces and modules. Fields. Linear transformations. Selected topics. HIGHLY RECOMMENDED FOR BEGINNERS.

278. Jacobson, Nathan. **Lectures in Abstract Algebra.** New York: Van Nostrand-Reinhold.
Volume I. **Basic Concepts.** 1951. 217p. $7.75.
Contents: Introduction. Semi-groups and groups. Rings, internal domains and fields. Extensions of rings and fields. Elementary factorization theory. Groups with operators. Modules and ideals. Lattices.
Volume II. **Linear Algebra.** 1953. 292p. $9.00.
Contents: Finite dimensional vector spaces. Linear transformations. The theory of a single linear transformation. Sets of linear transformations. Bilinear forms.

Euclidean and unitary spaces. Products of vector spaces. The ring of linear transformations. Infinite dimensional vector spaces.
Volume III. **Theory of Fields and Galois Theory.** 1964. 323p. $11.25.
Contents: Finite dimensional extension fields. Galois theory of equations. Abelian extensions. Structure theory of fields. Valuation theory. Artin-Schreier theory.

279. Johnson, Richard E. **University Algebra.** Englewood Cliffs, N.J.: Prentice-Hall, 1966. 279p.
Contents: Basic concepts. The real number system. Abelian groups. Commutative rings. Integral domains and fields. Polynomial rings. Algebraic extensions of a field. Factorization in integral domains. Vector spaces. Groups. Rings. Linear equations and determinants. Lattices.

280. Kurosh, A. G. **Lectures in General Algebra.** Elmsford, N.Y.: Pergamon, 1965. 374p.
Contents: Relations. Groups and rings. Universal algebra. Groups with multi-operators. Lattices. Operator groups and rings. Modules. Linear algebras. Ordered and topological groups and rings. Normed rings.

281. Laatsch, Richard. **Basic Algebraic Systems: An Introduction to Abstract Algebra.** New York: McGraw-Hill, 1968. 240p. $8.50.
Contents: Fundamental concepts. Groups. Subgroups and homomorphisms. Two-operation systems. Polynomials. Topics in group theory. Vector spaces. Boolean algebra and mathematical proof.

282 Landin, Joseph. **An Introduction to Algebraic Structures.** Boston: Allyn & Bacon, 1969. 247p.
Contents: Sets and numbers. The theory of groups. Group isomorphism and homomorphism. The theory of rings. Polynomial rings.

283. Lang, Serge. **Algebraic Structures.** Reading, Mass.: Addison-Wesley, 1967. 173p. $6.95.
Contents: The integers. Groups. Rings. Polynomials. Vector spaces. Modules. Fields. The real numbers (via Cauchy sequences). The complex numbers. Weierstrass's theorem. The fundamental theorem of algebra. Zorn's lemma.

284. Lewis, Donald J. **Introduction to Algebra.** New York: Harper & Row, 1965. 318p. $8.50.
Contents: The integers. Mappings. Abstract groups. Rings, integral domains, fields. Vector spaces.

285. Lindstrum, Andrew O., Jr. **Abstract Algebra.** San Francisco: Holden-Day, 1967. 224p.
Contents: Sets. Mappings. Laws of composition and the natural numbers. Semi-groups, equivalence relations and the rational integers. Groups. Systems with more than one law of composition. Polynomials, factorization, ideals and extension of fields. Fields. Linear mappings and matrices.

286. MacLane, Saunders, and Garret Birkhoff. **Algebra**. New York: Macmillan, 1967. 598p. $11.95.
Contents: Sets. Functions. Universal elements. Rings. Integral domains. Ordered
fields. Modules. Vector spaces. Matrices. Determinants. Tensor products. Similar matrices. Quadratic forms. Affine and projective spaces. Sylow's theorems.
The Jordan-Hölder theorem. Lattices. Categories. Adjoint functors. Multilinear algebra. M.R.35 No. 5266.

287. McCoy, Neal H. **Introduction to Modern Algebra**. Rev. ed. Boston:
Allyn & Bacon, 1968. 394p.
Contents: Some fundamental concepts. Rings. Integral domains. Some proper-
ties of the integers. Fields and the rational numbers. Real and complex numbers.
Groups. Polynomials. Ideals and quotient rings. Vector spaces. Systems of linear
equations. Determinants. Linear transformations and matrices.

288. Mostow, George D., J. H. Sampson, and Jean-Pierre Meyer. **Fundamental
Structures of Algebra**. New York: McGraw-Hill, 1963. 585p. $12.50.
Contents: Binary operations and groups. Rings. Integral domains. The integers.
Fields—the rational numbers. The real number system. The field of complex num-
bers. Polynomials. Rational functions. Vector spaces and affine spaces. Linear
transformations and matrices. Groups and permutations. Determinants. Rings of
operators and differential equations. The Jordan normal form. Quadratic and
Hermitian forms. Quotient structures. Tensors.

289. Patterson, E. M., and D. E. Rutherford. **Elementary Abstract Algebra**.
New York: John Wiley, 1966. 211p. $3.50.
Contents: Binary operations. Rings, integral domains and fields. Ideals and
factor rings. Polynomial and Euclidean rings. Vector spaces. M.R.33 No. 5417.

290. Rosenfeld, Azriel. **An Introduction to Algebraic Structures**. San Francisco: Holden-Day, 1968. 296p.
Contents: Sets, functions, and numbers. Ordered sets. Lattices. Quotient sets,
product sets, and cardinal numbers. Groupoids. Semigroups and groups. Subgroups, factor groups, direct products of groups. Finiteness conditions on groups.
Abelian groups. Rings. Vector spaces. Finiteness conditions on rings. M.R.38
No. 954.

291. Sah, Chih-Han. **Abstract Algebra**. New York: Academic Press, 1967.
342p. $11.50.
Contents: Natural numbers. Integers and rational numbers. Groups, rings, integral
domains, fields. Elementary theory of groups. Elementary theory of rings.
Modules and associated algebras over communative rings. Vector spaces. Elementary theory of fields. Galois theory. Real and complex numbers. M.R.36
No. 6245.

292. Wren, F. Lynwood, and John W. Lindsay. **Basic Algebraic Concepts**. New
York: McGraw-Hill, 1969. 368p. $8.95.
Contents: Groups, rings, integral domains, fields, and vector spaces. Relations
and functions. Polynomials. Systems of equations.

51

293. Zariski, Oscar, and Pierre Samuel. **Commutative Algebra.** New York: Van Nostrand-Reinhold, 1958, 1960. Volume I, 329p., $8.75. Volume II, 432p., $9.50.
Contents: VOLUME I. Introductory concepts. Elements of field theory. Ideals and modules. Noetherian rings. Dedekind domains—classical ideal theory. VOLUME II. Valuation theory. Polynomial and power series rings. Local algebra.

CHAPTER 9

LINEAR AND MULTILINEAR ALGEBRA
AND MATRIX THEORY

294. Ames, Dennis B. **Fundamentals of Linear Algebra.** Scranton, Penn.: Intext Educational Publishers, 1970. 279p. $9.00.
Contents: Vector spaces. Linear transformations. Quotient spaces and direct sums. Linear transformations and matrices. Inner-product vector spaces and dual spaces. Matrices and determinants. Eigenvalues and the spectral theorem. Bilinear and quadratic forms. Canonical forms for linear transformations. The tensor product of vector spaces. The exterior algebra of a vector space. M.R.41 No. 217.

295. Bellman, Richard. **Introduction to Matrix Analysis.** New York: McGraw-Hill, 1960. 328p. $11.50.
Contents: Maximization, minimization, and motivation. Vectors and matrices. Diagonalization and canonical forms for symmetric matrices. Reduction of general symmetric matrices to diagonal form. Constrained maxima. Functions of matrices. Variational description of characteristic roots. Inequalities. Dynamic programming. Matrices and differential equations. Explicit solutions and canonical forms. Symmetric functions, Kronecker products and circulants. Stability theory. Markoff matrices and probability theory. Stochastic matrices. Positive matrices, Perron's theorem and mathematical economics.

296. Boullion, T. L., and P. L. Odell. **Generalized Inverse Matrices.** New York: John Wiley, 1971. 103p.
Contents: Definitions and fundamental properties. Pseudoinverses of sums and products. Generalized inverses with spectral properties. Special topics. Solving systems of linear equations. Applications. RECOMMENDED.

297. Boullion, T. L., and P. L. Odell, editors. **Theory and Application of Generalized Inverses of Matrices.** Lubbock, Texas: Texas Technological College, 1968. 315p.
Contents: Generalized inverses of differential and integral operators, W. T. Reid. Some new generalized inverses with spectral properties, T. N. E. Greville. Inverses of rank invariant powers of a matrix, R. E. Cline. On generalized inverses and interval linear programming, A. Ben-Israel, A. Charnes, and P. D. Robers. Characteristic vectors for rectangular matrices, H. W. Milnes. Spectral eigenvalue property of A^+ for rectangular matrices, H. W. Milnes, J. Amburgey, T. O. Lewis, and T. L. Boullion. Specification problems and regression analysis, J. S. Chipman. A characterization of the maximal subgroups of the semigroup of m x n complex matrices, H. P. Decell, Jr. On applications of generalized inverses in nonlinear analysis, A. Ben-Israel. Partial isometries and generalized inverses, I. Erdelyi. Remarks on a generalized

inverse epsilon-algorithm for matrices, L. D. Pyle. Special applications of the theory of generalized matrix inversion to statistics, C. A. Rohde. On computing generalized characteristic vectors and values for a rectangular matrix, J. K. Amburgey, T. O. Lewis, and T. L. Boullion. The concept of a p-q generalized inverse, T. G. Newman, P. L. Odell, and M. Meicler. A bibliography on generalized matrix inverses, T. O. Lewis, T. L. Boullion, and P. L. Odell.

298. Bowman, F. **An Introduction to Determinants and Matrices.** New York: Van Nostrand-Reinhold, 1962. 163p. $3.95.
Contents: Determinants of the second order. Determinants of the third order. Determinants of any order. Matrices. Rank and linear equations. Properties of square matrices. Quadratic forms. Matrices and differential equations. Approximations.

299. Bronson, R. **Matrix Methods: An Introduction.** New York: Academic Press, 1970. 284p. $10.00.
Contents: Matrices. Determinants. The inverse. Simultaneous linear equations. Eigenvalues and eigenvectors. Matrix calculus. Differential equations. Jordan canonical forms. Special matrices.

300. Campbell, Hugh G. **Linear Algebra with Applications: Including Linear Programming.** New York: Appleton-Century-Crofts, 1971. 390p. $10.95.
Contents: Systems of linear equations. Matrix multiplication. Determinants. Algebraic systems. Vector spaces. Linear transformations. Characteristic values and vectors. Transformation of matrices. The simplex method.

301. Cullen, C. G. **Matrices and Linear Transformations.** 2nd ed. Reading, Mass.: Addison-Wesley, 1967. 227p.
Contents: Matrices and linear systems. Vector spaces. Determinants. Linear transformations. Similarity (part I): Polynomials and polynomial matrices. Similarity (part II): Matrix analysis; numerical methods.

302. Curtis, Charles W. **Linear Algebra: An Introductory Approach.** 2nd ed. Boston: Allyn & Bacon, 1968. 250p.
Contents: The real number system. Vector spaces and systems of linear equations. Linear transformations and matrices. Vector spaces with an inner product. Determinants. Polynomials and complex numbers. Theory of a single linear transformation. Orthogonal, unitary and symmetric transformations. RECOMMENDED.

303. De Pillis, John. **Linear Algebra.** New York: Holt, Rinehart, and Winston, 1969. 510p. $9.75.
Contents: Vector spaces. Systems of linear equations. Linear programming in the plane. Linear transformations. Matrices. Linear equations. Determinants. Inner products. Diagonalization. The spectral theorem for self-adjoint operators. Hermitian matrices.

304. Dixon, Crist. **Linear Algebra.** New York: Van Nostrand-Reinhold, 1971. 200p. $9.75.

Contents: Vector spaces. Linear transformations. Matrices. Determinants. Polynomials. Characteristic and minimal polynomials. Functions of matrices. Inner product spaces. Linear programming.

305. Eves, Howard. **Elementary Matrix Theory.** Boston: Allyn & Bacon, 1966. 325p.
Contents: Fundamental concepts and operations. Equivalence. Determinants. Matrices with polynomial elements. M.R.33 No. 7347.

306. Ficken, F. A. **Linear Transformations and Matrices.** Englewood Cliffs, N.J.: Prentice-Hall, 1967. 398p.
Contents: The system of real numbers. Vectors in three-dimensional Euclidean space. Systems and structures. Linear spaces. Linear transformations. Linear functionals. Duality. Properties of matrices. Systems of linear algebraic equations. Equivalence of matrices. Bilinear and quadratic functionals and forms. Determinants. Unitary and Euclidean spaces. Similar operators.

307. Finkbeiner, Daniel T., II. **Elements of Linear Algebra.** San Francisco: Freeman, 1972. 268p. $9.50.
Contents: An overview of linear algebra. Vector spaces. Linear mappings and matrices. Systems of linear equations. Diagonalization. RECOMMENDED.

308. Finkbeiner, Daniel T., II. **Introduction to Matrices and Linear Transformations.** 2nd ed. San Francisco: W. H. Freeman, 1966. 297p. $8.25.
Contents: Abstract systems. Vector spaces. Linear transformations. Matrices. Linear equations and determinants. Equivalence relations on matrices. A canonical form for similarity. Metric concepts. Combinatorial equivalence. Functions of matrices.

309. Franklin, Joel N. **Matrix Theory.** Englewood Cliffs, N.J.: Prentice-Hall, 1968. 292p.
Contents: Determinants. The theory of linear equations. Matrix analysis of differential equations. Eigenvalues, eigenvectors and canonical forms. The Jordan canonical form. Variational principles and perturbation theory. Numerical methods. M.R.38 No. 5798.

310. Fuller, Leonard E. **Basic Matrix Theory.** Englewood Cliffs, N.J.: Prentice-Hall, 1962. 245p.
Contents: Basic properties of matrices. Elementary matrix operations. Vector spaces and linear transformations. Determinants. Characteristic roots and vectors. Inversion of matrices. Inversion of matrices by iteration. Homogeneous forms.

311. Gilbert, Jimmie D. **Elements of Linear Algebra.** Scranton, Penn.: Intext Educational Publishers, 1970. 328p. $8.50.
Contents: Real coordinate spaces. Elementary operations on vectors. Matrix multiplication. Vector spaces, matrices, and linear equations. Linear transformations. Determinants. Eigenvalues and eigenvectors. Functions of vectors. Inner product spaces. Spectral decompositions. Numerical methods. Linear programming.

312. Greub, W. H. **Linear Algebra.** 3rd ed. New York: Springer-Verlag, 1967. 434p. $9.80.
Contents: Vector spaces. Linear transformations. Matrices. Determinants. Algebras. Gradations. Homology. Inner product spaces. Symmetric bilinear functions. Quadrics. Unitary spaces. M.R.37 No. 221.

313. Greub, W. H. **Multilinear Algebra.** New York: Springer-Verlag, 1967. 225p. $8.00.
Contents: Tensor products. Tensor algebra. Skew symmetry and symmetry in tensor algebra. Exterior and the mixed exterior algebra. M.R.37 No. 222.

314. Halmos, Paul R. **Finite-Dimensional Vector Spaces.** 2nd ed. New York: Van Nostrand-Reinhold, 1968. 200p. $6.50.
Contents: Spaces. Transfomations. Orthogonality. Analysis. Appendix—Hilbert space.

315. Hoffman, Kenneth, and Ray Kunze. **Linear Algebra.** 2nd ed. Englewood Cliffs, N.J.: Prentice-Hall, 1971. 416p.
Contents: Linear equations. Vector spaces. Linear transformations. Polynomials. Determinants. Invariant direct sum decompositions. Rational and Jordan forms. Inner products. Bilinear forms. RECOMMENDED.

316. Jeger, M., and B. Eckmann. **Vector Geometry and Linear Algebra.** New York: John Wiley, 1968. 259p. $9.50.
Contents: Vector algebra. Application of vector algebra to three-dimensional analytical geometry. Determinants and linear equations. Orthogonal coordinate transformations. Surfaces of the second degree. Linear transformations. Tensors.

317. Kahn, Peter J. **Introduction to Linear Algebra.** New York: Harper & Row, 1967. 452p. $11.95.
Contents: Introduction to linearity. Sets and functions. Vector spaces. Linear transformations. Linear equations and determinants. Inner product spaces. Canonical forms. M.R.35 No. 197.

318. Kaplansky, Irving. **Linear Algebra and Geometry: A Second Course.** Boston: Allyn & Bacon, 1969. 139p.
Contents: Inner product spaces. Orthogonal similarity. Geometry. M.R.40 No. 2689.

319. Lancaster, Peter. **Theory of Matrices.** New York: Academic Press, 1969. 316p. $11.00.
Contents: Linear spaces, algebra of matrices, and linear algebraic equations. Eigenvalues and eignevectors. The variational method. The minimal polynomial and normal forms. Functions of matrices. Norms of vectors and matrices. Perturbation theory and bounds for eigenvalues. Direct products. Solution of matrix equations and stability problems. Nonnegative matrices. M.R.39 No. 6885.

320. Liebeck, H. **Algebra for Scientists and Engineers.** New York: John Wiley, 1969. 405p. $13.95.
Contents: Vector spaces. Linear equations. Theory of vector spaces. Coordinate geometry by vector methods. Inner product spaces. Symmetry transformations. Groups. Symmetry groups. Linear transformations and matrices. The matrix inverse. Determinants. Eigenvalue problems. Orthogonal matrices. Quadratic forms and symmetric matrices. Reduction of quadratic forms. The evaluation of eigenvalues.

321. Mal'cev, A. I. **Foundations of Linear Algebra.** Trans. from the Russian by Thomas Craig Brown. Edited by J. B. Roberts. San Francisco: W. H. Freeman, 1963. 304p. $8.00.
Contents: Matrices. Linear spaces. Linear transformations. Polynomial matrices. Unitary and Euclidean spaces. Bilinear and quadratic forms. Linear transformations of bilinear metric spaces. Multilinear functions. Tensors.

322. Martin, A. D., and V. J. Mizel. **Introduction to Linear Algebra.** New York: McGraw-Hill, 1966. 440p. $9.50.
Contents: Gauss reduction for linear systems. Homogeneous systems. Matrices. Vector spaces. Subspaces. The inner product. Linear independence. Dimension. Orthogonal bases. The Gram-Schmidt procedure. The linear functional. Transformations. Invariants. Symmetric operators. The spectral theorem. The adjoint operator. Quadratic forms. The trace. The Jordan decomposition. The Cayley-Hamilton theorem.

323. Moore, John T. **Elementary Linear and Matrix Algebra: The Viewpoint of Geometry.** New York: McGraw-Hill, 1972. 274p. $9.95.
Contents: Geometric vectors. Matrices and linear equations. Real vector spaces. Linear transformations and matrices. Inner product spaces.

324. Moore, John T. **Elements of Linear Algebra and Matrix Theory.** New York: McGraw-Hill, 1968. 352p. $8.95.
Contents: Finite-dimensional vector spaces. Linear transformations and matrices. Determinants and systems of linear equations. Inner-product spaces. Bilinear and quadratic forms. Similarity and normal operators.

325. Nef, Walter. **Linear Algebra.** New York: McGraw-Hill, 1967. 305p. $12.00.
Contents: Sets and mappings. Vector spaces. Bases. Determinants. Linear mappings. Matrices. Linear functionals. Systems of linear equations and inequalities. Linear programming. Tchebychev approximations. Game theory. Euclidean and unitary vector spaces. Eigenvalues and eigenvectors. Invariant subspaces. Canonical forms of matrices.

326. Nering, E. D. **Linear Algebra and Matrix Theory.** 2nd ed. New York: John Wiley, 1970. 352p. $10.95.
Contents: Vector spaces. Linear transformations and matrices. Determinants, eigenvalues, and similarity transformations. Linear functionals, bilinear forms,

quadratic forms. Orthogonal and unitary transformations, normal matrices. Selected applications of linear algebra. Appendix. RECOMMENDED.

327. Noble, Ben. **Applied Linear Algebra.** Englewood Cliffs, N.J.: Prentice-Hall, 1969. 523p.
Contents: Matrix algebra. Some simple applications of matrices. Simultaneous linear equations and elementary operations. Linear dependence and vector spaces. Ranks and inverses. Linear programming. Determinants and square matrices. The numerical solution of simultaneous linear equations. Eigenvalues and eigenvectors. Unitary transformations. Similarity transformations. Quadratic forms and variational principles. Norms and error estimates. Abstract vector spaces. M.R.40 No. 153.

328. Nomizu, Katsumi. **Fundamentals of Linear Algebra.** New York: McGraw-Hill, 1966. 325p. $9.50.
Contents: Vector spaces. Linear mappings and matrices. Determinants. Minimal polynomials. Inner product. Affine spaces. Euclidean space.

329. Pease, Marshall C., III. **Methods of Matrix Algebra.** New York: Academic Press, 1965. 406p. $13.75.
Contents: Vectors and matrices. The inner product. Eigenvalues and eigenvectors. Hermitian, unitary, and normal matrices. Change of basis. Diagonalization and the Jordan canonical form. Functions of a matrix. The matricant. Decomposition theorems and the Jordan canonical form. The improper inner product. The dyad expansion and its application. Projectors. Singular and rectangular operators. The commutator operator. The direct product and the Kronecker sum. Periodic systems. Application to electromagnetic theory. Sturm-Liouville systems. Markoff matrices and probability theory. Stability.

330. Pedoe, D. **A Geometric Introduction to Linear Algebra.** New York: John Wiley, 1963. 224p. $8.50.
Contents: Coordinate geometry of the plane. Vectors in two dimensions. Coordinate geometry of three dimensions. Vectors in three dimensions. Linear equations with a unique solution. Vector spaces. Matrices. The concept of rank. Linear mappings and matrices.

331. Pettofrezzo, Anthony J. **Matrices and Transformations.** Englewood Cliffs, N.J.: Prentice-Hall, 1966. 133p.
Contents: Matrices. Inverses and systems of matrices. Determinants. Rank of a matrix. Systems of linear equations. Transformations. Mappings. Rotations. Reflections. Dilations and magnifications. Orthogonal matrices. Translations. Eigenvalues and eigenvectors. Characteristic functions. Diagnoalization of matrices. The Hamilton-Cayley theorems. Quadratic forms.

332. Rao, C. R., and S. K. Mitra. **Generalized Inverse of Matrices and Its Applications.** New York: John Wiley, 1971. 240p.
Contents: Notations and preliminaries. Generalized inverse of a matrix. Three

basic types of g-inverses. Other special types of g-inverses. Projectors, idempotent matrices and partial isometry. Simultaneous reduction of a pair of Hermitian forms. Estimation of parameters in linear models. Conditions for optimality and validity of least-squares theory. Distribution of quadratic forms. Miscellaneous applications of g-inverses. Computational methods.

333. Schwartz, Jacob T. **Introduction to Matrices and Vectors.** New York: McGraw-Hill, 1961. 163p. $6.50.
Contents: Matrices. Vectors and linear equations. Algebra of matrices and vectors. Eigenvalues and eigenvectors. Infinite series of matrices.

334. Shephard, G. C. **Vector Spaces of Finite Dimension.** New York: John Wiley, 1966. 200p. $3.25.
Contents: Set theory and algebra. Vector spaces and subspaces. Linear transformations. Dual vector spaces. Multilinear algebra. Norms and inner products. Coordinates and matrices. M.R.36 No. 5148.

335. Shilov, Georgi E. **An Introduction to the Theory of Linear Spaces.** Trans. by Richard A. Silverman. Englewood Cliffs, N.J.: Prentice-Hall, 1961. 310p.
Contents: Determinants. Linear spaces. Systems of linear equations. Linear functions of a vector argument. Coordinate transformations. Bi-linear and quadratic forms. Euclidean spaces. Orthogonalisation. Invariant subspaces and eigenvectors. Quadratic forms in a Euclidean space. Quadric surfaces. Infinite-dimensional Euclidean space.

336. Stewart, Frank M. **Introduction to Linear Algebra.** New York: Van Nostrand-Reinhold, 1963. 304p. $8.25.
Contents: The plane. Linear independence, dimension, bases, subspaces. Linear transformation. Simultaneous linear equations. The dual space. Alternative forms. Determinants. Inner product spaces.

337. Stoll, R. R., and E. T. Wong. **Linear Algebra.** New York: Academic Press, 1968. 326p. $9.25.
Contents: Vector spaces. Inner-product spaces. Linear transformations. Matrices. Algebraic properties of linear transformations. Bilinear forms and quadratic forms. Decomposition theorems for normal transformations. Several applications of linear algebra.

338. Suprunenko, D. A., and R. I. Tyshkevich. **Commutative Matrices.** Trans. by Scripta Technica, Inc. New York: Academic Press, 1968. 158p. $7.00; $3.95pa.
Contents: Elementary properties of commutative matrices. Commutative subgroups of $GL(n,P)$ and commutative subalgebras of P_n. Commutative nilpotent algebras of matrices over the field of complex numbers.

339. Varga, Richard. **Matrix Iterative Analysis.** Englewood Cliffs, N.J.: Prentice-Hall, 1962. 322p.

Contents: Basic properties of matrices. Elementary matrix operations. Vector spaces and linear transformations. Determinants. Characteristic roots and vectors. Inversion of matrices. Inversion of matrices by iteration. Homogeneous forms.

340. Vinograde, Bernard. **Linear and Matrix Algebra.** Lexington, Mass.: D. C. Heath, 1967. 252p.
Contents: Coordinates and vectors in the plane. Linear transformations and matrices in the plane. Similarity. Vector spaces. Invertibility. Linear transformations. Invariant subspaces and canonical matrices.

341. Wilcox, H. J. **Elementary Linear Algebra.** New York: Bogden & Quigley, 1970. 128p.
Contents: Preliminaries. Examples of vector spaces. Vector spaces. Bases. Systems of linear equations. Systems of homogeneous linear equations. Non-homogeneous systems. Determinants. Linear transformations. Geometry of vectors.

342. Yefimov, N. V. **Quadratic Forms and Matrices.** Trans. from the Russian and edited by A. Shenitzer. New York: Academic Press, 1964. 164p. $5.50; $2.45pa.
Contents: General theory of quadratic curves. General theory of quadratic surfaces. Linear transformations and matrices.

343. Zelinsky, Daniel. **A First Course in Linear Algebra.** New York: Academic Press, 1968. 266p. $7.50.
Contents: Vectors. Planes and lines. Linear functions. Solution of equations. Dimension. Determinants and transposes. Eigenvalues. Quadratic forms and change of basis.

344. Zuckerberg, Hyam L. **Linear Algebra.** Columbus, Ohio: Charles E. Merrill, 1972. 448p. $10.95.
Contents: Vector spaces. Linear operators. Matrices. Inner product spaces. Determinants. Polynomials. Diagonalization and reduction to Jordan canonical form. Real quadratic forms.

CHAPTER 10

ALGEBRAIC STRUCTURES
AND ORDERED ALGEBRAIC STRUCTURES

345. Abbott, James C. **Sets, Lattices, and Boolean Algebras.** Boston: Allyn & Bacon, 1969. 282p.
Contents: The axiomatics of set theory. The algebra of sets. Transfinite numbers. Partially ordered sets and lattice theory. Distributive and modular lattices. Boolean algebra. Semi-Boolean algebra and implication algebra. M.R. 39 No. 4052.

346. Adamson, I. T. **Introduction to Field Theory.** New York: John Wiley, 1964. 180p. $5.00.
Contents: Elementary definitions. Extension of fields. Galois theory. Applications.

347. Atiyah, M. F., and I. G. Macdonald. **Introduction to Commutative Algebra.** Reading, Mass.: Addison-Wesley, 1969. 128p. $7.50.
Contents: Rings. Ideals. Radicals. Extensions. Contractions. Modules. Primary decomposition theorems. Integral dependence. Valuations. The going-up and going-down theorems. Noetherian rings. Artin rings. Discrete valuation rings. Dedekind domains. Topologies. Filtrations. Graded rings. Dimension theory. M.R.39 No. 4129.

348. Barshay, Jacob. **Topics in Ring Theory.** New York: W. A. Benjamin, 1969. 145p. $15.00; $6.95pa.
Contents: Ideals. Residue rings. Quotient rings. Exact sequences. Noetherian domains. Dedekind domains. Artin rings. Semi-simple Artin rings. M.R.41 No. 1772.

349. Behrens, Ernst-August. **Ring Theory.** New York: Academic Press, 1971. 328p. $16.50.
Contents: Primitive rings. Rings with a faithful family of irreducible modules. Completely reducible modules. Tensor products, fields, and matrix representations. Separable algebras. Rings with identity. Frobenius algebras. Distributively representable rings. Noetherian ideal theory in nonassociative rings. Orders in semisimple Artinian rings. Rings of continuous functions.

350. Birkhoff, Garret. **Lattice Theory.** 3rd rev. ed. Providence, R.I.: American Mathematical Society, 1967. 418p. $11.80.
Contents: Lattices. Structure representation theory. Geometric lattices. Free algebras. Group theory. Structure lattice. Applications to analysis. Theory of sets. Topology and measure theory. Transfinite induction. Algebraic lattices. Continuous lattices. Lattices of open and closed sets. Stone's representation

theorem. Metric and topological lattices. Von Neumann lattices. Borel and measure algebras. Dimension theory. Lattice-ordered groups. Lattice-ordered monoids. Lattice-ordered rings. 166 unsolved problems. M.R.37 No. 2638.

351. Burton, David M. **A First Course in Rings and Ideals.** Reading, Mass.: Addison-Wesley, 1970. 309p. $10.50.
Contents: Ideals. The classical isomorphism theorems. Integral domains. Fields. Maximal, prime and primary ideals. Wedderburn theorem. Direct sums of rings and rings with chain conditions. M.R.41 No. 3509.

352. Cohn, R. M. **Difference Algebra.** New York: John Wiley, 1965. 355p. $14.95.
Contents: Difference rings and difference fields. Relations among difference rings. Ritt difference rings. Varieties of difference polynomials. Extensions of difference fields. Systems of difference equations. Isomorphisms of difference fields. The variety of a difference polynomial.

353. Dieudonné, Jean. **Topics in Local Algebra.** Edited by M. Borelli. Notre Dame, Ind.: University of Notre Dame, 1967. 122p. $2.25.
Contents: Commutative and local algebra. Krull dimension theory. Hilbert polynomial. Regularity. Normality. Depth. Flat ring homomorphisms. Complete Noetherian local rings. Depth. Excellent ring. M.R.39 No. 2748.

354. Donnellan, Thomas. **Lattice Theory.** Elmsford, N.Y.: Pergamon, 1968. 296p.
Contents: Sets, relations, and operations. Lattice as an algebra. Lattice as a partially ordered set. Properties of lattices. Modular and semi-modular lattices. Distributive lattices. Boolean algebras. M.R.38 No. 2059.

355. Faith, Carl. **Lectures on Injective Modules and Quotient Rings.** New York: Springer-Verlag, 1967. 140p. $3.00.
Contents: Injective modules. Essential extensions. The injective hull. Quasi-injectivity. Structures theory for rings. Rational extensions of modules. Maximal rational extensions. Prime rings. Semi-prime rings. Quotient rings. Goldie's theorem. Transitivity. M.R.37 No. 2791.

356. Fuchs, L. **Partially Ordered Algebraic Systems.** Elmsford, N.Y.: Pergamon, 1963. 228p.
Contents: Partially ordered groups, semigroups, and rings. Fully ordered groups, semigroups, and rings. Lattice-ordered groups, semigroups, and rings.

357. Gilmer, Robert W. **Multiplicative Ideal Theory.** Kingston, Ontario: Queen's University, 1968. 400p. $11.00.
Contents: Basic concepts. Integral extension. Valuation theory. Prüfer domains. Polynomial rings. Domains of classical ideal theory. M.R.37 No. 5198.

358. Grätzer, George. **Universal Algebra.** New York: Van Nostrand-Reinhold, 1968. 368p. $12.50.

Contents: Sub-algebras and homomorphisms. Partial algebras, constructions of algebras. Free algebras. Independence. Elements of model theory. Elementary properties of algebraic constructions. Free Σ-structures. M.R.40 No. 1320.

359. Gray, Mary. **A Radical Approach to Algebra.** Reading, Mass.: Addison-Wesley, 1970. 232p. $10.50.
Contents: Elementary ring theory. The radical of an ideal. Primary decompositions. Unique factorization domains. Polynomial rings. The Hilbert basis theorem. Wedderburn theorem. Nilpotent and idempotent elements. Modules. Tensor products. The Jacobson radical. The density theorem. The upper and lower radicals. The Levitzki radical. The Brown-McCoy radical. Banach algebras. Group algebras. Lie algebras. The Jordan algebras. Category theory. Radical subcategories. Sheaf theory.

360. Gulliksen, T. H., and G. Levin. **Homology of Local Rings.** Kingston, Ontario: Queen's University, 1969. 192p.
Contents: Differential graded algebras. The structure of Tor R(k, k). Poincare series. Derivation of local rings. M.R.41 No. 6837.

361. Gunning, R. C. **Lectures on Complex Analytic Varieties: The Local Parametrization Theorem.** Princeton, N.J.: Princeton University Press, 1970. 167p.
Contents: Analytic varieties and subvarieties. Local rings. The local parametrization theorem for varieties and subvarieties.

362. Halmos, P. R. **Algebraic Logic.** New York: Chelsea, 1962. 271p. $4.50.
Contents: Part I, Monadic algebras. Part II, Polyadic algebras. Part III, Summary.

363. Herstein, I. N. **Noncommutative Rings.** New York: John Wiley, 1968. 199p. $6.00.
Contents: The Jacobson radical. Semisimple rings. Commutativity theorems. Simple algebras. Representations of finite groups. Polynomial identities. Goldie's theorem. The Golod-Shafarevitch theorem. M.R.37 No. 2790.

364. Jacobson, Nathan. **Structure and Representations of Jordan Algebras.** Providence, R.I.: American Mathematical Society, 1968. 453p. $10.80.
Contents: Foundations. Elements of representation theory. Pierce decompositions and Jordan matrix algebra. Jordan algebras with minimum conditions on quadratic ideals. Structure theory for finite dimensional Jordan algebras. Generic minimum polynomials. Traces and norms. Representation theory for separable Jordan algebras. Connections with Lie algebras. Exceptional Jordan algebras. HIGHLY RECOMMENDED.

365. Jategaonkar, A. V. **Left Principal Ideal Rings.** New York: Springer-Verlag, 1970. 145p. $3.30.
Contents: Goldie's and Small's theorems on orders in Artinian rings. Structure theory. The ideal theory of a pli-ring. M.R.41 No. 8449.

366. Kaplansky, Irving. **Commutative Rings.** Boston: Allyn & Bacon, 1970. 180p.
Contents: Prime ideals and integral extensions. Noetherian rings. Macaulay rings and regular rings. Homological aspects of ring theory. M.R.40 No. 7234.

367. Kruse, Robert L., and David T. Price. **Nilpotent Rings.** New York: Gordon & Breach, 1969. 136p.
Contents: Definition and examples of nilpotent rings. The subring structure of nilpotent rings.

368. Lambek, Joachim. **Lectures on Rings and Modules**. With an Appendix by I. G. Connell. Waltham, Mass.: Blaisdell, 1966. 84p. $8.50.
Contents: Lattices. Boolean algebras. Rings. Modules. Isomorphism. Direct sums and products. Commutative rings. Prime radicals. Primitive and semi-primitive rings. The Jacobson radical. The density theorem. Reducible rings and modules. The Wedderburn-Artin theorems. Artin rings. Noetherian rings. The complete ring of quotients. Homological algebra. Tensor product. Functional representations of rings. Group rings. Semi-prime group rings. M.R.34 No. 5857.

369. Larsen, Max D., and Paul J. McCarthy. **Multiplicative Theory of Ideals.** New York: Academic Press, 1971. 320p. $17.00.
Contents: Modules. Primary decompositions. Noetherian rings. Rings and modules of quotients. Integral dependence. Valuation rings. Prüfer and Dedekind domains. Dimension of commutative rings. Krull domains. Generalizations of Dedekind domains. Prüfer rings.

370. Northcott, D. G. **Lessons on Rings, Modules and Multiplicities.** New York: Cambridge University Press, 1968. 458p. $14.50.
Contents: Prime ideals and primary submodules. Rings and modules of fractions. Noetherian rings and modules. The theory of grade. Hilbert rings and the zeros theorem. Multiplicity theory. The Koszul complex. Filtered rings and modules.

371. Pierce, Richard S. **Introduction to the Theory of Abstract Algebras.** New York: Holt, Rinehart, and Winston, 1968. 148p. $5.50.
Contents: Basic concepts of abstract algebras. Direct and subdirect decomposition theorems. Ore's uniqueness theorems. Free products. Free extensions. Characterization of varieties of algebras. M.R.37 No. 2655.

372. Pierce, Richard S. **Modules Over Commutative Regular Rings.** Providence, R.I.: American Mathematical Society, 1967. 112p. $1.80.
Contents: Sheaves. Ringed spaces. Finitely generated modules over commutative regular rings. The Grothendieck group.

373. Postnikov, M. M. **Foundations of Galois Theory.** Elmsford, N.Y.: Pergamon, 1962. 115p.
Contents: Basic definitions. Group theory. Field theory. Galois theory of equations. Unsolvability by radicals of the general equation of degree larger than five.

374. Ribenboim, P. **Rings and Modules**. New York: John Wiley, 1969. 162p. $12.95.
Contents: Modules. Structure of modules and rings. Artinian and Noetherian modules. Radicals of modules and rings. Centralizer and double centralizer. Finitely generated modules. Von Neumann regular rings. The group-ring. M.R.39 No. 4204.

375. Rickart, Charles E. **General Theory of Banach Algebras**. New York: Van Nostrand-Reinhold, 1960. 400p. $11.50.
Contents: The radical, semi-simplicity and the structure spaces. Commutative Banach algebras. Algebras with an involution. Structure of ideals and representations of B*-algebras.

376. Samuel, P. **Lectures on Unique Factorization Domains**. Bombay: Tata Institute of Fundamental Research, 1964. 84p. $2.00.
Contents: Krull rings and divisors. Graded Krull rings. Homological characterization of regularity. Unique factorization in regular local rings.

377. Schafer, Richard D. **An Introduction to Nonassociative Algebras**. New York: Academic Press, 1966. 166p. $7.95.
Contents: Arbitrary nonassociative algebras. Alternative algebras. Jordan algebras. Power-associative algebras.

378. Seligman, G. B. **Modular Lie Algebras**. New York: Springer-Verlag, 1967. 165p. $9.75.
Contents: Fundamentals. Classical semi-simple Lie algebras. Automorphisms of the classical algebras. Forms of the classical algebras. Comparison of the modular and non-modular cases. M.R.39 No. 6933.

379. Simis, Aron. **When Are Projective Modules Free?** Kingston, Ontario: Queen's University, 1969. 254p.
Contents: Hereditary and semi-hereditary rings. Von Neumann regular rings. Commutative local rings and semilocal rings. The rank function of projective module. Dedekind domains. The group of fractional ideals. Graded rings and modules. Vector bundles. Exterior algebras and outer products. Bounds on the number of generators of a finitely generated module. M.R.41 No. 260.

380. Stewart, Ian. **Lie Algebras**. New York: Springer-Verlag, 1970. 97p. $2.80.
Contents: Basic definitions. Representations of nilpotent algebars. Cartan subalgebras. The Killing form. The Cartan decompositions. Systems of fundamental roots. Dynkin diagrams. Some astronomical observations. Algebras with a given star. Subideals. Derivations. Automorphisms. The Baer radical. Other radicals. Lie algebras in which every subalgebra is a subideal. The minimal conditions for subideals. M.R.41 No. 8483.

381. Sweedler, Moss E. **Hopf Algebras**. New York: W. A. Benjamin, 1969. 336p.
Contents: Rational modules. The fundamental theorem of co-algebras.

Maschke's theorem. Application of Hopf algebars to Galois theory. Lie algebras. Linear algebras. M.R.40 No. 5705.

382. Szasz, Gabor. **Introduction to Lattice Theory**. 3rd rev. ed. Trans. from the Hungarian. New York: Academic Press, 1964. 229p. $8.50.
Contents: Partially ordered sets. Lattices in general. Complete lattices. Distributive and modular lattices. Special subclasses of the class of modular lattices. Boolean algebras. Semimodular lattices. Ideals of lattices. Congruence relations. Direct and subdirect decompositions.

CHAPTER 11

GROUP THEORY AND GENERALIZATIONS

383. Arbib, Michael A., editor. **The Algebraic Theory of Machines, Languages, and Semigroups.** New York: Academic Press, 1968. 359p. $16.00.
Contents: Semigroups: elementary definitions and examples, John L. Rhodes and Bret R. Tilson. Algebraic machine theory and logical design, E. F. Assmus, Jr., and J. J. Florentin. Automation decompositions and semigroup extensions, Michael A. Arbib. Cascade decomposition of automata using covers, H. Paul Zeiger. The prime decomposition theorem of the algebraic theory of machines, Kenneth Krohn, John L. Rhodes, and Bret R. Tilson. Complexity and group complexity of finite-state machines and finite semi-groups, Michale A. Arbib, John L. Rhodes, and Bret R. Tilson. Local structure theorems for finite semi-groups, John L. Rhodes and Bret R. Tilson. Homomorphisms and semilocal theory, Kenneth Krohn, John L. Rhodes, and Bret R. Tilson. Axioms for complexity of finite semigroups, Kenneth Krohn, John L. Rhodes, and Bret R. Tilson. Expository lectures on topological semigroups, Jane M. Day. The syntactic monoid of a regular event, Robert McNaughton and Seymour Papert. Lectures on context-free power languages, Seymour Ginsburg. Algebraic, rational, and context-free power series in noncommuting variables, Eliahu Shamir. M.R.38 No. 1198.

384. Bhagavantam, S., and T. Vankatarayuda. **Theory of Groups and Its Application to Physical Problems.** New York: Academic Press, 1969. 279p. $6.50.
Contents: Groups. One dimensional lattice. Lattices in two dimensions. Some properties of groups. Matrix groups. The wave equation and its properties. Vibrations of dynamical system. Vibrational Raman effect and infra-red absorption. Molecular structure and normal modes. Molecular structure and normal frequencies. Lattices in three dimensions. Raman and infra-red spectra of crystals. Crystal symmetry and physical properties. Rotation groups. Application to problems of atomic spectra.

385. Boerner, H. **Representations of Groups.** Amsterdam: North-Holland, 1970. 341p. $16.00.
Contents: Matrices. Groups. General representation theory. Representations of the symmetric group. Representations of the full linear, unimodular and unitary groups. Characters of the linear and permutation groups. The alternating group. Characters and single-valued representations of the rotation group. Spin representations. Infinitesimal ring. Ordinary rotation group. The Lorentz group.

386. Burrow, Martin. **Representation Theory of Finite Groups.** New York: Academic Press, 1965. 185p. $6.50; $3.45pa.
Contents: Foundation. Representation theory of rings with identity. The representation theory of finite groups. Applications of the theory of characters. The construction of irreducible representations. Modular representations.

387. Clifford, A. H., and G. B. Preston. **The Algebraic Theory of Semigroups.** Vol. II. Providence, R.I.: American Mathematical Society, 1967. 350p. $13.70. Contents: Minimal ideals and minimal conditions. Inverse semigroups. Simple semigroups. Finite representations of semigroups and free products with amalgamations. Congruences. Representation by transformations of a set. Embedding of a semigroup in a group. M.R.24 No. A2627.

388. Folley, Karl, editor. **Semigroups.** New York: Academic Press, 1969. 277p. $14.00. Contents: Remarks on the algebraic subsemigroups of certain compact semigroups, L. W. Anderson and R. P. Hunter. Semigroups and amenability, Mahlon M. Day. Ternary operations and semigroups, Edwin Hewitt and Herbert S. Zuckerman. Problems about compact semigroups, Karl Heinrich Hofmann and Paul S. Mostert. A survey of results on threads, R. J. Koch. Some recent results on the structure of inverse semigroups, W. D. Munn. Structure numbers and structure theorems for finite semigroups, John Rhodes. On the p length of p solvable semigroups: preliminary results, Bret R. Tilson. Stationary measures for random walks on semigroups, M. Rosenblatt. The study of closets and free contents related to semilattice decomposition of semigroups, Takayuki Tamura. Recent results on binary topological algebra, Alexander Doniphan Wallace. M.R.41 No. 3637.

389. Fuchs, Laszlo. **Abelian Groups.** Elmsford, N.Y.: Pergamon, 1960. 367p. Contents: Basic concepts. Direct sum of cyclic groups. Divisible groups. Direct summands and pure subgroups. Basic subgroups. The structure of P-groups. Torsion free groups. Mixed groups. Homomorphism groups and endomorphism rings. Group extensions. Tensor products. The additive group of rings. The multiplicative group of fields. The lattice of subgroups. Decompositions into direct sums of subsets.

390. Fuchs, Laszlo. **Infinite Abelian Groups.** New York: Academic Press, 1970. Vol. 1, 288p. $15.00. Contents: Direct sums. Direct sums of cyclic groups. Divisible groups. Pure subgroups. Basic subgroups. Algebraically compact groups. Homomorphism groups. Groups of extensions. Tensor and torsion products. M.R.41 No. 333.

391. Gelfand, I. M., R. A. Minlos, and Z. Ya. Shapiro. **Representations of the Rotation and Lorentz Groups and Their Applications.** Elmsford, N.Y.: Pergamon, 1963. 384p. Contents: Representations of the group of rotations of three-dimensional space. Infinitesimal rotations. The irreducible representations of the group of rotations. Tensors and tensor representations. Spinors and spinor representations. Representations of the Lorentz group. The Lorentz group. Infinitesimal operators and representations of the Lorentz group. Spinors and spinor representations of the Lorentz group.

392. Gorenstein, Daniel. **Finite Groups.** New York: Harper & Row, 1968. 527p. $15.50.

Contents: Representations of groups. Character theory. Groups of prime power order. Solvable and Π-solvable groups. Fusion, transfer, and p-factor groups. p-constrained and p-stable groups. Groups of even order. Fixed-point-free automorphisms. Zassenhaus groups. General classification problems.

393. Gruenberg, K. W. **Some Cohomologic Topics in Group Theory**. London: Queen Mary College, 1967. 117p. $2.00.
Contents: Fixed point free action. The homology and cohomology groups. Free groups. Classical extension theory. Finite p-groups.

394. Hall, G. G. **Applied Group Theory**. New York: American Elsevier, 1967. 128p. $6.00.
Contents: Theory of groups. Group algebra. Characters. Group representations. Symmetry adapted vectors. Translation groups. Rotation groups. Continuous groups. Product representations. M.R.36 No. 2672.

395. Heine, Volke. **Group Theory in Quantum Mechanics: An Introduction to Its Present Usage**. 2nd rev. ed. Elmsford, N.Y.: Pergamon, 1963. 468p.
Contents: Symmetry transformations. The quantum theory of a free atom. The representations of finite groups. Applications.

396. Jansen, L., and M. Boon. **Theory of Finite Groups: Applications in Physics (Symmetry Groups of Quantum Mechanical Systems)**. Amsterdam: North-Holland, 1967. 367p. $16.80.
Contents: The theory of finite groups. Groups with operators. Group representations. Theory of co-representations. Applications. M.R.36 No. 6490.

397. Macdonald, Ian D. **The Theory of Groups**. Oxford: Clarendon Press, 1968. 254p. $4.50.
Contents: Properties of groups. Homomorphisms and isomorphisms. Finitely generated Abelian groups. The Sylow theorems. Normal and Sylow structure. Finite soluble groups. Free groups. Nilpotent groups.

398. Magnus, W., A. Karrass, and D. Solitar. **Combinatorial Group Theory**. New York: John Wiley, 1966. 444p. $19.95.
Contents: Basic concepts. Interchange. Nielsen transformations. Free products and free products with amalgamations. Commutator calculus. Introduction to some recent developments. M.R.34 No. 7617.

399. Naimark, M. A. **Linear Representations of the Lorentz Group**. Elmsford, N.Y.: Pergamon, 1964. 464p.
Contents: The three-dimensional rotation group and the Lorentz group. The representations of the three-dimensional rotation group. Irreducible linear representations of the proper and full Lorentz groups.

400. Neumann, H. **Varieties of Groups**. New York: Springer-Verlag, 1967. 192p. $11.50.
Contents: Free groups. Laws. Verbal subgroups. Variety of groups. Product

varieties. Nilpotent varieties. Relatively free groups. The laws of a finite group. M.R.35 No. 6734.

401. Passman, Donald. **Permutation Groups.** New York: W. A. Benjamin, 1968. 310p. $9.50; $4.95pa.
Contents: Basic results on permutation groups. The Burnside theorem. Frobenius groups. Classification of Frobenius complements. Nilpotence of Frobenius kernels. Solvable doubly transitive groups. Sharply transitive permutation groups. Witt's construction of Mathieu groups. M.R.38 No. 5908.

402. Rédei, L. **The Theory of Finitely Generated Commutative Semigroups.** Elmsford, N.Y.: Pergamon, 1966. 368p.
Contents: Kernel functions. Properties of the kernel functions. Ideal theory of free semimodules of finite rank. Equivalent kernel functions. The cases of semigroups without a unity element.

403. Rotman, Joseph J. **The Theory of Groups: An Introduction.** Boston: Allyn & Bacon, 1965. 305p.
Contents: Groups, homomorphisms, isomorphisms. Permutation groups. Direct products. Sylow theorems. Normal and subnormal series. Infinite Abelian groups. Free groups and free products.

404. Schenkman, Eugene. **Group Theory.** New York: Van Nostrand-Reinhold, 1965. 304p. $10.50.
Contents: Abelian groups. Subgroups, mappings. Families of groups. Sylow theorems. Free groups and free products. Nilpotent groups. Solvable and supersolvable groups. Group representations. Recent developments.

405. Schmidt, O. U. **Abstract Theory of Groups.** Trans. from the Russian by Fred Holling and J. B. Roberts. Edited by J. B. Roberts. San Francisco: W. H. Freeman, 1966. 174p. $6.00.
Contents: Conjugation and invariance. Homomorphism and automorphism. The theory of finite groups. Fundamental theorems. Abelian groups and direct products. Groups of order p^m and their direct products. The theory of characters. M.R.35 No. 238.

406. Scott, W. R. **Group Theory.** Englewood Cliffs, N.J.: Prentice-Hall, 1964. 479p.
Contents: Isomorphism theorems. Transformations and subgroups. Direct sums. Abelian groups. p-Groups and p-subgroups. Supersolvable groups. Free groups and free products. Extensions. Permutation groups. Symmetric and alternating groups representations. Products of subgroups. The multiplicative group of a division ring. Topics in infinite groups.

407. Wielandt, Helmut. **Finite Permutation Groups.** Trans. from the German by R. Bercov. New York: Academic Press, 1964. 114p. $5.50; $2.95pa.
Contents: Fundamental concepts. Transitive groups. The transitive constituents of G_N. The method of Schur. Relationship with representation theory.

408. Zak, J., A. Casher, M. Gluck, and Y. Gur. **The Irreducible Representations of Space Groups**. New York: W. A. Benjamin, 1969. 271p. $35.00. Contents: The irreducible representations by the induction method and the ray representation method. Part II: Tables.

CHAPTER 12

CATEGORY THEORY
AND HOMOLOGICAL ALGEBRA

409. Bucur, I., and A. Deleanu. **Introduction to the Theory of Categories and Functors.** New York: John Wiley, 1968. 224p. $13.50.
Contents: Basic concepts. Sums and products. Inductive and projective limits. Structures on the objects of a category. General theory of Abelian categories. Injective and projective objects in Abelian categories. Elements of homological algebra. M.R.38 No. 4534.

410. Hu, Sze-Tsen. **Introduction to Homological Algebra.** San Francisco: Holden-Day, 1968. 211p.
Contents: Modules. Categories and functors. Tor_n and Ext^n. M.R.38 No. 2190.

411. MacLane, S., editor. **Reports of the Midwest Category Seminar II.**
New York: Springer-Verlag, 1968. 91p. $2.40.
Contents: On the vanishing of the second cohomology group of a commutative algebra, M. Andre. Homology and universality relative to a functor, David A. Buchsbaum. Some algebraic problems in the context of functorial semantics of algebraic theories. F. W. Lawvere. An application of categories of fractions to homotopy theory, R. L. Knighten. Adjoint triangles, Eduardo Dubuc. M.R.37 No. 6341.

412. Mitchell, Barry. **Theory of Categories.** New York: Academic Press, 1965. 273p. $13.75.
Contents: Diagrams and functors. Complete categories. Group. Valued functors. Adjoint functors. Applications of adjoint functors. Extensions. Satellites. Global dimension. Sheaves.

413. Moss, R. M. F., and C. B. Thomas, editors. **Algebraic K-Theory and Its Applications.** New York: Springer-Verlag, 1969. 86p.
Contents: K_2 and symbols, H. Bass. Foundations of equivariant algebraic K-theory, A. Fröhlich and C. T. C. Wall. Triangulated categories and algebraic K-theory, I. Bucur. On modules with quadratic forms, A. Bak. Normal subgroups of integral orthogonal groups, M. Knesser. A splitting theorem and the Kunneth formula in algebraic K-theory, W. C. Hsiang. Obstructions for group actions on S^{2n-1}, C. B. Thomas. M.R.40 No. 1450.

414. Northcott, D. G. **An Introduction to Homological Algebra.** New York: Cambridge University Press, 1960. 292p. $9.50.
Contents: Generalities concerning modules. Tensor products and groups of homomorphisms. Categories and functors. Homology functors. Projective and injective modules. Derived functors. Torsion and extension functors. Some useful identities. Commutative and Noetherian rings of finite global dimension. Homology and cohomology theories of groups and monoids.

415. Pareigis, Bodo. **Categories and Functors**. New York: Academic Press, 1970. 273p. $13.00.
Contents: Adjoint functors and limits. Universal algebra. Abelian categories. Finitely generated objects.

416. Swan, R. G. **Algebraic K-Theory**. New York: Springer-Verlag, 1968. 262p. $4.50.
Contents: The theory of Abelian categories. Krull-Schmidt theorem. Heller's theorem. Hilbert's syzygy theorem. Grothendieck's theorem. Serre's theorem. M.R.41 No. 6940.

417. Weiss, Edwin. **Cohomology of Groups**. New York: Academic Press, 1969. 276p. $15.00.
Contents: Cohomology groups of G in A. Mappings of cohomology groups. Some properties of cohomology groups. The cup product. Group extensions. Abstract class field theory. M.R.41 No. 8499.

CHAPTER 13

TOPOLOGICAL GROUPS AND LIE THEORY

418. Adams, J. F. **Lectures on Lie Groups.** New York: W. A. Benjamin, 1969. 182p.
Contents: Prerequisite algebraic topology. Lie groups. Smooth manifolds. The representation theory of compact groups. Haar measure. Peter-Weyl's theorem. The conjugacy theorem for maximal tori of a compact group. Stiefel diagram. Weyl integral and character formulae. RECOMMENDED. M.R.40 No. 5780.

419. Burckel, R. B. **Weakly Almost Periodic Functions on Semigroups.** New York: Gordon & Breach, 1970. 118p. $12.50; $6.00pa.
Contents: The compactification and existence of the mean. The structure theorem the direct sum decomposition. The extent of W(S). Conditions on W(G) implying compactness of G. Approximation by semi-characters and coefficients of representations. M.R.41 No. 8562.

420. Freudenthal, H., and H. de Vries. **Linear Lie Groups.** New York: Academic Press, 1969. 547p. $27.50.
Contents: The connection between local linear Lie groups and Lie algebras. Solvability and semisimplicity. Dressings and classification of semisimple complex Lie algebras. Topological and integration methods. The algebraic approach to linear representations. Reality in Lie groups and algebras and their linear representations. Symmetric spaces. Tits' geometries. Betti numbers of semi-simple Lie groups and regular subalgebras of semisimple Lie algebras. M.R.41 No. 5546.

421. Greenleaf, Frederick P. **Invariant Means of Topological Groups.** New York: Van Nostrand-Reinhold, 1969. 113p.
Contents: Invariant means on discrete groups. Semigroups and locally compact groups. Applications of invariant means. M.R.40 No. 4776.

422. Hausner, Melvin, and J. T. Schwartz. **Lie Groups; Lie Algebras.** New York: Gordon & Breach, 1968. 229p. $9.50; $4.50pa.
Contents: Complex simple Lie algebras. Real semi-simple Lie algebras. Representation of real and complex Lie algebras. The Campbell-Hausdorf formula. Levi decomposition. Ado's theorem. Compact connected Lie groups. M.R.38 No. 3377.

423. Hermann, Robert. **Lie Groups for Physicists.** New York: W. A. Benjamin, 1966. 193p. $12.50.
Contents: Lie groups and Lie algebras. Compact and non-compact semi-simple Lie algebras. The Cartan decomposition. Symmetric subalgebras. Symmetric spaces. Dual symmetric spaces. Decomposition of semi-simple Lie groups.

Representation theory. Vector bundles over homogeneous spaces. Induced representations. The Poincare group. The Dirac equation. Tensor products of induced representations. Applications to physics. HIGHLY RECOMMENDED. M.R.35 No. 4327.

424. Hochschild, G. **Introduction to Affine Algebraic Groups**. San Francisco: Holden-Day, 1971. 120p.
Contents: Group representations and Hopf algebras. Affine algebraic groups. Decomposition into components. Polynomial maps and algebraic subgroups. Factor groups. The Lie algebra. Lie subalgebras and algebraic subgroups. Polynomial representations. Unipotent groups. Abelian groups. Semisimple representations.

425. Hochschild, G. **The Structure of Lie Groups**. San Francisco: Holden-Day, 1965. 240p.
Contents: Topological groups. Compact groups. Elementary structure theory. Coverings. Power series maps. Analytic manifolds. Analytic groups and their Lie algebras. Closed subgroups of Lie groups. Automorphism groups and semi-direct products. The Campbell-Hausdorff formula. Elementary theory of Lie algebras. Simply connected analytic groups. Compact analytic groups. Cartan subalgebras. Compact subgroups of Lie groups. Centers of analytic groups and closures of analytic subgroups. Complex analytic groups. Faithful representations.

426. Hofmann, Karl H., and P. S. Mostert. **Elements of Compact Semigroups**. Columbus, Ohio: Charles E. Merrill, 1966. 384p. $15.00.
Contents: Idempotents. Completely simple semigroups. Maximal ideals. Green's relations. Compact semigroups. Centralizers. Solenoidal semigroups. Monothetic semigroups. Cylindrical semigroups. One-parameter semigroups. Connected semigroups. Irreducibility. Totally ordered decompositions. One dimensional semigroups. Examples. M.R.35 No. 285.

427. Husseini, S. Y. **The Topology of Classical Groups and Related Topics**. New York: Gordon & Breach, 1969. 136p.
Contents: Fibration and their classification. Cohomology and homology of the classical group. The homology and cohomology of the classifying spaces and loopspaces of the classical Lie groups and Bott periodicity. K-theory.

428. Jacobson, N. **Lie Algebras**. 3rd ed. New York: John Wiley, 1966. 331p. $11.95.
Contents: Basic concepts. Solvable and nilpotent Lie algebras. Cartan's criterion and its consequences. Split semi-simple Lie algebras. Universal enveloping algebras. The theorem of Ado-Iwasawa. Classification of irreducible modules. Characters of the irreducible modules. Automorphisms. Simple Lie algebras over an arbitrary field.

429. Kaplansky, Irving. **Lie Algebras and Locally Compact Groups**. Chicago: University of Chicago Press, 1971. 148p.

Contents: Lie Algebras. Lie groups. Hilbert's fifth problem. Locally compact groups. FOR RESEARCHERS.

430. Koszul, J. L. **Lectures on Groups of Transformations.** Bombay: Tata Institute of Fundamental Research, 1965. 97p. $2.00.
Contents: Proper actions of groups of transformations. Slices. Presentation by finitely many generators. Weil's theorem on discrete subgroups of Lie groups.

431. Loos, Ottmar. **Symmetric Spaces. I: General Theory. II: Compact Spaces and Classification.** New York: W. A. Benjamin, 1969. Vol. I: 198p.; $12.50; $3.95pa. Vol. II: 183p.; $12.50; $3.95pa.
Contents: VOLUME I. Tangent vectors. Affine connections. Lie transformation groups. Symmetric spaces. Riemannian symmetric spaces. VOLUME II. Compact Lie groups. Root systems. Compact symmetric spaces. Classification of compact symmetric spaces. Hermitian symmetric spaces.

432. McCarty, George. **Topology: An Introduction with Application to Topological Groups.** New York: McGraw-Hill, 1967. 270p. $8.95.
Contents: Sets and functions. Groups. Metric spaces. Topologies. Topological groups. Compactness and connectedness. Function spaces. The fundamental group. The fundamental group of the circle. Locally isomorphic groups. RECOMMENDED.

433. Pontryagin, L. S. **Topological Groups.** Trans. from the Russian by A. Brown. 2nd ed. New York: Gordon & Breach, 1966. 543p.
Contents: Groups. Topological spaces. Topological groups. Topological division rings. Linear representation of compact topological groups. Locally compact commutative groups. The concept of a Lie group. The structure of compact groups. Locally isomorphic groups. Lie groups and Lie algebras. The structure of compact Lie groups.

434. Simms, D. J. **Lie Groups and Quantum Mechanics.** New York: Springer-Verlag, 1968. 90p. $2.00.
Contents: Group theory and physical applications. The Lorentz group. Induced representations. The Dirac equation.

CHAPTER 14

TOPOLOGY

435. American Mathematical Society. **Eleven Papers on Topology.** A.M.S. Translations: Series II, Vol. 78. Providence, R.I.: American Mathematical Society, 1968. 251p. $13.30.
Contents: Closed mappings, bicompact sets and a problem of P. S. Aleksandrov, A. V. Arhangel'skii. On the Cebysev point of a system of sets, P. K. Belobrov. Homeomorphisms of Euclidean space and topological imbeddings of polyhedra in Euclidean spaces, I, A. V. Cernavskii. Homeomorphisms of R^n are k-stable for $k \leq n - 3$, A. V. Cernavskii. Regular polyhedra on a closed surface whose Euler characteristic is $X = 3$, A. S. Grek. Periodic transformations and fixed point theorem, I, Liao Shan-dao. On the structure of point sets in three-dimensional space, F. I. Smidov. Monotonically-open mappings of a sphere, A. B. Sosinskii. H-closed topological spaces, N. V. Velicko. On the realization of complexes in Euclidean spaces, I and II, Wu Wen-jun.

436. Baum, John D. **Elements of Point Set Topology.** Englewood Cliffs, N.J.: Prentice-Hall, 1964. 150p.
Contents: Topological spaces. Basic definitions and theorems. Continuous functions and homomorphisms. Various spatial types of topological spaces. Metric spaces.

437. Blackett, Donald W. **Elementary Topology: A Combinatorial and Algebraic Approach.** New York: Academic Press, 1967. 224p. $9.50.
Contents: Sphere. Torus. Cylinder. The Mobius strip. The projective plane. Complex conics. Covering surfaces. The winding number. Vector fields. Network topology. Manifolds. Orientability. Fibre bundles. M.R.35 No. 951.

438. Borsuk, Karol. **Theory of Retracts.** Warsaw: Polish Scientific Publishers, 1967. 245p.
Contents: Point set topology. Algebraic topology. Retracts. Absolute retracts in metric spaces. M.R.35 No. 7306.

439. Bourbaki, Nicolas. **Elements of Mathematics: General Topology.** Trans. from the French. Reading, Mass.: Addison-Wesley, 1966. 437p.
Contents: Part I. Topological structures. Uniform structures. Topological groups. Real numbers. Part II. One parameter groups. Real number spaces and projective spaces. Function spaces.

440. Bourgin, D. G. **Modern Algebraic Topology.** New York: Macmillan, 1963. 544p. $11.50.
Contents: Absolute homology groups. Relative omology modules. Manifolds. Fixed cells. Omology exact sequences. Gratings. Homological algebra. Homotopy. Spectral sequences. M.R.28 No. 3415.

441. Bredon, Glen E. **Sheaf Theory**. New York: McGraw-Hill, 1967. 272p.
Contents: Presheaves. Sheaves. Sheaf cohomology. Spectral sequence of a
differential sheaf. Application of spectral sequences and Borel-Moore homology.
M.R.36 No. 4552.

442. Brown, Ronald. **Elements of Modern Topology**. New York: McGraw-
Hill, 1968. 351p. $11.95.
Contents: Topology of the real line. Continuity. Connectedness. Compactness.
Adjunction spaces. Finite cell complexes. Projective spaces. The fundamental
groupoid. Homotopy. M.R.37 No. 3563.

443. Burgess, D. C. J. **Analytical Topology**. New York: Van Nostrand-
Reinhold, 1967. 190p. $7.50.
Contents: List of special symbols. Sets, functions and orderings. Basic concepts
for metric spaces. Types of metric spaces. Properties of any topological space.
Separation axioms.

444. Cech, Edward. **Point Sets**. New York: Academic Press, 1969. 271p.
$12.00.
Contents: General metric spaces. Special metric spaces. Connectedness. Local
connectedness. Mappings of a space into the circle. Topology of the plane.

445. Cech, Edward. **Topological Spaces**. Trans. by Charles O. Junge. Rev. ed.
by Z. Frolik and M. Katetov. New York: John Wiley, 1966. 893p. $24.00.
Contents: Classes and relations. Algebraic structure and order. Topological
spaces. Uniform and proximity spaces. Separation. Generation of topological
spaces. Generation of uniform and proximity spaces.

446. Choquet, G. **Topology**. Trans. from the French by A. Feinstein. New
York: Academic Press, 1966. 337p. $12.50.
Contents: Topological spaces. Metric spaces. Numerical functions. The Stone-
Weierstrass theorem. Topological vector spaces.

447. Conner, Pierre E. **Lectures on the Action of a Finite Group**. New York:
Springer-Verlag, 1968. 123p. $2.50.
Contents: Complex line bundles with a group of operators. The bordism theory
of orientation-preserving involutions on closed oriented manifolds. M.R.41
No. 2670.

448. Cooke, George E., and R. L. Finney. **Homology of Cell Complexes**.
Princeton: Princeton University Press, 1967. 256p. $3.75.
Contents: Definition and examples of complexes. Regular complexes and their
homology groups. Compactly generated spaces. The Kunneth formula. Singular
theory. The homotopy extension problem. Regular quasi-complexes. Skeletal
homology. M.R.36 No. 2142.

449. Copson, E. T. **Metric Spaces**. New York: Cambridge University Press,
1968. 152p. $5.00.

Contents: Metric spaces. Open and closed sets. Complete metric spaces. Connected sets. Compactness. Functions and mappings. Some applications. Further developments. M.R.37 No. 877.

450. Crowell, R. H., and R. H. Fox. **Introduction to Knot Theory**. Boston: Ginn & Co., 1963. 182p.
Contents: Knots. The fundamental group. The calculation of the fundamental group. Free groups. The knot group. Knot polynomials and their properties.

451. Cullen, Helen F. **Introduction to General Topology**. Lexington, Mass.: D. C. Heath, 1968. 427p.
Contents: The general topological structure. Generation of spaces. Separation axioms. Metrizable spaces and uniformizable spaces. Connectedness. Compactness. Homotopy. Notations and conventions. RECOMMENDED. M.R.36 No. 4507.

452. Dugundji, James. **Topology**. Boston: Allyn & Bacon, 1966. 447p.
Contents: Elementary set theory. Ordinals and cardinals. Topological spaces. Cartesian products. Connectedness. Identification topology. Weak topology. Separation axioms. Covering axioms. Metric spaces. Convergence. Compactness. Function spaces. Complete spaces. Homotopy. Maps into spheres. Homotopy type. Path spaces. H-spaces. Fiber spaces.

453. Franz, Wolfgang. **Algebraic Topology**. Trans. from the German by Leo Boron. New York: Frederick Ungar, 1968. 169p.
Contents: Simplicial complexes. Chain complexes. Cell complexes.

454. Gaal, Steven. **Point Set Topology**. New York: Academic Press, 1964. 317p. $13.00.
Contents: Set theory. Topological spaces. Separation properties. Compactness and uniformization. Continuity. Theory of convergence.

455. Gabriel, P., and M. Zisman. **Calculus of Fractions and Homotopy Theory**. New York: Springer-Verlag, 1967. 168p. $9.50.
Contents: Localization of categories. Semi-simplicial sets. The Poincare groupoids. Geometric realization functors. Homotopy category. Semi-simplicial theory. Minimal fibrations. M.R.35 No. 1019.

456. Gemignani, Michael C. **Elementary Topology**. Reading, Mass.: Addison-Wesley, 1967. 258p.
Contents: Metric spaces. Topological spaces. Separation axioms. Covering. Compactness. Connectedness. Metrizability. Homotopy theory. M.R.36 No. 838.

457. Gillman, Leonard, and Meyer Jerison. **Rings of Continuous Functions**. New York: Van Nostrand-Reinhold, 1960. 304p. $9.75.
Contents: Functions on a topological space. Ideals and z-filters. Completely regular spaces. Fixed ideals, compact spaces. Ordered residue class rings. The Stone-Cech compactification. Characterization of maximal ideals. Real compact spaces. Cardinals of classed sets in Beta x. Homomorphisms and continuous

mappings. Embedding in products of real lines. Discrete spaces—nonmeasurable cardinals. Hyper-real residue class fields. Prime ideals. Uniform spaces. Dimension.

458. Greenberg, Marvin J. **Lectures on Algebraic Topology**. New York: W. A. Benjamin, 1967. 235p. $10.00; $5.95pa.
Contents: The fundamental group. Covering spaces. Singular homology. The Eilenberg-Steenrod axioms. The Poincare theorem. The Jordan-Brower separation theorem. Manifolds. Duality. Cup- and Cap-products. The universal coefficient theorem. Lefschetz fixed point theorem. M.R.35 No. 6137.

459. Greever, John. **Theory and Examples of Point-Set Topology**. Belmont, Calif.: Brooks/Cole, 1967. 130p. $9.25.
Contents: Sets. Functions. Relations. Linear ordering. Well-ordering. The real numbers. Topological spaces. Closed sets. Open sets. Continuity. Separable spaces. Local compactness. Connectedness. Separations axioms. Compactness. Product spaces. Local connectedness. Metric spaces. Complete metric spaces. Finite and infinite products of metric spaces. M.R.36 No. 5876.

460. Hilton, Peter J. **Homotopy Theory and Duality**. New York: Gordon & Breach, 1965. 224p.
Contents: Exact sequences. Universal coefficient theorem for homotopy groups. Induced fibre and cofibre maps. Homotopy theory of modules. Cohomology and homotopy products.

461. Hilton, Peter J., editor. **Studies in Modern Topology**. Englewood Cliffs, N.J.: Prentice-Hall, 1968. 212p.
Contents: Modern topology, P. J. Hilton. What is a curve?, G. T. Whyburn. Some results on surfaces in 3-manifolds, Wolfgang Haken. Semi-simplicial homotopy theory, V. K. A. M. Gugenheim. The functions of algebraic topology, Eldon Dyer. On the geometry of differential manifolds, Valentin Poénaru.

462. Hilton, Peter J., and S. Wylie. **Homology Theory**. New York: Cambridge University Press, 1960. 500p. $13.50; $3.95pa.
Contents: The topology of polyhedra. Homology theory of a simplicial complex. Chain complexes. The contrahomology ring for polyhedra. Abelian groups and homological algebra. The fundamental group and covering spaces. Contrahomology and maps. Singular homology theory. The singular contrahomology ring. Special homology theory and homology theory of groups.

463. Hocking, J. G., and G. S. Young. **Topology**. Reading, Mass.: Addison-Wesley, 1961. 374p.
Contents: Topological spaces and functions. The elements of point-set topology. The elements of homotopy theory. Polytopes and triangulated spaces. Simplicial homology theory. Developments in algebraic topology. General homology theories.

464. Hu, Sze-Tsen. **Cohomology Theory**. Chicago: Markham, 1968. 149p. $9.50.

Contents: The Eilenberg-Steenrod axioms. Uniqueness on finite cell complexes. Singular cohomology groups. Alexander-Kolmogorov-Spanier-Wallace cohomology theory. The Kunneth formula. Cross-products. Cup-products. Cap-products. Oriented topological manifolds. Direct limit of groups. Poincare duality theorems. M.R.38 No. 2765.

465. Hu, Sze-Tsen. **Homology Theory: A First Course in Algebraic Topology.** San Francisco: Holden-Day, 1966. 240p.
Contents: Axioms and uniqueness. Further consequences of the axioms. Computation of homology groups. Elementary applications. Cellular homology groups. Singular homology theory.

466. Hu, Sze-Tsen. **Introduction to General Topology.** San Francisco: Holden-Day, 1966. 240p.
Contents: Sets and functions. Spaces and maps. Properties of spaces and maps. Metric spaces. Uniformity and boundedness. Topological linear spaces.

467. Husemoller, Dale. **Fibre Bundles.** New York: McGraw-Hill, 1966. 291p. $14.50.
Contents: Homotopy theory. Vector bundles. General fibre bundles. Local coordinate description of fibre bundles. Change of structure group in fibre bundles. Calculations involving the classical groups. Elements of K-theory. Characteristic classes. Chern classes and Stiefel-Whitney classes. Differentiable manifolds. General theory of characteristic classes.

468. Kasriel, Robert H. **Undergraduate Topology.** Philadelphia: W. B. Saunders, 1971. 285p.
Contents: Sets. Relations. Functions. The real line. Metric spaces. Functions on metric spaces. Topological spaces. Compactness. Connectedness. Net and filter convergence. Quotient spaces. Product spaces.

469. Kelley, J. L., Isaac Namioka, et al. **Linear Topological Spaces.** New York: Van Nostrand-Reinhold, 1963. 256p. $8.00.
Contents: Foreword. Linear spaces. Linear topological spaces. The category theorems. Convexity in linear topological spaces. Duality.

470. Kuratowski, K. **Topology.** Trans. from the French by J. Jaworowski. New York: Academic Press, 1966, 1969. Vol. 1, 560p., $18.50. Vol. 2, 608p., $19.75.
Contents: VOLUME 1. Topological spaces. Metric spaces. Complete spaces. VOLUME 2. Compact spaces. Connected spaces. Locally connected spaces. Absolute retracts.

471. Lundell, Albert T., and Stephen Weingram. **The Topology of CW Complexes.** New York: Van Nostrand-Reinhold, 1969. 224p. $12.95.
Contents: Combinatorial cell complexes. CW complexes. Regular and semi-simplicial CW complexes. Homotopy type of CW complexes. Singular homology of CW complexes.

472. Mansfield, Maynard J. **Introduction to Topology.** New York: Van Nostrand-Reinhold, 1963. 116p. $5.75.
Contents: Topological spaces. Continuity. Homeomorphisms. Connectedness. Compactness. T_1-spaces. Regular spaces. Normal spaces. T_3-spaces. Tychonoff spaces. Metric spaces.

473. Massey, William S. **Algebraic Topology: An Introduction.** New York: Harcourt, 1967. 261p. $9.25.
Contents: Two dimensional manifolds. Free product of groups. Covering spaces. The fundamental group. M.R.35 No. 2271.

474. Maunder, C. R. F. **Algebraic Topology.** New York: Van Nostrand-Reinhold, 1970. 375p.
Contents: Homotopy and simplicial complexes. The fundamental group. Homotopy theory. Homology theory.

475. Maunder, C. R. F. **On Stable Homotopy Theory.** Aarhus: Mathematisk Institut, Aarhus Universitet, 1968. 55p. $1.00.
Contents: Classes of Abelian groups. The EHP sequence. Stable homotopy theory. The Adams spectral sequence for S^O. Applications of K-theory. M.R. 41 No. 1051.

476. May, J. P. **Simplicial Objects in Algebraic Topology.** Princeton, N.J.: D. Van Nostrand, 1967. 161p. $2.95.
Contents: Simplicial objects and homotopy. Fibrations. Postnikov systems. Minimal complexes. Geometric realization. Twisted Cartesian products and fibre bundles. Eilenberg-MacLane complexes and Postnikov systems. Loop groups. Acyclic models. Twisted tensor products.

477. Mendelson, Bert. **Introduction to Topology.** 2nd ed. Boston: Allyn & Bacon, 1968. 202p.
Contents: Theory of sets. Metric spaces. Topological spaces. Connectedness. Compactness.

478. Moore, Theral O. **Elementary General Topology.** Englewood Cliffs, N.J.: Prentice-Hall, 1964. 174p.
Contents: Set theory. Topological spaces. Mappings. Compactness. Product spaces. Metric spaces. Nets and convergence. Peano spaces.

479. Mosher, Robert E., and Martin C. Tangora. **Cohomology Operations and Applications in Homotopy Theory.** New York: Harper & Row, 1968. 214p. $13.50.
Contents: Cohomology operations. Construction of the Steenrod squares. The Steenrod algebra. Exact couples and spectral sequences. Fibre spaces. Homotopy. M.R.37 No. 2223.

480. Nachbin, Leopoldo. **Topology and Order.** Princeton, N.J.: D. Van Nostrand, 1965. 122p. $2.50.

Contents: Topological ordered spaces. Uniform ordered spaces. Locally convex ordered vector spaces.

481. Nagami, Keio. **Dimension Theory.** New York: Academic Press, 1970. 256p. $13.50.
Contents: Theory of open coverings. Dimension of normal spaces. Dimension of metric spaces. Gaps between dimension functions. Dimension-changing closed mappings. Product theorem and expansion theorem. Metric-dependent dimension functions.

482. Naimpally, S. A., and B. D. Warrack. **Proximity Spaces.** Cambridge: Cambridge University Press, 1970. 128p.
Contents: Definition and properties of proximity spaces. Compactification of proximity spaces clusters. Ultra filters. Proximity. Uniformity. Sequential proximity. Generalized proximities.

483. Pervin, William J. **Foundations of General Topology.** New York: Academic Press, 1964. 209p. $10.50.
Contents: Algebra of sets. Cardinal and ordinal numbers. Topological spaces. Connectedness, compactness, and continuity. Separation and countability axioms. Metric spaces. Complete metric spaces. Product spaces. Function and quotient spaces. Metrization and paracompactness. Uniform spaces.

484. Schaefer, Helmut H. **Topological Vector Spaces.** New York: Macmillan, 1966. 294p.
Contents: Topological vector spaces. Locally convex topological vector spaces. Linear mappings. Duality. Order structures. With an appendix on spectral properties of positive operators.

485. Schubert, Horst. **Topology.** Trans. from the German by Siegfried Moran. Boston: Allyn & Bacon, 1968. 358p.
Contents: Topological spaces. Uniform spaces. Homotopy. Singular homology theory.

486. Simmons, George F. **Introduction to Topology and Modern Analysis.** New York: McGraw-Hill, 1963. 372p. $9.95.
Contents: Sets and functions. Metric spaces. Topological spaces. Compactness. Separation. Connectedness. Approximation. Operators. Algebraic systems. Banach spaces. Hilbert spaces. Finite-dimensional spectral theory. Algebras of operators. Banach algebras. The structure of commutative Banach algebras. RECOMMENDED.

487. Spanier, Edwin H. **Algebraic Topology.** New York: McGraw-Hill, 1966. 528p. $15.00.
Contents: Set theory. General topology. Group theory. Modules. Euclidean spaces. Homotopy and the fundamental group. Covering spaces and fibrations. Polyhedra. Homology. Products. General cohomology theory and duality. Homotopy theory. Obstruction theory. Spectral sequences and homotopy groups of spheres. M.R.35 No. 1007.

488. Steen, L. A., and J. A. Seebach, Jr. **Counterexamples in Topology.** New York: Holt, Rinehart, and Winston, 1970. 210p.
Contents: General introduction. Separation axioms. Compactness. Connectedness. Metric spaces. Counterexamples.

489. Steenrod, N. **The Topology of Fibre Bundles.** 5th printing. Princeton, N.J.: Princeton University Press, 1965. 229p. $5.00.
Contents: Part I: The general theory of bundles. Part II: The homotopy theory of bundles. Part III: The cohomology theory of bundles.

490. Swan, Richard G. **The Theory of Sheaves.** Chicago: University of Chicago Press, 1964. 150p.
Contents: Sheaves. Homological algebra. Homology of sheaves. Sections of sheaves. Duality theorems. Cup and cap products.

491. Wallace, Andrew H. **Algebraic Topology, Homology and Cohomology.** New York: W. A. Benjamin, 1970. 272p.
Contents: Homology and cohomology thoery. Cech homology theory. Cech homology theory. Homotopy. The fundamental group. M.R.43 No. 4023.

492. Whitehead, George W. **Homotopy Theory.** Cambridge, Mass.: M.I.T. Press, 1966. 124p.
Contents: Homotopy. Homology. Fibre spaces.

493. Willard, Stephen. **General Topology.** Reading, Mass.: Addison-Wesley, 1970. 369p. $13.50.
Contents: Set theory. Metric spaces. Subspaces. Continuous functions. Product spaces. Quotient spaces. Weak topologies. Nets. Filters. The separation axioms. Regularity. Normality. Compactness. Local compactness. Paracompactness. The Baire theorem. Connectedness. Pathwise connectedness. Local connectedness. Total disconnectedness. The fundamental group. Uniform subspaces. Pointwise and uniform convergence. M.R.41 No. 9173.

CHAPTER 15

TOPOLOGY AND GEOMETRY OF MANIFOLDS

494. Abraham, Ralph, and Joel W. Robbin. **Transversal Mappings and Flows.** New York: W. A. Benjamin, 1967. 161p. Contents: Calculus in Banach spaces. Banach manifolds. Vector bundles. Jet bundles. Transversality theory. Representation of manifolds. Density theorems. Isotopy theorems. Vector fields in a manifold. HIGHLY RECOMMENDED. M.R.39 No. 2181.

495. Bishop, R., and R. Crittenden. **Geometry of Manifolds.** New York: Academic Press, 1964. 273p. $12.50. Contents: Manifolds. Lie groups. Fibre bundles. Differential forms. Connexions. Affine connexions. Riemannian manifolds. Geodesics and complete Riemannian manifolds. Riemannian curvature. Immersions and the second fundamental form. Second variation of arc length.

496. Chern, S. S. **Complex Manifolds without Potential Theory.** Princeton, N.J.: D. Van Nostrand, 1967. 92p. $2.75. Contents: Complex manifolds. Complex vector spaces. Hermitian inner products. Exterior derivatives. Sheaves. Cohomology. Vector bundles. Whitney sums. Chern classes. Holomorphic vector line bundles. Hermitian geometry. Kahlerian geometry. Projective spaces. Tori. Imbeddings. Immersions. Grassmanians. FOR RESEARCH WORKERS. M.R.37 No. 940.

497. Chern, S. S., editor. **Studies in Global Geometry and Analysis.** Englewood Cliffs, N.J.: Prentice-Hall, 1967. 197p. Contents: What is analysis in the large, M. Morse. Curves and surfaces in Euclidean space, S. S. Chern. Differential forms, H. Flanders. On conjugate and cut loci, S. Kobayashi. Surface area, L. Cesari. Integral geometry, L. A. Santalo. M.R.41 No. 7686; M.R.35 No. 1429.

498. Conner, P. E., and E. E. Floyd. **The Relation of Cobordism to K-Theories.** New York: Springer-Verlag, 1966. 112p. Contents: K-theory. Homomorphism from cobordism to K-theory. Theory of cobordism characteristic classes. The Stong-Hattori theorem. Chern numbers.

499. Hu, S. T. **Differentiable Manifolds.** New York: Holt, Rinehart and Winston, 1969. 177p. $11.50. Contents: Differentiable manifolds and their properties. Differential forms. Riemannian manifolds. De Rham's theorem. M.R.39 No. 6343.

500. Hudson, J. F. P. **Piecewise Linear Topology.** New York: W. A. Benjamin, 1969. 282p. $12.50; $4.95pa. Contents: Polyhedra. Complexes. Subdivisions. Simplicial maps. Combinatorial

manifolds. Existence and uniqueness of regular neighborhoods. Abstract piecewise linear spaces and manifolds. Approximation of continuous maps by piecewise linear maps. Sunny collapsing and Zeeman's theorems. Relations between concordance, isotopy, ambient isotopy and isotopy by moves. Engulfing theorems. Handlebody theory. M.R.40 No. 2094.

501. Lang, Serge. **Introduction to Differentiable Manifolds.** 2nd ed. New York: John Wiley, 1966. 126p.
Contents: Differential calculus. Manifolds. Vector fields and differential equations. Differential forms. The theorem of Frobenius. Riemannian metrics. The spectral theorem. Local coordinates.

502. Liulevicius, Arunas. **On Characteristic Classes.** Aarhus: Mathematisk Institut, Aarhus Universitet, 1968. 99p. $2.00.
Contents: Basic facts about fibre bundles. Serre spectral sequence. Cohomology of classifying spaces. Thom spectra. Cobordism. Structure of cobordism rings.

503. Morse, Marston, and S. S. Cairns. **Critical Point Theory in Global Analysis and Differential Topology: An Introduction.** New York: Academic Press, 1969. 389p. $18.00.
Contents: Analysis of nondegenerate functions. Abstract differentiable manifolds. Singular homology theory. Other applications of critical point theory. M.R.39 No. 6358.

504. Palais, Richard S. **Foundations of Global Non-Linear Analysis.** New York: W. A. Benjamin, 1968. 131p. $9.50; $3.95pa.
Contents: Banach spaces of sections of vector bundles. Non-linear differential operators. Polynomial differential operators. Linearization and symbols of operators. Elliptic operators. The calculus of variations in fibre bundles. Existence and smoothness theorems for elliptic variational problems. HIGHLY RECOMMENDED.

505. Porteous, Ian R. **Topological Geometry.** New York: Van Nostrand-Reinhold, 1970. 457p. $12.95.
Contents: Quaternions. Correlations. Quadrics. Clifford algebras. The Cayley algebra. Normed linear spaces. Topological spaces. Homogeneous spaces. Affine approximation. The inverse function theorem. Smooth manifolds. M.R.40 No. 8059.

506. Singer, I. M., and J. A. Thorpe. **Lecture Notes on Elementary Topology and Geometry.** Glenview, Ill.: Scott, Foresman, 1967. 214p. $6.75.
Contents: Point set topology. Fundamental group and covering spaces. Simplicial complexes. Manifolds. Homology theory and the de Rham theory. Intrinsic Riemannian geometry of surfaces. Imbedded manifolds in R^3.

507. Sorani, Giuliano. **An Introduction to Real and Complex Manifolds.** New York: Gordon & Breach, 1969. 212p.
Contents: Multilinear algebra. Euclidean spaces. Real manifolds. Complex manifolds. Lie groups. Fiber bundles. Sheaves. Cohomology with coefficients in a sheaf. Kähler manifolds.

508. Stallings, John R. **Lectures on Polyhedral Topology.** Bombay: Tata Institute of Fundamental Research, 1967. 260p. $2.00.
Contents: Basic concepts of polyhedral category. The simplicial approximation theorem. Links. Stars. Dual complexes. Manifolds. Joins. Cells. Position maps. Regular neighborhoods. Regular collapsing. The s-cobordism theorem. M.R.38 No. 6605.

509. Wallace, Andrew H. **Differential Topology: First Steps.** New York: W. A. Benjamin, 1968. 130p. $9.50.
Contents: Point set topology. Differentiable manifolds. Critical values theory. Classification of 2-manifolds.

CHAPTER 16

GEOMETRY

510. Abram, J. **Tensor Calculus through Differential Geometry**. London: Butterworths, 1965. 170p. $7.50.
Contents: Index operations. Summation. Permutation symbols. Riemannian geometry. Applications to dynamics of continuous media.

511. Adler, Claire F. **Modern Geometry**. 2nd ed. New York: McGraw-Hill, 1967. 302p. $8.75.
Contents: Non-Euclidean geometry. Pure, nonmetric projective geometry. Algebraic projective geometry and linear algebra.

512. American Mathematical Society. **Seven Papers on Algebra, Algebraic Geometry and Algebraic Topology**. A.M.S. Translations: Series II, Vol. 63. Providence, R.I.: American Mathematical Society, 1967. 280p. $14.70.
Contents: A remark on Lie p-algebras, Ju. I. Manin. Normed ideals of intermediate type, B. S. Mitjagin. A criterion for projectivity of complete algebraic abstract varieties, B. G. Moisezon. On n-dimensional compact varieties with n algebraically independent meromorphic functions, I, II, and III, B. G. Moisezon. Differentiable sphere bundles, S. P. Novikov. Classification of vector bundles over an algebraic curve of arbitrary genus, A. N. Tjurin. Duality on elliptic curves over local fields, I, O. N. Vedenskii.

513. American Mathematical Society. **Seventeen Papers on Topology and Differential Geometry**. A.M.S. Translations: Series II, Vol. 92. Providence, R.I.: American Mathematical Society, 1970. 284p. $14.40.
Contents: An invariant tensor criterion for conformally reducible Riemannian spaces, A. M. Ancikov. On a class of spaces containing all metric spaces and all locally bicompact spaces, A. V. Arhangel'skii. Conformally Euclidean generalized Riemannian spaces, S. M. Bahrah. On simple systems of paths on complete pretzels, H. Cisang (Heiner Zieschang). Parallel and normal correspondence of two-dimensional surfaces in four-dimensional Euclidean space E_4, L. N. Krivonosov. Connections in homogeneous bundles, Ju. G. Lumiste. A method for determining a quaternary differential quadratic form with prescribed characteristics, P. I. Petrov. Characterization of Riemannian manifolds, P. I. Petrov. Theory of the curvature of X_2 and E_4, K. S. Ramazanova. Some remarks on limiting spaces of Postnikov systems, E. G. Skljarenko. On Demoulin transforms of projective minimal surfaces, I, Su Buchin. On Demoulin transforms of projective minimal surfaces, II, Su Buchin. Completely reducible connections, V. I. Vedernikov. On extension of mappings of topological spaces, N. Velicko. On Pontrjagin classes, I, Wu Wen-jun. On Pontrjagin classes, II, Wu Wen-jun. On Pontrjagin classes, III, Wu Wen-jun.

514. Auslander, Louis. **Differential Geometry**. New York: Harper & Row, 1967. 271p. $11.95.
Contents: Algebraic preliminaries. Differentiable structures. Matrix Lie groups and frame bundles. Differential invariants of surfaces and curves. Local theory of surfaces. Global study of surfaces. Integration of forms and the Gauss-Bonnet theorem. M.R.35 No. 2208.

515. Auslander, Louis, and Robert E. Mackenzie. **Introduction to Differentiable Manifolds**. New York: McGraw-Hill, 1963. 219p. $11.50.
Contents: Euclidean, affine, and differentiable structure on R^n. Differentiable manifolds. Projective spaces and projective algebraic varieties. The tangent bundle of a differentiable manifold. Submanifolds and Riemann metrics. The Whitney imbedding theorem. Lie groups and their one-parameter subgroups. Integral manifolds and Lie subgroups. Fiber bundles. Multilinear algebra.

516. Benson, Russel V. **Euclidean Geometry and Convexity**. New York: McGraw-Hill, 1966. 265p. $9.50.
Contents: Basic concepts. Convex bodies. Transformations. Extremum problems. Euclidean n-dimensional space. Minkowski geometry.

517. Blattner, John W. **Projective Plane Geometry**. San Francisco: Holden-Day, 1968. 307p. $10.95.
Contents: Introduction to projective geometry. Transformations of projective planes. Desarguesian planes. Coordinates in a Desarguesian plane. Pappian planes. M.R.39 No. 2054.

518. Blumenthal, Leonard M. **A Modern View of Geometry**. San Francisco: W. H. Freeman, 1961. 191p. $2.25pa.
Contents: Historical development of the modern view. Sets and propositions. Postulational systems. Coordinates in an affine plane. Coordinates in an affine plane with Desargues and Pappus properties. Coordinatizing projective planes. Metric postulates for the Euclidean plane. Postulates for the non-Euclidean plane.

519. Blumenthal, Leonard M., and Karl Menger. **Studies in Geometry**. San Francisco: W. H. Freeman, 1970. 512p. $15.00.
Contents: Lattice geometries. Projective geometry. Metric geometry. Curve theory.

520. Borel, Armand. **Linear Algebraic Groups**. New York: W. A. Benjamin, 1969. 398p.
Contents: Elements of algebraic geometry. Algebraic groups. Homogeneous spaces. Solvable groups. Borel subgroups. M.R.40 No. 4273.

521. Bumcrot, Robert J. **Modern Projective Geometry**. New York: Holt, Rinehart, and Winston, 1969. 149p. $8.95.
Contents: Projective spaces. The axioms of Veblen and Young. Collineations. Affine planes. Subspaces. Hall's coordinatization. Desargue's configuration. Pappian planes. Algebraic curves. Projective planes. M.R.38 No. 6450.

522. Burago, Yu. D. **Isoperimetric Inequalities in the Theory of Surfaces of Bounded External Curvature.** Trans. from the Russian. New York: Plenum Press, 1971. 99p.
Contents: Area of polyhedral surface (closed simply connected or any topological structure).

523. Byrne, J. Richard. **Modern Elementary Geometry.** New York: McGraw-Hill, 1970. 344p. $9.50.
Contents: Points, lines, and planes. Congruence and similarity.

524. Coxeter, H. S. M. **Introduction to Geometry.** 2nd ed. New York: John Wiley, 1969. 469p. $10.95.
Contents: Triangles. Regular polygons. Isometry in the Euclidean plane. Two-dimensional crystallography. Similarity in the Euclidean plane. Circles and spheres. Isometry and similarity in Euclidean space. Coordinates. Complex numbers. The five Platonic solids. The golden section and Phyllotaxis. Ordered geometry. Affine geometry. Projective geometry. Absolute geometry. Hyperbolic geometry. Differential geometry of curves. The tensor notation. Differential geometry of surfaces. Geodesics. Topology of surfaces. Four-dimensional geometry. Tables. RECOMMENDED.

525. Dembowski, P. **Finite Geometries.** New York: Springer-Verlag, 1968. 375p. $17.00.
Contents: Basic concepts. Designs. Projective and affine planes. Collineations of finite planes. Constructions of finite planes. Inversive planes. M.R.38 No. 1597.

526. Eves, Howard. **Fundamentals of Geometry.** Boston: Allyn & Bacon, 1969. 237p.
Contents: Modern elementary geometry. Elementary transformations. Euclidean constructions. Projective geometry. M.R.39 No. 843.

527. Eves, Howard. **A Survey of Geometry.** Boston: Allyn & Bacon, 1963, 1965. Vol. I, 489p. Vol. II, 496p.
Contents: VOLUME I. The fountainhead. Modern elementary geometry. Elementary transformations. Euclidean constructions. Dissection theory. Projective geometry. Non-Euclidean geometry. The foundations of geometry. VOLUME II. Analytic geometry I. Analytic geometry II. Geometry and the theory of groups. Geometry of complex numbers. Limit operations in geometry. Differential geometry. Combinatorial topology. n-Dimensional geometry and abstract spaces.

528. Fishback, W. T. **Projective and Euclidean Geometry.** 2nd ed. New York: John Wiley, 1969. 298p. $10.95.
Contents: Euclid reexamined. Hilbert's axioms. The growth of geometry. Projective geometry. Fundamentals of synthetic projective geometry. Natural homogeneous coordinates. Vectors and matrices. Fundamentals of analytic projective geometry. Conics. Axiomatic projective geometry. Affine and Euclidean geometry. Hyperbolic geometry. Elliptic geometry. M.R.39 No. 6152.

529. Flanders, Harley. **Differential Forms: With Applications to the Physical Sciences**. New York: Academic Press, 1963. 203p. $7.50.
Contents: Exterior algebra. The exterior derivative. Applications. Manifolds and integration. Applications in Euclidean space. Applications to differential equations. Applications in differential geometry. Applications to group theory. Applications to physics. RECOMMENDED.

530. Gruenberg, K. W., and Alan J. Weir. **Linear Geometry**. New York: Van Nostrand-Reinhold, 1967. 200p. $6.00.
Contents: Vector spaces. Affine and projective geometry. Isomorphisms. Linear mappings. Bilinear forms. Euclidean geometry. Modules. M.R.36 No. 6423.

531. Grunbaum, B. **Convex Polytopes**. New York: John Wiley, 1967. 456p. $18.75.
Contents: Notation and prerequisites. Convex sets. Polytopes. Examples. Fundamental properties and constructions. Polytopes with few vertices. Neighborly polytopes. Euler's relation. Analogues of Euler's relation. Extremal problems concerning numbers of faces. Properties of boundary complexes. k-Equivalence of polytopes. 3-Polytopes. Angle-sums relations. The Steiner point. Addition and decomposition of polytopes. Diameters of polytopes. Long paths and circuits on polytopes. Arrangements of hyperplanes. Concluding remarks. RECOMMENDED. M.R.37 No. 2085.

532. Guggenheimer, Heinrich W. **Differential Geometry**. New York: McGraw-Hill, 1963. 378p. $14.00.
Contents: Elementary differential geometry. Curvature. Evolutes and involutes. Calculus of variations. Transformation groups. Lie group germs. Space curves. Tensors. Surfaces. Inner geometry of surfaces. Affine geometry of surfaces. Riemannian geometry.

533. Guggenheimer, Heinrich W. **Plane Geometry and Its Groups**. San Francisco: Holden-Day, 1967. 300p.
Contents: Axioms. Products of reflections. Groups. Circles. Metric geometry. Similitude. Geometric inequalities. Circular transformations. Hyperbolic geometry. M.R.35 No. 4796.

534. Hartshorne, Robin. **Local Cohomology**. New York: Springer-Verlag, 1967. 106p. $2.50.
Contents: Elementary properties of local cohomology groups. Applications to preschemes. Functors on modules. Local duality. M.R.37 No. 219.

535. Hartshorne, Robin. **Residues and Duality**. New York: Springer-Verlag, 1966. 423p.
Contents: The derived category. Derived functors. Duality for projective morphisms. Local cohomology. Dualizing complexes and local duality. Noetherian schemes. Residual complexes. The duality theorem. M.R.36 No. 5145.

536. Hawkins, George A. **Multilinear Analysis for Students in Engineering and Science.** New York: John Wiley, 1963. 219p.
Contents: Calculus of tensors and vector analysis.

537. Helgason, Sigurdur. **Differential Geometry and Symmetric Spaces.** New York: Academic Press, 1962. 486p. $14.50.
Contents: Elementary differential geometry. Lie groups and Lie algebras. Structure of semi-simple Lie algebras. Symmetric spaces. Decomposition of symmetric spaces. Symmetric spaces of the non-compact type. Symmetric spaces of the compact type. Hermitian symmetric spaces. On the classification of symmetric spaces. Functions on symmetric spaces.

538. Hemmerling, E. M. **Fundamentals of College Geometry.** 2nd ed. New York: John Wiley, 1970. 464p. $8.95.
Contents: Basic elements of geometry. Elementary logic. Deductive reasoning. Congruence—congruent triangles. Parallel and perpendicular lines. Polygons—parallelograms. Circles. Proportion—similar polygons. Inequalities. Geometric constructions. Geometric loci. Areas of polygons. Coordinate geometry. Areas and volumes of solids.

539. Hermann, Robert. **Differential Geometry and the Calculus of Variations.** New York: Academic Press, 1968. 440p. $18.50.
Contents: Differential and integral calculus on manifolds. The Hamilton-Jacobi theory and calculus of variations. Global Riemannian geometry. Differential geometry and the calculus of variations. Additional topics in differential geometry.

540. Hicks, N. J. **Notes on Differential Geometry.** New York: Van Nostrand-Reinhold, 1965. 183p. $3.95.
Contents: Manifolds. Hypersurfaces in R^n. Surfaces in R^3. Tensors and forms. Connexions. Riemannian manifolds and submanifolds. Operators on forms and integration. Gauss-Bonnet theory and rigidity. Existence theory. Topics in Riemannian geometry.

541. Hutchinson, M. W. **Geometry: An Intuitive Approach.** Columbus, Ohio: Charles E. Merrill, 1972. 352p. $9.95.
Contents: Set theory and logic. Congruence and measure. Geometric constructions. Parallels and parallelograms. Area. Space figures and volumes. Coordinate geometry.

542. Karamcheti, K. **Vector Analysis and Cartesian Tensors with Selected Applications.** San Francisco: Holden-Day, 1967. 268p.
Contents: Vector algebra. Differentiation. Integration. Integral definitions and relations. Properties of certain vector fields. Expressions in orthogonal curvilinear coordinates. Cartesian tensors.

543. Kobayashi, S., and K. Nomizu. **Foundations of Differential Geometry.** New York: John Wiley, 1963, 1969. Vol. I, 329p., $16.95; Vol. II, 470p. $17.50.

Contents: VOLUME I. Differentiable manifolds. Theory of connections. Linear and affine connections. Riemannian connections. Curvature and space forms. Transformations. VOLUME II. Submanifolds. Variations of the length integral. Complex manifolds. Homogeneous spaces.

544. Laugwitz, Detlef. **Differential and Riemannian Geometry.** Trans. from the German by Fritz Steinhardt. New York: Academic Press, 1965. 238p. $10.75. Contents: Local differential geometry of space curves and surfaces. Tensor calculus. Riemannian geometry.

545. Meschkowski, Herbert. **Noneuclidean Geometry.** Trans. from the German by A. Shenitzer. New York: Academic Press, 1964. 104p. $5.50; $2.45pa. Contents: Hilbert's system of axioms. From the history of the parallel postulate. The Poincare model. Elementary theorems of hyperbolic geometry. Constructions. Trigonometry. Elliptic geometry.

546. Modenov, P. S., and A. S. Parkhomenko. **Geometric Transformations.** Trans. from the Russian by M. B. P. Slater. New York: Academic Press, 1965. Vol. I, 160p.; $5.50; $2.45pa. Vol. II, 136p.; $5.50; $2.45pa. Contents: VOLUME I. **Euclidean and Affine Transformations.** Orthogonal transformations. Similarity transformations. Affine transformations. VOLUME II. **Projective Transformations.** Projective transformations. Topology of the projective plane. Principle of duality. M.R.33 No. 1777.

547. O'Neill, Barrett. **Elementary Differential Geometry.** New York: Academic Press, 1966. 411p. $11.50. Contents: Calculus on Euclidean space. Frame fields. Euclidean geometry. Calculus on a surface. Shape operators. Geometry of surfaces in E^3. Riemannian geometry.

548. Oort, F. **Commutative Group Schemes.** New York: Springer-Verlag, 1966. 133p. $9.80. Contents: Finite group schemes. Grothendieck's theorems on the Picard shceme. Abelian schemes. Pro-algebraic groups. Unipotent group schemes. Duality theorem. M.R.35 No. 4229.

549. Pedoe, D. **An Introduction to Projective Geometry.** Elmsford, N.Y.: Pergamon, 1963. 232p. Contents: The Desargues and Pappus theorems. The cross-axis, united points and involutions. Collineations in projective planes. The introduction of coordinates in a projective plane. Homogeneous coordinates and collineations. Cross-ratio. Conics. Projective space of n-dimensions.

550. Perfect, Hazel. **Topics in Geometry.** Elmsford, N.Y.: Pergamon, 1963. 153p. Contents: Point transformations. Some properties of triangles. Theorems of Ceva and Menelaus. Ptolemy's theorem, Apollonius' circle and Simpson's line. Properties of incidence. Homogeneous Cartesian co-ordinates and the representa-

tion of points at infinity. Projection. Inversion. Coaxial circles. Cross ratios. Involutions and conic sections. Polar properties of conics. Projective coordinates.

551. Petrov, A. **Einstein Spaces**. Trans. from the Russian by R. F. Kelleher. Elmsford, N.Y.: Pergamon, 1969. 411p. $12.00.
Contents: Tensor analysis. Einstein spaces. Classification of gravitation fields. Conformal mapping of Einstein spaces. The Cauchy problem for the Einstein field equations. M.R.25 No. 4897; M.R.28 No. 5792; M.R.39 No. 6225.

552. Rainich, G. Y., and S. Dowdy. **Geometry for Teachers: An Introduction to Geometrical Theories**. New York: John Wiley, 1968. 228p. $7.95.
Contents: Euclidean geometry. Projective geometry. Inversive geometry. Hyperbolic geometry. Foundations of geometry. Axiom systems. Groups. Numbers in geometries.

553. Ringenberg, L. A. **College Geometry**. New York: John Wiley, 1968. 308p. $9.50.
Contents: Elementary geometry. Incidence. Distance and coordinates on a line. Convexity and separation. Angles. Polygons. Parallelism and similarity. Affine geometry. Coordinates in a plane. Space geometry. Further geometry of triangles. Cross ratio. Further geometry of circles. Construction with Euclidean ruler and compasses.

554. Rosskopf, Myron F., Joan L. Levine, and Bruce R. Vogell. **Geometry: A Perspective View**. New York: McGraw-Hill, 1969. 306p. $8.95.
Contents: Measurement of sets of points. Congruence, parallelism, and similarity of sets of points. Geometry of transformations.

555. Samuel, P. **Lectures on Old and New Results on Algebraic Curves**. Notes by S. Anantharaman. Bombay: Tata Institute of Fundamental Research, 1966. 127p. $2.00.
Contents: Fundamentals of algebraic geometry. Some classical results on algebraic curves. The Riemann-Roch theorem. Some new results of Grauert and Samuel.

556. Schreier, O., and E. Sperner. **Projective Geometry of n Dimensions**. Trans. from the German by Calvin A. Rogers. Bronx, N.Y.: Chelsea, 1961. 208p. $6.00.
Contents: n-Dimensional projective space. General projective coordinates. Hyperplane coordinates. The duality principle. The cross ratio. Projectivities. Linear projectivities of P_n onto itself. Correlations. Hypersurfaces of the second order. Projective classification of hypersurfaces of the second order. Projective properties of hypersurfaces of the second order. The affine classification of hypersurfaces of the second order. The metric classification of hypersurfaces of the second order.

557. Schwartz, Jacob T. **Differential Geometry and Topology**. New York: Gordon & Breach, 1969. 180p.

Contents: General theory of manifolds. Degree of a map and intersection theory. Applications. Theory of fiber bundles. Spectral sequences. Theory of characteristic classes. Riemannian geometry. An application of characteristic classes. Generalized cohomology theories. Continuation of K-theory. Vector fields on spheres.

558. Seidenberg, A. **Elements of the Theory of Algebraic Curves.** Reading, Mass.: Addison-Wesley, 1968. 216p. $11.00.
Contents: Plane curve. Intersection multiplicity. Genus. Branch. Ground fields of positive characteristic. Infinitely near points.

559. Seidenberg, A. **Lectures in Projective Geometry.** New York: Van Nostrand-Reinhold, 1962. 230p. $7.75.
Contents: The axiomatic foundation. Establishing co-ordinates in a plane. Axiomatic introduction of higher-dimensional space. Conics. Co-ordinate systems and linear transformations. Co-ordinates on a conic. Quadric surfaces. The Jordan canonical form.

560. Serre, Jean-Pierre. **Abelian ℓ-Adic Representations and Elliptic Curves.** New York: W. A. Benjamin, 1968. 177p.
Contents: Definition and representations of ℓ-adic representations. Construction of some Abelian ℓ-adic representations. M.R.41 No. 8422.

561. Snapper, Ernest, and Robert J. Troyer. **Metric Affine Geometry.** New York: Academic Press, 1971. 435p. $13.50.
Contents: Affine geometry. Metric vector spaces. Metric affine spaces.

562. Sokolnikoff, I. S. **Tensor Analysis: Theory and Applications to Geometry and Mechanics of Continua.** 2nd ed. New York: John Wiley, 1964. 361p. $10.95.
Contents: Linear vector spaces. Matrices. Tensor theory. Geometry. Analytical mechanics. Relativistic mechanics. Mechanics of continuous media.

563. Springer, C. E. **Geometry and Analysis of Projective Spaces.** San Francisco: W. H. Freeman, 1964. 299p. $8.00.
Contents: Homographies in one dimension. Geometry of points on a line. Geometry of points on a line. Geometry and invariants. Homogeneous coordinates in two dimensions. Homography in two dimensions. Geometry in the projective plane. Collineations and correlations in the plane. The conic. Non-euclidean geometries. Higher-dimensional geometry.

564. Sternberg, Shlomo. **Lectures on Differential Geometry.** Englewood Cliffs, N.J.: Prentice-Hall, 1964. 390p.
Contents: Algebraic preliminaries. Differentiable manifolds. Integral calculus on manifolds. The calculus of variations. Lie groups. Differential geometry of Euclidean space. The geometry of G-structures.

565. Stoker, J. J. **Differential Geometry.** New York: John Wiley, 1969. 404p. $14.95.

Contents: Operations with vectors. Plane curves. Space curves. The basic elements of surface theory. Some special surfaces. The partial differential equations of surface theory. Inner differential geometry in the small from the extrinsic point of view. Differential geometry in the large. Intrinsic differential geometry of manifold. Relativity. The wedge product and the exterior derivative of differential forms, with applications to surface theory. M.R.39 No. 2072.

566. Thomas, T. Y. **Concepts from Tensor Analysis and Differential Geometry.** 2nd ed. New York: Academic Press, 1965. 178p. $7.50.
Contents: Elements of tensor algebra. Metric and affinely connected spaces. Covariant differentiation and extension. Curves and surfaces in space. Einstein-Riemann spaces.

567. Tuller, Annita. **A Modern Introduction to Geometries.** New York: Van Nostrand-Reinhold, 1967. 214p. $7.50.
Contents: Non-Euclidean geometry. Axiom systems. Plane projective geometry. Klein's Erlanger program. Linear transformations. Projective geometry. Subgeometries of projective geometry. Projective metric geometries. Circular transformations. Inversion geometry. Axiom sets for Euclidean geometry. M.R.34 No. 6607.

568. Valentine, Frederick. **Convex Sets.** New York: McGraw-Hill, 1964. 238p. $12.50.
Contents: Hyperplanes and the separation theorem. The Minkowski metric. Some characterizations of convex sets and local convexity. The support function and the dual cone. Helly-type theorems. Further characterizations of convex sets. Properties of S determined by conditions on each set of r points of S. Maximal convex subsets and upper semicontinuous decompositions. Convex functions. Special boundary points. Properties of convex sets.

569. Verdina, Joseph. **Projective Geometry and Point Transformations.** Boston: Allyn & Bacon, 1971. 206p.
Contents: Projective spaces. Projectivities between lines. Collineations and correlations between projective planes. Conics and polarities. Contacts between plane curves and between point transformations. Rational algebraic curves. Quadratic transformations. Point transformations between two projective planes. Cubic transformations.

570. Wolf, Joseph A. **Spaces of Constant Curvature.** New York: McGraw-Hill, 1967. 408p. $12.50.
Contents: Riemannian geometry. Affine differential geometry. The Euclidian space form problem. The spherical space form problem. Space form problems on symmetric spaces and on indefinite metric manifolds. M.R.36 No. 829.

571. Wylie, Clarence R., Jr. **Foundations of Geometry.** New York: McGraw-Hill, 1964. 338p. $8.95.
Contents: The axiomatic method. Euclidean geometry. The geometry of four dimensions. Plane hyperbolic geometry. A Euclidean model of the hyperbolic plane.

572. Wylie, Clarence R., Jr. **Introduction to Projective Geometry.** New York: McGraw-Hill, 1970. 512p. $11.50.
Contents: The elements of perspective. Linear algebra. The extended Euclidean plane. Affine, Euclidean, similarity, and equiareal transformations. The axiomatic foundation. Conics. The introduction of coordinates. The introduction of a metric. Singular metric gauges.

573. Yale, Paul B. **Geometry and Symmetry.** San Francisco: Holden-Day, 1968. 300p.
Contents: Algebraic and combinatoric preliminaries. Isometrics and similarities. An introduction to crystallography. Fields and vector spaces.Affine spaces. Projective spaces.

574. Yano, K. **Differential Geometry on Complex and Almost Complex Spaces.** Elmsford, N.Y.: Pergamon, 1963. 340p.
Contents: Complex manifolds. Kahler spaces. Almost complex spaces. Almost Hermite spaces. Locally product spaces. Almost product spaces. F-connections.

575. Zalgaller, V. A., editor. **Convex Polyhedra with Regular Faces.** Trans. from the Russian. New York: Plenum, 1969. 130p. $12.50.
Contents: Fundamental results. Tools for the proof. Trihedral vertices. Faces with more than five sides. Tetrahedral vertices with one triangular face. Preliminary investigations for tetrahedral vertices with two triangular faces. Two triangular faces in a row. Two triangular faces in opposition. The remaining tetrahedral vertices. Pentahedral vertices. M.R.39 No. 2064.

576. Zariski, Oscar. **An Introduction to the Theory of Algebraic Surfaces.** New York: Springer-Verlag, 1969. 100p. $2.00.
Contents: Projective varieties. Affine varieties. Intersection theory. Differentials. Arithmetic genus. Hilbert characteristic polynomial. The Riemann-Roch theorem. M.R. 41 No. 8418.

CHAPTER 17

REAL ANALYSIS

577. American Mathematical Society. **Four Papers on Functions of Real Variables.** A.M.S. Translations: Series II, Vol. 81. Providence, R.I.: American Mathematical Society, 1969. 280p. $14.70.
Contents: On equivalent norms for fractional spaces, K. K. Golovkin. Some inequalities in function spaces and their applications to the study of the convergence of variational processes, V. P. Il'in. On some properties of differentiable functions of several variables, V. P. Il'in and V. A. Solonnikov. The properties of some classes of differentiable functions of several variables defined in an n-dimensional region, V. P. Il'in.

578. American Mathematical Society. **Thirteen Papers on Functions of Real and Complex Variables.** A.M.S. Translations: Series II, Vol. 80. Providence, R.I.: American Mathematical Society, 1969. 278p. $14.70.
Contents: Generalization of a theorem of Hayman on subharmonic functions in an m-dimensional cone, V. S. Azarin. On the representation of continuous positive-definite kernels, N. N. Caus. On completeness of a set of analytic functions, Ju. F. Korobeinik. Integration of almost periodic functions with values in a Banach space, B. M. Levitan. Estimate of the Cauchy integral along an analytic curve, M. S. Mel'nikov. Inequalities for entire functions of finite degree and their application to the theory of differentiable functions of several variables, S. M. Nikol'skii. Properties of certain classes of functions of several variables on differentiable manifolds, S. M. Nikol'skii. On total differentials, G. H. Sindalovskii. Differentiation and integration in abstract spaces, G. P. Tolstov. Estimates of the Cauchy integral, A. G. Vituskin. On bounds of convexity for starlike functions of order a in the circle $|z| < 1$ and in the circular region $0 < |z| < 1$, V. A. Zmorovic. On a class of extremal problems associated with regular functions with positive real part in the circle $|z| < 1$, by V. A. Zmorovic. On the bounds of starlikeness and on univalence in certain classes of functions regular in the circle $|z| < 1$, V. A. Zmorovic.

579. American Mathematical Society. **Twelve Papers on Real and Complex Function Theory.** A.M.S. Translations: Series II, Vol. 88. Providence, R.I.: American Mathematical Society, 1970. 325p. $16.50.
Contents: On systems of inequalities with convex functions in the left sides, I. I. Eremin. Inequalities with convex functions, E. K. Godunova. An integral with respect to a semiadditive measure and its application to the theory of entire functions I, II, III, IV, A. A. Gol'dberg. On uniform convergence of convex functions in a closed domain, I. Ja. Guberman. On some extremal problems in the theory of univalent functions, Ja. S. Mirosnicenko. Some properties of differentiable functions defined in an n-dimensional open set, S. M. Nikol'skii. On coefficients of bounded univalent functions, M. I. Red'kov. On the growth of entire functions of several complex variables, L. I. Ronkin. On the connection

between the growth of the maximum modulus of an entire function and the absolute values of the coefficients of its power series expansion, M. N. Seremeta.

580. Anderson, Johnston A. **Real Analysis.** New York: Gordon & Breach, 1970. 356p.
Contents: Sets, logic and mathematical argument. Raw materials. The real numbers, functions. The limit process (convergence). Sequences. Series. Continuous functions. The calculus. The derivative. Integration. The elementary functions.

581. Anderson, Kenneth W., and Dick Wick Hall. **Elementary Real Analysis.** New York: McGraw-Hill, 1972. 178p. $8.95.
Contents: Set theory. Mappings. Continuity. Sequences. Series. Compactness. The derivative. FOR BEGINNERS.

582. Anderson, Kenneth W., and Dick Wick Hall. **Sets, Sequences, and Mappings: The Basic Concepts of Analysis.** New York: John Wiley, 1963. 191p. $6.50.
Contents: Introduction to sets and mappings. Sequences. Countable, connected, open and closed sets. Convergence. Continuity and uniform continuity. Metric spaces.

583. Bartle, R. G. **The Elements of Real Analysis.** New York: John Wiley, 1964. 447p. $12.95.
Contents: Set theory. The real numbers. The topology of Cartesian spaces. Convergence. Continuous functions. Differentiation. Integration. Infinite series.

584. Buck, R. Creighton, editor. **Studies in Modern Analysis.** Englewood Cliffs, N.J.: Prentice-Hall, 1962. 182p.
Contents: Preface, R. P. Dilworth. Introduction, R. C. Buck. A theory of limits, E. J. McShane. A generalized Weierstrass approximation theorem, M. H. Stone. The spectral theorem, E. R. Lorch. Preliminaries to functional analysis, Casper Goffman.

585. Burkill, J. C., and H. Burkill. **A Second Course in Mathematical Analysis.** New York: Cambridge University Press, 1970. 526p. $12.50.
Contents: Sets. Functions. Abstract metric spaces. Continuous functions. Real and complex limits. Real and complex series. Uniform convergence. Riemann-Stieltjes integrals. Differential and integral calculus. Fourier series. Cauchy's theorem. Laurent expansions. Calculus of residues. Asymptotic expansions. RECOMMENDED. M.R.41 No. 3197.

586. Burrill, Claude W. **Foundations of Real Numbers.** New York: McGraw-Hill, 1967. 163p. $6.95.
Contents: Inductive sets. Natural numbers. The integers. Properties of integers. The real numbers. Uniqueness of the real number system. Dedekind's definition. Cantor's definition. M.R.36 No. 1585.

587. Burrill, Claude W., and John R. Knudsen. **Real Variables.** New York: Holt, Rinehart and Winston, 1969. 419p.
Contents: The system of integers. Finite and infinite sets. The real number system. Topological concepts. Sequences of real numbers. Series of constants. Limit of a function. Continuous functions. Derivatives of functions. The Riemann integral. Sequences of functions. Series of functions. Monotone functions and bounded variation. The Stieltjes integral. Measure of a set of real numbers. The Lebesgue integral. Generalized measure and the Lebesgue-Stieltjes integral.

588. Byrne, J. Richard. **Number Systems: An Elementary Approach.** New York: McGraw-Hill, 1967. 291p. $6.95.
Contents: Sets. Cardinal numbers. Numerals. Addition and multiplication algorithms. Subtraction and division. Topics from number theory. Rational numbers and fractions. Properties of the nonnegative rational numbers. Negative numbers. Real numbers.

589. Cohen, Leon W., and Gertrude Ehrlich. **The Structure of the Real Number System.** New York: Van Nostrand-Reinhold, 1963. 116p. $5.25.
Contents: The natural numbers. The integers. The rational numbers. Ordered fields. The real numbers. The complex numbers.

590. Cooper, R. **Functions of Real Variables.** New York: Van Nostrand-Reinhold, 1966. 228p. $8.50; $4.95pa.
Contents: Numbers and sets of points. Sequences and series. Infinite products. Functions. Integration. The elementary transcendental functions. Curvilinear, surface, and volume integrals. Differential equations. Legendre polynomials. Fourier series and series of orthogonal functions.

591. De Branges, Louis, and James Rounyak. **Square Summable Series.** New York: Holt, Rinehart, and Winston, 1966.104p. $3.95.
Contents: Hilbert spaces. The space of power series with square summable coefficients. The Beurling invariant subspace theorem. The shift operator. Cauchy's theorem. Poisson's formula. Herglotz representation of harmonic functions. The Helly selection theorem.

592. Dieudonné, J. **Treatise on Analysis.** New York: Academic Press, 1969, 1970. Vol. I, **Foundations of Modern Analysis.** Enlarged and corrected printing. 387p. $12.50. Vol. II, trans. from the French by I. G. Macdonald. 422p. $16.00. Vol. III, in preparation.
Contents: VOLUME I. Theory of sets. Real numbers. Metric spaces. Normed spaces. Hilbert spaces. Spaces of continuous functions. Differential calculus. Analytic functions. Applications of analytic functions to plane topology. Existence theorems. Elementary spectral theory. VOLUME II. Topology and topological algebra. Integration. Integration in locally compact groups. Normed algebras and spectral theory. M.R.41 No. 3198.

593. Eggleston, H. G. **Elementary Real Analysis.** New York: Cambridge University Press, 1962. 290p. $7.50.

Contents: Enumerability and sequences. Bounds for sets of numbers. Bounds for functions and sequences. The O notation. Limits of sequences; the o, ∼ notation. Monotonic sequences. Upper and lower limits of real sequences. Convergence of series. Absolute convergence. Conditional convergence. Rearrangement and multiplication of absolutely convergent series. Double series. Power series. The Bolzano-Weierstrass, Cantor and Heine-Borel theorems. Functions. Limits and continuity. Monotonic functions. Functions of bounded variation. Differentiation. Mean-value theorems. Taylor's theorem. Convex and concave functions. The elementary transcendental functions. Inequalities. The Riemann integral. Integration and differentiation. The Riemann-Stieltjes integral. Improper integrals. Convergence of integrals. Uniform convergence. Continuity and differentiability.

594. Fichtenholz, G. M. **Functional Series.** Trans. and adapted by Richard A. Silverman. New York: Gordon & Breach, 1971. 176p.
Contents: Uniform convergence. Functional properties of the sum of a series. Application. More on power series. Enveloping and asymptotic series.

595. Fichtenholz, G. M. **Infinite Series: Ramifications.** Trans. and adapted by Richard A. Silverman. New York: Gordon & Breach, 1970. 138p.
Contents: Operations on series. Iterated and double series. Computations involving series. Summation of divergent series.

596. Fichtenholz, G. M. **Infinite Series: Rudiments.** Trans. from the Russian and adapted by Richard Silverman. New York: Gordon & Breach, 1970. 146p.
Contents: Positive series. Arbitrary series. Infinite products. Series and product expansions of elementary functions.

597. Fleming, Wendell H. **Functions of Several Variables.** Reading, Mass.: Addison-Wesley, 1965. 337p. $9.75.
Contents: Euclidean spaces. Convexity. Differentiation of real-valued functions. Vector-valued functions of one and several variables. Integration. Exterior algebra and differential calculus. Integration on manifolds.

598. Gelbaum, Bernard, and John Olmsted. **Counterexamples in Analysis.** San Francisco: Holden-Day, 1964. 220p.
Contents: Functions and limits. Differentiation. Riemann integration. Sequences and series. Sets and measures.

599. Gemignani, Michael C. **Introduction to Real Analysis.** Philadelphia: W. B. Saunders, 1971. 160p. $8.00.
Contents: Preliminaries. The real numbers. Some basic topology of the real numbers. Sequences and series. The derivative. The Riemann integral. Sequences and series of functions.

600. Goffman, Casper. **Introduction to Real Analysis.** New York: Harper & Row, 1966. 160p. $9.95.
Contents: The real numbers. Topology of the reals. Infinite series. Continuous

functions. Special functions. Sequences and series of functions. Differentiation. Integration. Power series. Fourier series.

601. Goldberg, R. R. **Methods of Real Analysis.** Waltham, Mass.: Blaisdell, 1964. 359p. $12.25.
Contents: Sets and functions. Sequences of real numbers. Series of real numbers. Limits and metric spaces. Continuous functions on metric spaces. Connectedness, completeness and compactness. Calculus. The elementary functions. Taylor series. Sequences and series of functions. Three famous theorems. The Lebesgue integral. Fourier series. HIGHLY RECOMMENDED FOR BEGINNERS.

602. Hartnett, W. E. **An Introduction to the Fundamental Concepts of Analysis.** New York: John Wiley, 1964. 154p. $6.95.
Contents: Real sequences. Real functions. Continuous real functions. Differentiable real functions. Integration of functions.

603. Hayes, C. A. **Concepts of Real Analysis.** New York: John Wiley, 1964. 190p. $7.95.
Contents: Elements of set theory. The real number system. Finite and infinite sets. Sequences and convergences of real-valued sequences. Sequential limit theory in the extended real number system. Definition by induction. Functions of a real variable. Limits and continuity.

604. Hewitt, E., and K. Stromberg. **Real and Abstract Analysis.** New York, Springer-Verlag, 1965. 476p. $9.50.
Contents: Set theory and algebra. Topology and continuous functions. The Lebesgue integral. Function spaces and Banach spaces. Differentiation. Integration on product spaces.

605. Hirschman, I. I., Jr., editor. **Studies in Real and Complex Analysis.** Englewood Cliffs, N.J.: Prentice-Hall, 1965. 213p.
Contents: Introduction, I. I. Hirschman, Jr. Several complex variables, H. J. Bremermann. Nonlinear mappings between Banach spaces, Lawrence M. Graves. What is a semi-group?, Einar Hille. The Laplace transform, the Stieltjes transform and their generalizations, I. I. Hirschman, Jr., and D. V. Widder. A brief introduction to the Lebesgue-Stieltjes integral, H. H. Schaefer. Harmonic analysis, Guido Weiss. Toeplitz matrices, Harold Widom.

606. Hu, Sze-Tsen. **Elements of Real Analysis.** San Francisco: Holden-Day, 1967. 377p.
Contents: Sets, functions and relations. The real line. Topological spaces. Metric spaces. Topological linear spaces. Measures and integrals. Differentiation. M.R.38 No. 3389.

607. Isaacs, G. L. **Real Numbers: A Development of Real Numbers in an Axiomatic Set Theory.** New York: McGraw-Hill, 1968. 128p. $8.50.
Contents: Fundamental concepts. Natural numbers. Integers. Rational numbers. Cuts. Totally ordered fields. The axiom of completeness. The elementary

functions of analysis. The recursion and course of values. Recursion theorems. M.R.37 No. 6138.

608. Kripke, Bernard. **Introduction to Analysis.** San Francisco: W. H. Freeman, 1968. 274p. $8.50.
Contents: Axioms for the real number system and their consequences. Vector spaces, inner products, and linear maps. Metric and normed spaces. Introduction to complex numbers. Some applications. Compactness. Connectedness. M.R. 38 No. 1218.

609. Littlewood, John E. **Some Problems in Real and Complex Analysis.** Lexington, Mass.: D. C. Heath, 1968. 57p.
Contents: A collection of problems, questions, conjectures and strategies for extending conjectures.

610. Randolph, John F. **Basic Real and Abstract Analysis.** New York: Academic Press, 1968. 515p. $15.50.
Contents: Orientation. Sets and spaces. Sequences and series. Measure and integration. Measure theory. Continuity. Derivatives. Stieltjes integrals.

611. Royden, H. L. **Real Analysis.** 2nd ed. New York: Macmillan, 1971 (fifth printing). 349p. $12.95.
Contents: Set theory. The real number system. Lebesgue measure. The Lebesgue integral. Differentiation and integration. The classical Banach spaces. Metric spaces. Topological spaces. Compact spaces. Banach spaces. Measure and integration. Measure and outer measure. The Daniell integral. Mappings of measure spaces. HIGHLY RECOMMENDED.

612. Rudin, Walter. **Principles of Mathematical Analysis.** 2nd ed. New York: McGraw-Hill, 1964. 270p. $8.95.
Contents: The real and complex number systems. Elements of set theory. Numerical sequences and series. Continuity. Differentiation. The Riemann-Stieltjes integral. Sequences and series of functions. Functions of several variables. The Lebesgue theory.

613. Rudin, Walter. **Real and Complex Analysis.** New York: McGraw-Hill, 1966. 412p. $11.50.
Contents: The exponential function. Abstract integration. Positive Borel measures. L^P-spaces. Elementary Hilbert space theory. Complex measures. Integration on product spaces. Differentiation. Fourier transforms. Holomorphic functions. Harmonic functions. The maximum modulus principle. Approximation by rational functions. Conformal mapping. Zeroes of holomorphic functions. Analytic continuation. H^P-spaces. Banach algebras. Holomorphic Fourier transforms. Uniform approximation by polynomials. M.R.35 No. 1420.

614. Shilov, G. Ye. **Mathematical Analysis.** Elmsford, N.Y.: Pergamon, 1965. 485p.
Contents: Sets. Metric spaces. The calculus of variations. Theory of the integral.

Geometry of Hilbert space. Differentiation and integration. The Fourier transform.

615. Sion, Maurice. **Introduction to the Methods of Real Analysis.** New York: Holt, Rinehart, and Winston, 1968. 134p. $8.95.
Contents: Point set topology. Convergence of sequences. Accumulation points. Continuous functions. Ascoli's theorem. Measure theory. Lebesgue measures. Lebesgue-Stieltje's measures. Absolute integration. The Radon-Nykodym derivative. Vitali covering. The Riesz representation. M.R.37 No. 5332.

616. Sprecher, David A. **Elements of Real Analysis.** New York: Academic Press, 1970. 341p. $11.50.
Contents: Sets and functions. Rational numbers. The real number system. Sequences and series of numbers. The structure of point sets. Continuity. Differentiability. Spaces of continuous functions. Measure and integration. Fourier series.

CHAPTER 18

MEASURE AND INTEGRATION

617. Bartle, R. G. **The Elements of Integration.** New York: John Wiley, 1966. 129p. $8.95.
Contents: Measurable functions. Measures. The integral. Integrable functions. The Lebesgue spaces L_p modes of convergence. Decomposition of measures. Generation of measures.

618. Burrill, Claude W. **Measure, Integration, and Probability.** New York: McGraw-Hill, 1972. 464p. $15.95.
Contents: Metric spaces. σ-Fields. Measure. Integration. Differentiation. Types of convergence. Hilbert space. Probability. Characteristic functions. Central limit problem. Conditional probability and expectation. Martingales. Stochastic processes.

619. Dinculeanu, N. **Vector Measures.** Elmsford, N.Y.: Pergamon, 1967. 432p. $17.50.
Contents: Classes of sets. Set functions. Variation of set functions. Semivariation of set functions. Extension of set functions. Measurable functions. Integration of step functions. Integrable functions. Measures defined by densities. Borel sets. Borel measures. Regular measures. Construction and extension of regular measures. Stieltjes measures. Reduction of the integration on abstract spaces to the integration on locally compact spaces. Disintegration of measure. M.R.34 No. 6011a.

620. Federer, Herbert. **Geometric Measure Theory.** New York: Springer-Verlag, 1969. 676p. $29.50.
Contents: Measure theory. The approach of Caratheodory and the approach of Riesz and Daniell. Suslin sets. Haar measures. Tangential and rectifiability properties of sets. Theory of currents.

621. Fichtenholz, G. M. **The Indefinite Integral.** New York: Gordon & Breach, 1971. 150p.
Contents: Integration of rational expressions. Integration of irrational expressions. Integration of trigonometric expressions.

622. Hartman, S., and J. Mikusinski. **The Theory of Lebesgue Measure and Integration.** Elmsford, N.Y.: Pergamon, 1961. 176p.
Contents: Lebesgue measure of linear sets. Measurable functions. The definite Lebesgue integral. Convergence in measure and equi-integrability. The Stieltjes integral.

623. Nachbin, Leopoldo. **The Haar Integral.** Trans. by Lulu Bechtolsheim. New York: Van Nostrand-Reinhold, 1965. 168p. $7.00.

Contents: Integration on locally compact spaces. Integration on locally compact groups. Integration on locally compact homogeneous spaces.

624. Parry, William. **Entropy and Generators in Ergodic Theory.** New York: W. A. Benjamin, 1969. 124p. $10.50; $3.95pa.
Contents: Basic convergence theorems. Entropy of a transformation. Lebesgue spaces and measurable partitions. α-generators. Generators for ergodic automorphisms. Zero entropy and finite generators. σ-finite endomorphisms and automorphisms. HIGHLY RECOMMENDED FOR BEGINNERS. M.R.41 No. 7071.

625. Pesin, I. N. **Classical and Modern Integration Theories.** New York: Academic Press, 1970. 195p. $12.50.
Contents: From Cauchy to Riemann. Development of integration ideas in the second half of the 19th century. The Borel measure. Lebesgue's measure and integration. Young's integral. Other definitions related to the definition of Lebesgue's integral. Stieltjes' integral. M.R.41 No. 8614.

626. Segal, Irving E., and R. A. Kunze. **Integrals and Operators.** New York: McGraw-Hill, 1968. 308p. $10.50.
Contents: Basic integrals. Measurable functions. Convergence differentiation. Euclidean spaces. Function spaces. Invariant integrals. Applications in commutative Banach algebras. Commutative set of normal operators and locally Abelian groups.

627. Shilov, G. E., and B. L. Guervich. **Integral, Measure and Derivative: A Unified Approach.** Trans. from the Russian by R. A. Silverman. Englewood Cliffs, N.J.: Prentice-Hall, 1966. 233p.
Contents: Riemann integration. Lebesgue integration in n-space. Riemann-Stieltjes integration. Lebesgue-Stieltjes integration. Measure. Constructive measure theory. Axiomatic measure theory. The derivative. M.R.30 No. 3188; M.R.33 No. 2781; M.R.36 No. 2765.

CHAPTER 19

COMPLEX ANALYSIS

628. Ahlfors, Lars V. **Complex Analysis: An Introduction to the Theory of Analytic Functions of One Complex Variable.** 2nd ed. New York: McGraw-Hill, 1966. 317p. $9.50.
Contents: Complex numbers. Analytic functions. Power series. Conformal mappings. Complex integration. Cauchy's integral formula. The calculus of residues. Harmonic functions. Power series expansion. Entire functions. Dirichlet's problem. The Riemann mapping theorem. Elliptic functions. The Weierstrass theory. Analytic continuation. RECOMMENDED.

629. Ash, R. B. **Complex Variables.** New York: Academic Press, 1970. 255p. $9.50.
Contents: The elementary theory. The general Cauchy theorem. Applications of the Cauchy theory. Entire functions. Families of analytic functions. The prime number theorem.

630. Carrier, George F., Max Krook, and Carl E. Pearson. **Functions of a Complex Variable: Theory and Technique.** New York: McGraw-Hill, 1966. 438p. $11.00.
Contents: Complex numbers. Analytic functions. Contour integration. Conformal mapping. Special functions. Asymptotic methods. Transform methods.

631. Churchill, Ruel V. **Complex Variables and Applications.** 2nd ed. New York: McGraw-Hill, 1960. 297p. $9.50.
Contents: Complex numbers. Analytic functions. Integrals. Power series. Residues and poles. Conformal mapping. The Schwarz-Christoffel transformation. Integral formulas of Poisson type.

632. Cohn, Harvey. **Conformal Mapping on Riemann Surfaces.** New York: McGraw-Hill, 1967. 319p. $12.95.
Contents: Review of complex analysis. Riemann manifolds. Elliptic functions. Derivation of existence theorems. Real existence proofs. Algebraic applications. M.R.36 No. 3974.

633. Collingwood, Edward, and A. J. Lohwater. **The Theory of Cluster Sets.** New York: Cambridge University Press, 1966. 222p. $8.50.
Contents: Functions analytic in a circular disc. Topics in the theory of conformal mapping. Intrinsic properties of cluster sets. Cluster sets of functions analytic in the unit disc. Boundary properties of functions meromorphic in the disc. Classification of singularities.

634. Depree, John D., and C. C. Oehring. **Elements of Complex Analysis.** Reading, Mass.: Addison-Wesley, 1969. 390p. $10.95.

Contents: Some point set topology. Arcs. Curves. Angles. Orientation. Differentiation. Riemann integration and Riemann surfaces. Cauchy's theorem. Cauchy-Goursat theorem. Isolated singularities and residues. Sequences. Series. Infinite products. Mittag-Leffler theorem. Weierstrass factorization theorem. Geometric function theory. Conformal mappings. The Riemann mapping theorem. Harmonic functions. Dirichlet problem. Green's functions. Entire functions. Hadamard's factorization theorem. Analytic continuation. Monodromy theorem. Picard theorem. M.R.40 No. 309.

635. Duncan, J. **The Elements of Complex Analysis.** New York: John Wiley, 1958 (1968pa.). 313p. $11.50; $5.75pa.
Contents: Metric spaces. The complex numbers. Continuous and differentiable complex functions. Power series functions. Arcs, contours and complex integration. Cauchy's theorem for starlike domains. Local analysis. Global analysis. Conformal mapping. Analytic continuation.

636. Fuchs, B. A., and B. V. Shabat. **Functions of a Complex Variable and Some of Their Applications.** Elmsford, N.Y.: Pergamon, 1964 (Vol. I), 1962 (Vol. II). Vol. I, 452p. Vol. II (by B. A. Fuchs and V. I. Levin), 296p.
Contents: VOLUME I. Complex analysis. Conformal mappings. Integral representation of regular functions. Harmonic functions. Power series. VOLUME II. Algebraic functions. The Laplace transform. Contour integration. Asymptotic expansions.

637. Goodstein, R. L. **Complex Functions: A First Course in the Theory of Functions of a Single Complex Variable.** New York: McGraw-Hill, 1965. 218p. $6.95.
Contents: The real numbers. The complex numbers. Convergence. Continuity. The Cauchy-Riemann equations. Cauchy's integral theorem. Cauchy's formula. Taylor's series. Poles. Residue theory. The Mittag-Leffler's theorem. Rouché's theorem. The Weierstrass factorization theorem.

638. Greenleaf, Frederick P. **Introduction to Complex Variables.** Philadelphia: W. B. Saunders, 1972. 588p.
Contents: The complex number system. Limits and functions of a complex variable. Power series and analytic functions. Holomorphic functions. Integration in the complex plane. Singularities and residues. Harmonic functions and boundary value problems. Physical applications of potential theory. Conformal mapping problems. RECOMMENDED.

639. Gunning, Robert C., and Hugo Rossi. **Analytic Functions of Several Complex Variables.** Englewood Cliffs, N.J.: Prentice-Hall, 1965. 317p.
Contents: Holomorphic functions. Local rings of holomorphic functions. Varieties. Analytic sheaves. Analytic spaces. Cohomology theory. Stein spaces. Geometric theory. Sheaf theory.

640. Heins, Maurice. **Complex Function Theory.** New York: Academic Press, 1968. 416p. $12.00.

Contents: Complex differentiation. Cauchy theory. Laurent expansion. Meromorphic functions. Zeros and poles. Power series. Harmonic functions. The
modular functions. Riemann surfaces.

641. Heins, Maurice. **Hardy Classes on Riemann Surfaces.** New York:
Springer-Verlag, 1969. 106p. $2.50.
Contents: Results on certain classes of analytic and harmonic functions on
Riemann surfaces. M.R.40 No. 338.

642. Hille, E. **Analytic Function Theory.** New York: Blaisdell, 1965 (Vol. I,
4th printing); 1962 (Vol. II.). Vol. I, 308p.; Vol. II, 496p.
Contents: VOLUME I. Number systems. The complex plane. Fractions, powers,
and roots. Holomorphic functions. Power series. Some elementary functions.
Complex integration. Representation theorems. The calculus of residues.
VOLUME II. Analytic continuation. Singularities and representation of analytic
functions. Algebraic functions. Elliptic functions. Entire and meromorphic
functions. Normal families. Lemniscates. Conformal mapping. Majorization.
Functions holomorphic in a half-plane. RECOMMENDED.

643. Hörmander, Lars. **An Introduction to Complex Analysis in Several
Variables.** New York: Van Nostrand-Reinhold, 1966. 208p. $6.95.
Contents: Cauchy's integral formula. The theorems of Runge, Mittag-Leffler, and
Weierstrass. Reinhardt domains. Applications to Banach algebras. Inhomogeneous
Cauchy-Riemann equations. Stein manifolds. Weierstrass preparation theorem.
RECOMMENDED. M.R.34 No. 2933.

644. Levinson, Norman, and Ray Redheffer. **Introduction to Complex
Variables.** San Francisco: Holden-Day, 1970. 350p.
Contents: The complex derivative. Complex integration. Residue theory. Con-
formal mapping. Infinite sequences. Singular integrals. RECOMMENDED.

645. Marden, Morris. **Geometry of Polynomials.** 2nd rev. ed. Providence,
R.I.: American Mathematical Society, 1966. 243p. $12.70.
Contents: A study of the location of the zeros of polynomials in the complex
plane.

646. Markushevich, A. I. **Theory of Functions of a Complex Variable.** Trans.
and ed. by Richard A. Silverman. Englewood Cliffs, N.J.: Prentice-Hall. Vol. I,
1965; 459p. Vol. II, 1965; 333p. Vol. III, 1967; 360p.
Contents: VOLUME I. Differentiation. The Cauchy-Riemann equations. Meromorphic functions. Integration. Cauchy's theorem. Infinite products. VOLUME
II. Calculus of residues. Harmonic and subharmonic functions. Entire and meromorphic functions. VOLUME III. Conformal mapping. Periodic and elliptic
functions. Riemann surfaces.

647. Nachbin, Leopoldo. **Holomorphic Functions, Domains of Holomorphy
and Local Properties.** Amsterdam: North-Holland, 1971. 129p. $4.95pa.
Contents: Differentiation of holomorphic functions and the Cauchy inequali-

ties. Taylor series and unique analytic continuation. Maximum modulus theorem. Removable singularities. The Cartan-Thullen theorem.

648. Pennisi, Louis L. **Elements of Complex Variables.** New York: Holt, Rinehart, and Winston, 1966. 459p. $8.50.
Contents: Complex numbers and their geometric representation. Point sets. Sequences and mappings. Single-valued analytic functions. Integration. Power series. The calculus of residues. Conformal representation.

649. Phillips, E. G. **Some Topics in Complex Analysis.** New York: Pergamon, 1966. 139p. $5.50.
Contents: Elliptic functions. The Jacobian elliptic functions. Conformal transformations. Schlicht functions. The maximum-modulus principle. Integral functions. Expansions in infinite series. Contour integrals defining some special functions. M.R.36 No. 360.

650. Pyatetskii-Shapiro, I. I. **Automorphic Functions and the Geometry of Classical Domains.** Trans. from the Russian. New York: Gordon & Breach, 1969. 268p.
Contents: Siegel domains. The geometry of classical domains. Normal discrete groups of analytical automorphisms of classical domains. Automorphic forms. Bounded homogeneous domains. M.R.40 No. 5908.

651. Rodin, Burton, and Leo Sario. **Principal Functions.** New York: Van Nostrand-Reinhold, 1968. 360p. $10.50.
Contents: Principal functions. Riemann surface theory. The normal operator method. Conformal mapping. The theorems of Riemann-Roch and Abel. Capacity, stability, and extremal length. Principal functions in harmonic spaces. M.R.37 No. 5378.

652. Rudin, Walter. **Function Theory in Polydiscs.** New York: W. A. Benjamin, 1969. 188p. $12.50; $3.95pa.
Contents: Elements of complex analysis in several variables. The Poisson integral. The Hardy spaces. The Nevannlinna spaces. Zero sets. Inner functions. Zeros and interpolations on the torus. Embeddings. M.R.41 No. 501.

653. Sario, Leo, and M. Nakai. **Classification Theory of Riemann Surfaces.** New York: Springer-Verlag, 1970. 446p. $27.00.
Contents: Riemann surfaces. Holomorphic functions. Harmonic functions with finite Dirichlet integral. Harmonic functions satisfying certain boundedness conditions. Harmonic and holomorphic functions with a simple logarithmic singularity. Meromorphic functions. Soilow compactification. M.R.41 No. 8660.

654. Sario, Leo, and K. Noshiro. **Value Distribution Theory.** Princeton, N.J.: D. Van Nostrand, 1966. 236p. $7.50.
Contents: Mappings into closed Riemann surfaces. Mappings into open Riemann surfaces. Functions of bounded characteristic. Functions on parabolic Riemann surfaces. Picard sets. Riemann images. Appendices: basic properties of Riemann surfaces, Gaussian mapping of arbitrary minimal surfaces.

655. Schwartz, M. H. **Lectures on Stratification of Complex Analytic Sets.**
Bombay: Tata Institute of Fundamental Research, 1966. 77p. $2.00.
Contents: Pseudo-bundles. Regular stratifications of analytic sets.

656. Siegel, C. L. **Topics in Complex Function Theory. Vol. I: Elliptic
Functions and Uniformization Theory.** Trans. from the German. New York:
John Wiley, 1969. 186p. $9.95.
Contents: Elliptic functions. Doubling the arc of the lemniscate. Riemann
surfaces. Elliptic integrals. Uniformization theory for algebraic functions.

657. Veech, William A. **A Second Course in Complex Analysis.** New York:
W. A. Benjamin, 1967. 246p. $8.75.
Contents: Analytic continuation. Koebe and Riemann's mapping theorems.
The modular functions. Hadamard's product theorem. The prime number theorem.

658. Wayland, Harold. **Complex Variables Applied in Science and Engineering.**
New York: Van Nostrand-Reinhold, 1970. 350p. $9.50.
Contents: Complex numbers. Functions. Integration in the complex plane.
Series of functions. Analytic functions. Residue theory. Applications to dif-
ferential equations and potential theory.

CHAPTER 20

POTENTIAL THEORY

659. Burago, Yu. D., and V. G. Maz'ya. **Potential Theory and Function Theory for Irregular Regions.** Trans. from the Russian. New York: Plenum, 1969. 68p. $12.50.
Contents: Multidimensional potential theory and the solution of boundary value problems for regions with irregular boundaries. Space of functions whose derivatives are measures. M.R.39 No. 1633.

660. Du Plessis, Nicolas. **An Introduction to Potential Theory.** Edinburgh: Oliver and Boyd, 1970. 177p.
Contents: Superharmonic, subharmonic, and harmonic functions in R^n. The conductor problem and capacity. The Dirichlet problem.

661. Helms, L. L. **Introduction to Potential Theory.** New York: John Wiley, 1969. 282p. $14.95.
Contents: Harmonic functions. Functions harmonic on a ball. Boundary limit theorems. Superharmonic functions. Green functions. Green potentials. Capacity. The generalized Dirichlet problem. Dirichlet problem for unbounded regions. Fine topology. Energy. Martin boundary. M.R.41 No. 5638.

662. Sneddon, Ian N. **Mixed Boundary Value Problems in Potential Theory.** New York: John Wiley, 1966. 283p.
Contents: Mixed boundary value problems in potential theory. Electrostatic problems. Heat conduction problems. Special functions. Integral equations. Operators of fractional integration. Electrified circular disk. Dual integral equations. Dual series relations. Triple integral equations. Integral representations of harmonic functions.

CHAPTER 21

SPECIAL FUNCTIONS

663. Babister, A. W. **Transcendental Functions Satisfying Nonhomogeneous Linear Differential Equations.** New York: Macmillan, 1967. 414p. $14.95. Contents: The Bessel functions. The hypergeometric functions. The confluent hypergeometric functions. The Legendre functions. The generalized hypegeometric functions. The Heun functions. The Lame functions. The Mathieu functions. The Laplace equation. The wave equation. The Lorentz equation. Series expansions. Asymptotic expansions.

664. Lebedev, N. N. **Special Functions and Their Applications.** Trans. by Richard A. Silverman. Englewood Cliffs, N.J.: Prentice-Hall, 1965. 308p. Contents: The Gamma function. The probability integral and related functions. The exponential integral and related functions. Orthogonal polynomials. Cylinder function, theory. Cylinder functions, applications. Spherical harmonics, theory. Spherical harmonics, applications. Hypergeometric functions. Parabolic cylinder functions.

665. Luke, Yudell L. **The Special Functions and Their Approximations.** New York: Academic Press, 1969. Vol. I, 349p.; $19.50. Vol. 2, 487p.; $24.50. Contents: VOLUME I. Asymptotic expansions. The gamma function and related functions. Hypergeometric functions. Confluent hypergeometric functions. The generalized hypergeometric function and the G-function. Identification of the F and G-functions with the special functions of mathematical physics. Asymptotic expansions of F for large parameters. Orthogonal polynomials. M.R.39 No. 3039. VOLUME II. Expansions of generalized hypergeometric functions in series of functions of the same kind. The τ-method. Polynomial and rational approximations to generalized hypergeometric functions. Recursion formula for polynomials and functions which occur in infinite series and rational approximations to generalized hypergeometric functions. Polynomial and rational approximations for $E(z) = {}_2F_1(1, \sigma p + 1; -1/z)$. Polynomial and rational approximations for the incomplete gamma function. Trapezoidal rule integration formulas.

666. MacRobert, T. M. **Spherical Harmonics: An Elementary Treatise on Harmonic Functions with Applications.** Elmsford, N.Y.: Pergamon, 1967. 368p. Contents: Fourier series. Spherical harmonics. The hypergeometric function. The Legendre polynomials and functions. Applications to potential theory and to electrostatics. Ellipsoids of revolution. Clerk Maxwell's theory of spherical harmonics. Bessel functions. Asymptotic expansions. Fourier-Bessel expansion. Applications. M.R.36 No. 4037.

667. Magnus, W., F. Oberhettinger, and R. P. Soni. **Formulas and Theorems for the Special Functions of Mathematical Physics.** 3rd rev. ed. New York: Springer-Verlag, 1966. 508p. $16.50.

Contents: Gamma functions. Hypergeometric functions. Bessel functions. Legendre functions. Orthogonal polynomials. Kummer's functions. Whittaker functions. Parabolic cylinder functions. Parabolic functions. Incomplete gamma functions. Elliptic integrals. Theta functions. Elliptic functions. Integral transforms. Transformation of systems of coordinates.

668. Miller, Willard, Jr. **Lie Theory and Special Functions.** New York: Academic Press, 1968. 338p. $16.50.
Contents: Resume of Lie theory. Representations and realizations of Lie algebras. Lie theory and Bessel functions. Lie theory and confluent hypergeometric functions. Lie theory and hypergeometric functions. Special functions related to the Euclidean group in 3-space. The factorization method. Generalized Lie derivatives.

669. Slater, Lucy Joan. **Confluent Hypergeometric Functions.** New York: Cambridge University Press, 1960. 260p. $15.00.
Contents: Differential equations satisfied by confluent hypergeometric functions. Differential properties. Integral properties. Asymptotic expansion. Related differential equations and particular cases of the functions. Descriptive properties.

670. Slater, Lucy Joan. **Generalized Hypergeometric Functions.** New York: Cambridge University Press, 1966. 288p. $11.50.
Contents: The Gauss function. The generalized Gauss function. Basic hypergeometric functions. Hypergeometric integrals. Basic hypergeometric integrals. Bilateral series. Basic bilateral series. Appel series. Basic Appel series.

671. Talman, James D. **Special Functions: A Group Theoretic Approach.** New York: W. A. Benjamin, 1968. 260p.
Contents: Lie algebras and Lie groups and their representations. Orthogonal groups. Euclidean groups. The coefficients of irreducible representations of such groups expressed in terms of special functions. M.R.39 No. 511.

672. Tranter, C. J. **Bessel Functions with Some Physical Applications.** New York: Hart, 1969. 148p. $10.00.
Contents: The solution of Bessel's and associated equations. Some indefinite integrals. Expansions and additions theorems. Zeros of Bessel functions. Fourier-Bessel series and Hankel transforms. Some finite and infinite definite integrals containing Bessel functions. Dual integrals and dual series equations. The equations of mathematical physics. Solution by separation of variables. The equations of mathematical physics. Solution by integral transforms.

CHAPTER 22

FUNCTIONAL ANALYSIS

673. American Mathematical Society. **Five Papers on Functional Analysis.**
A.M.S. Translations: Series II, Vol. 62. Providence, R.I.: American Mathematical
Society, 1967. 284p. $14.90.
Contents: Generalized extensions of nonsymmetric operators, E. R. Cekanovskii.
On the Friedrichs model in the theory of perturbations of a continuous spectrum,
L. D. Faddeev. The existence of spectral functions of generalized second-order
differential systems with a boundary condition at the singular end, I. S. Kac.
Fundamental concepts of the spectral theory of Hermitian operators, A. I.
Plesner. Spectral theory of linear operators, A. I. Plesner and V. A. Rohlin.

674. American Mathematical Society. **Nine Papers on Functional Analysis.**
A.M.S. Translations: Series II, Vol. 93. Providence, R.I.: American Mathematical
Society, 1970. 253p. $13.00.
Contents: On an eigenfunction expansion for selfadjoint operators, Ju. M.
Berezanskii. On regular bases in nuclear spaces, M. M. Dragilev. Positive-
definite functionals on nuclear spaces, A. G. Kostjucenko and B. S. Mitjagin.
Introduction to the geometry of indefinite J-spaces and to the theory of opera-
tors in those spaces, M. G. Krein. On some new Banach algebras and Wiener-
Lévy type theorems for Fourier series and integrals, M. G. Krein. Nuclearity and
other properties of spaces of type S, B. S. Mitjagin. Exponential law for spaces
of continuous linear mappings, D. A. Raikov. Theorems on positive solutions of
equations of the second kind with nonlinear operators, V. Ja. Stecenko and A. R.
Esajan. Bessel and Hilbert systems in Banach spaces and questions of stability,
B. E. Veic.

675. American Mathematical Society. **Thirteen Papers on Functional Analysis.**
A.M.S. Translations: Series II, Vol. 90. Providence, R.I.: American Mathematical
Society, 1970. 253p. $12.90.
Contents: An evolutionary equation and an expansion of open systems, Z. S.
Agranovic. On estimation of the spectrum of a class of positive linear operators,
I. A. Bahtin. On continuous branches of semieigenvectors of nonlinear opera-
tors, I. A. Bahtin. On factorization of operators relative to a discrete chain of
projectors in Banach space, M. A. Barkar' and I. C. Gohberg. On the factoriza-
tion of operators in Banach space, M. A. Barkar' and I. C. Gohberg. On multiplicative operators in Banach algebras, I: General propositions, M. S. Budjanu and
I. C. Gohberg. Unconditional bases and semiorderedness, Ja. M. Ceitlin. Analytic
problems and results in the theory of linear operators in Hilbert space, M. G.
Krein. The characteristic operator-function for an arbitrary bounded operator,
O. B. Kuzel'. Imbedding theorems for functions with partial derivatives considered
in various metrics, S. M. Nikol'skii. On some questions of analysis in Hilbert
space, II, G. E. Silov. On one-parameter families of extensions of a symmetric
operator, A. V. Straus. Imbedding theorems for the generalized Sobolev
classes W_p^r, S. V. Uspenskii.

676. Bachman, G., and L. Narici. **Functional Analysis.** New York: Academic Press, 1966. 530p. $16.25.
Contents: Inner product spaces. Orthogonal projections. Normed spaces. Metric spaces. Isometrics and completion of a metric space. Compactness. Separable spaces. Topological spaces. Banach spaces. Equivalent norms. Factor spaces. Hilbert spaces. Bessel's inequality. Complete orthonormal sets. The Hahn-Banach theorem. Weak convergence. Bounded linear transformations. The principle of uniform boundedness. Closed transformations. The closed graph theorem. Conjugate transformations. Banach algebras. Adjoints and sesquilinear functionals. Orthogonal projections. Positive definite operators. Square roots and a spectral decomposition theorem. The spectral theorem. RECOMMENDED. M.R.36 No. 638.

677. Beals, Richard. **Topics in Operator Theory.** Chicago: University of Chicago Press, 1971. 130p.
Contents: Bounded and unbound operators in Hilbert spaces. Self-adjoint operators. Spectral decompositions of self-adjoint and unitary operators. Shift operators. Dissipative operators. Spectral measures. Positive measures. Unitary groups.

678. Bonic, Robert A. **Linear Functional Analysis.** New York: Gordon & Breach, 1969. 136p.
Contents: Vector spaces. Metric spaces. Banach spaces. Some special Banach spaces. Some important classes of operators. Chains and nuclear spaces.

679. Browder, Felix E., editor. **Functional Analysis and Related Fields.** New York: Springer-Verlag, 1970. 241p.
Contents: Nonlinear eigenvalue problems and group invariance, F. E. Browder. Minimal submanifolds of a sphere with second fundamental form of constant length, S. S. Chern, M. do Carmo, and S. Kobayashi. Eisenstein series over finite fields, Harish-Chandra. Lp transforms on compact groups, E. Hewitt. Theory of simple scattering and eigenfunction expansion, T. Kato and S. T. Kuruda. Induced representations of locally compact groups and applications, G. W. Mackey. Convolution operators in spaces of nuclearly entire functions on a Banach space, L. Nachbin. Operants, a functional calculus for non-commuting operators, E. Nelson. Local non-linear functions of quantum fields, I. Segal. On the analogue of the modular group in characteristic "p", A. Weil. A theorem on the formal multiplication of trigonometric series, A. Zygmund. The influence of M. H. Stone on the origins of category theory, S. MacLane.

680. Collatz, L. **Functional Analysis and Numerical Mathematics.** Trans. by H. Oser. New York: Academic Press, 1966. 473p. $18.50.
Contents: Foundations of functional analysis and applications. Iterative methods. Monotonicity, inequalities, and other topics. Remarks on Schauder's fixed-point theorem. RECOMMENDED. M.R.34 No. 4961.

681. Dauns, John, and Karl H. Hofmann. **Representation of Rings by Sections.** Providence, R.I.: American Mathematical Society, 1968. 180p. $2.20.

Contents: Point-set topological theory of fields. The stalks of fields. Representation of topological algebras. M.R.40 No. 752.

682. Davis, Martin. **A First Course in Functional Analysis.** New York: Gordon & Breach, 1966. 110p.
Contents: Set theoretic preliminaries. Normed linear spaces and algebras. Functions on Banach spaces. Homomorphisms on normed linear spaces. Analytic functions into a Banach space. RECOMMENDED FOR BEGINNERS.

683. Dunford, N., and J. T. Schwartz. **Linear Operators.** New York: John Wiley. Part I: General Theory; 1958; 872p.; $24.95. Part II: Spectral Theory, Self Adjoint Operators in Hilbert Space; 1963; 1072p.; $37.50. Part III: Spectral Operators; 1971; 667p.; $32.50.
Contents: PART I. Three basic principles of linear analysis. Integration and set functions. Special spaces. Convex sets and weak topologies. Operators and their adjoints. General spectral theory. Applications. PART II. B-algebras. Bounded normal operators in Hilbert space. Miscellaneous applications. Unbounded operators in Hilbert space. Ordinary differential operators. Linear partial differential equations and operators. PART III. Spectral operators. Spectral operators, sufficient conditions. Algebras of spectral operators. Unbounded spectral operators. Perturbations of spectral operators with discrete spectra. Spectral operators with continuous spectra. Applications of the general theory.

684. Duren, Peter L. **Theory of H^p Spaces.** New York: Academic Press, 1970. 251p. $14.50.
Contents: Harmonic and subharmonic functions. Basic structure of H^p functions. Applications. Conjugate functions. Mean growth and smoothness. Taylor coefficients. H^p as a linear space. External problems. Interpolation theory. H^p spaces over general domains. H^p spaces over a half-plane. The Corona theorem.

685. Edwards, R. E. **Functional Analysis: Theory and Applications.** New York: Holt, Rinehart, and Winston, 1965. 781p. $25.00.
Contents: Set theory and topology. Vector spaces. Topological spaces. The Hahn-Banach theorem. Fixed point theorems. Duals of spaces. Radon measures. Distributions and differential equations. The open mapping theorem. The closed graph theorem. Boundedness principles. Duality theory. Compact operators. The Krein-Milman theorem.

686. Epstein, Bernard. **Linear Functional Analysis.** Philadelphia: W. B. Saunders, 1970. 230p. $9.50.
Contents: Metric spaces. Lebesgue measure and integration. The L^p- and l^p-spaces. Normed linear spaces. Linear functionals. Operators. Operators on finite-dimensional spaces. Elements of spectral theory in infinite-dimensional Hilbert spaces.

687. Fano, Guido. **Mathematical Methods of Quantum Mechanics.** Trans. from the Italian. New York: McGraw-Hill, 1971. 428p.
Contents: Elements of quantum mechanics. Linear spaces. Topological spaces. Hilbert spaces. Measure theory.

688. Friedman, A. **Foundations of Modern Analysis**. New York: Holt, Rinehart, and Winston, 1970. 250p.
Contents: Measure theory. Integration. Metric spaces. Elements of functional analysis in Banach spaces. Completely continuous operators. Hilbert spaces and spectral theory.

689. Friedrichs, K. O. **Perturbation of Spectra in Hilbert Space**. Providence, R.I.: American Mathematical Society, 1965. 178p.
Contents: Justification of the computational methods used in quantum mechanics.

690. Goffman, Casper, and George Pedrick. **A First Course in Functional Analysis**. Englewood Cliffs, N.J.: Prentice-Hall, 1965. 282p.
Contents: Metric space. Banach spaces. Measure and integration. Lp spaces. Hilbert space. Topological vector spaces. Banach algebras. RECOMMENDED.

691. Gohberg, I. C., and M. G. Krein. **Theory and Applications of Volterra Operators in Hilbert Space**. Trans. from the Russian by A. Feinstein. Providence, R.I.: American Mathematical Society, 1970. 430p. $31.10.
Contents: The abstract triangular representation of completely continuous operators. General theorems on transformers. The triangular truncation transformer. Relations between the spectra of the Hermitian components of Volterra operators. Factorization of operators close to the identity. Triangular models for Volterra operators. Self-adjoint boundary value problems for canonical equations. Tests for the stable boundedness of solutions of canonical equations with periodic coefficients. Fundamental theorem on the density of the spectrum of the real component of a Volterra operator with nuclear imaginary component. Unicellular operators and connections with analytical problems. M.R.36 No. 2007; M.R.39 No. 7447.

692. Goldberg, Seymour. **Unbounded Linear Operators: Theory and Applications**. New York: McGraw-Hill, 1966. 199p. $11.50.
Contents: Introduction to normed linear spaces. Linear operators and their conjugates. Strictly singular operators. Operators with closed range. Perturbation theory. Applications to ordinary differential operators. The Dirichlet operator. RECOMMENDED. M.R.34 No. 580.

693. Goldstein, Allen A. **Constructive Real Analysis**. New York: Harper & Row, 1967. 178p. $9.25.
Contents: Solutions of non-linear operator equations. Newton's method. Gradient techniques. Non-linear programming. Polyhedral and infinite convex programming. M.R.36 No. 705.

694. Halmos, Paul R. **A Hilbert Space Problem Book**. New York: Van Nostrand-Reinhold, 1967. 384p. $11.95.
Contents: Vectors and spaces. Weak topology. Analytic functions. Infinite matrices. Boundedness and invertibility. Operator matrices. Properties of spectra. Examples of spectra. Spectral radius. Norm topology. Strong and weak topologies. Partial isometries. Unilateral shift. Compact operators. Subnormal

operators. Numerical range. Unitary dilations. Commutators of operators. Toeplitz operators. M.R.34 No. 8178.

695. Helmberg, G. **Introduction to Spectral Theory in Hilbert Space.** Amsterdam: North-Holland, 1969. 360p. $19.50.
Contents: The concept of Hilbert space. Specific geometry of Hilbert space. Bounded linear operators. General theory of linear operators. Spectral analysis of compact linear operators. Spectral analysis of bounded linear operators. Spectral analysis of unbounded selfadjoint operators. HIGHLY RECOMMENDED. M.R.39 No. 4689.

696. Hille, E., and R. S. Phillips. **Functional Analysis and Semi-Groups.** Rev. ed. Providence, R.I.: American Mathematical Society, 1968. 808p.
Contents: Abstract spaces. Linear transformations. Banach algebras. Analysis in a Banach algebra. Laplace integrals and binomial series. Subadditive functions. Semi-modules. Addition theorems in a Banach algebra. Semi-groups in the strong topology. Generator and resolvent. Generation of semi-groups. Perturbation theory. Adjoint theory. Operational calculus. Spectral theory. Translations and powers. Trigonometric semi-groups. Semi-groups in Hilbert space. RECOM-MENDED.

697. Horvath, J. **Topological Vector Spaces and Distributions.** Vol. I. Reading, Mass.: Addison-Wesley, 1966. 449p.
Contents: Banach spaces. Locally convex spaces. Duality. Distributions.

698. Kantorovich, L. V., and G. P. Akhilov. **Functional Analysis in Normed Spaces.** Elmsford, N.Y.: Pergamon, 1964. 784p.
Contents: Metric spaces. Normed spaces. Linear operations and functionals. The extension of linear operations and functionals. Spaces of operations and functionals. The analytic representation of functionals. Sequences of linear operations. The weak convergence of functionals and elements. Completely continuous and adjoint operations. The analytic representation of certain classes of linear operations. Linear topological spaces. Adjoint equations. The general theory of approximation methods. The method of steepest descent. The fixed point principle. The differentiation of non-linear operations. Newton's method. RECOMMENDED.

699. Kato, Tosio. **Perturbation Theory for Linear Operators.** New York: Springer-Verlag, 1966. 592p. $19.80.
Contents: Operator theory in finite-dimensional vector spaces. Perturbation in a finite-dimensional space. Introduction to the theory of operators in Banach spaces. Stability theorems. Operators in Hilbert spaces. Sesquilinear forms in Hilbert spaces and associated operators. Analytic perturbation theory. Asymptotic perturbation theory. Perturbation theory for semigroups of operators. Perturbation of continuous spectra and unitary equivalence. RECOMMENDED. M.R.34 No. 3324.

700. Lorch, Edgar Raymond. **Spectral Theory**. New York: Oxford University Press, 1962. 158p.
Contents: Banach spaces. Linear transformations. Hilbert space. Spectral theory of linear transformations. The structure of self-adjoint transformations. Commutative Banach algebras.

701. Maddox, I. J. **Elements of Functional Analysis**. New York: Cambridge University Press, 1970. 208p. $6.95.
Contents: Basic set theory and analysis. Metric and topological spaces. Linear metric spaces. Normed linear spaces. Banach algebras. Hilbert space. Matrix transformations in sequence spaces. HIGHLY RECOMMENDED FOR BEGINNERS.

702. Malgrange, B. **Ideals of Differentiable Functions**. London: Oxford University Press, 1967. 106p. $6.00.
Contents: Whitney's extension theorem. Closed ideals. Analytic and differentiable algebras. Metric and differential properties of analytic sets. The preparation theorem for differentiable functions. Ideals defined by analytic functions. Applications to the theory of distributions. M.R.35 No. 3446.

703. Nachbin, Leopoldo. **Lectures on the Theory of Distributions**. Recife: Instituto de Fisica e Matematica, Universidade do Recife, 1964. 280p.
Contents: Differentiable maps from one normed space into another. Taylor's formula for such maps. Properties of distributions. Convolutions. The Fourier transform. The space of tempered distributions. Analytic maps between normed spaces. Entire functions. Entire functions of exponential type. The Paley-Wiener-Schwartz theorem. Schwartz's kernel theorem.

704. Nachbin, Leopoldo. **Topology on Spaces of Holomorphic Mappings**. New York: Springer-Verlag, 1969. 66p. $4.50.
Contents: Description of a method of endowing some vector spaces of holomorphic mappings with locally convex topologies and some results. M.R.40 No. 7787.

705. Naimark, M. A. **Normed Rings**. Rev. ed. Trans. by L. F. Boron. The Netherlands: Wolters-Noordhoff, 1970. 572p. $18.50.
Contents: Basic ideas from topology and functional analysis. Fundamental concepts and propositions in the theory of normed rings. Commutative normed rings. Representations of symmetric rings. Some special rings. Group rings. Rings of operators in Hilbert space. Decomposition of a ring of operators into irreducible rings. RECOMMENDED.

706. Peressini, Anthony L. **Ordered Topological Vector Spaces**. New York: Harper & Row, 1967. 228p. $11.50.
Contents: Ordered vector spaces. The Choquet-Kendall theorem. Lattice ideals. Convergence in vector lattices. Ordered topological vector spaces. The order topology. The topology of uniform convergence on order bounded sets. M.R. 37 No. 3315.

707. Porcelli, Pasquale. **Linear Spaces of Analytic Functions.** New York: McGraw-Hill, 1967. 158p. $9.00.
Contents: Log mean. Functions with positive real part. Boundary values and H_p spaces. H_p algebras. The ring of analytic functions. Problems in approximation. Topology of pointwise convergence. The Fourier transform. Infinite systems of equations in Hilbert space. The moment problem. M.R.41 No. 4219.

708. Przeworska-Rolewicz, D., and S. Rolewicz. **Equations in Linear Spaces.** Warsaw: Polish Scientific Publ., 1968. 380p.
Contents: Operators with a finite and semifinite dimensional characteristic. Algebraic and almost algebraic operators. Linear operators in linear topological spaces. Compact operators in linear topological spaces. Linear operators in Banach spaces. Perturbations of operators. Fredholm alternative transform and similar transforms. M.R.40 No. 1845.

709. Putnam, C. R. **Commutation Properties of Hilbert Space Operators and Related Topics.** New York: Springer-Verlag, 1967. 167p. $8.50.
Contents: Commutators of bounded operators. Commutators and spectral theory. Semi-normal operators. Commutation relations in quantum mechanics. Wave operators and unitary equivalence of self-adjoint operators. Laurent and Toeplitz operators. Singular integral operators and Jacobi matrices. M.R.36 No. 707.

710. Rellich, Franz. **Perturbation Theory of Eigenvalue Problems.** New York: Gordon & Breach, 1969. 127p. $9.25; $6.25pa.
Contents: Small perturbation parameter and small perturbation. Character of perturbation. Non-analytic perturbation. Perturbation of the discrete spectrum. Power series in Hilbert space.

711. Riesz, F., and B. Sz.-Nagy. **Functional Analysis.** Trans. by L. F. Boron. New York: Ungar, 1971 (5th printing). 468p. $12.50.
Contents: Differentiation. The Lebesgue integral. The Stieltjes integral and its generalizations. Integral equations. Hilbert and Banach spaces. Completely continuous symmetric transformations of Hilbert spaces. Bounded symmetric, unitary, and normal transformations of Hilbert space. Unbounded linear transformations of Hilbert space. Self-adjoint transformations, functional calculus, spectrum, and perturbations. Groups and semi-groups of transformations. Spectral theories for linear transformations of general type.

712. Schechter, M. **Principles of Functional Analysis.** New York: Academic Press, 1971. 383p. $15.00.
Contents: Basic notions. Duality. Linear operators. The Riesz theory for compact operators. Fredholm operators. Spectral theory. Unbounded operators. Reflexive Banach spaces. Banach algebras. Semigroups. Hilbert space. Bilinear forms. Self-adjoint operators. Examples and applications. RECOMMENDED.

713. Schmeidler, Werner. **Linear Operators in Hilbert Space.** Trans. from the German by J. Strum; ed. by A. Shenitzer and D. Solitar. New York: Academic Press, 1965. 122p. $6.00; $2.95pa.

Contents: Abstract Hilbert spaces. Orthogonal systems. Bounded linear operators. Biorthogonal systems. Completely continuous operators. The Fredholm theorems. Self-adjoint and normal completely continuous operators. Positive bounded self-adjoint operators and their square roots. Spectral theory.
RECOMMENDED.

714. Schwartz, Jacob T. **Nonlinear Functional Analysis**. New York: Gordon & Breach, 1969. 244p.
Contents: Basic calculus. Implicit functional theorems. Degree theory and applications. Morse theory on Hilbert manifolds. Category. Applications of Morse theory to the calculus of variations in the large. Applications. Closed geodesics on topological spheres.

715. Schwartz, Jacob T. **W*-Algebras**. New York: Gordon & Breach, 1967. 256p. $13.50; $7.95pa.
Contents: Direct integral decompositions. Types of infinite-dimensional factors. M.R.38 No. 547.

716. Segal, Irving E., and Ray Kunze. **Integrals and Operators**. New York: McGraw-Hill, 1968. 308p. $12.00.
Contents: Basic integrals. Measurable functions and their integrals. Convergence and differentiation. Set functions. Locally compact and Euclidean spaces. Function spaces. Invariant integrals. Banach algebras and Hilbert space. Spectral analysis in Hilbert space. Group representations and unbounded operators.

717. Sz.-Nagy, Bela, and Ciprian Foias. **Harmonic Analysis of Operators on Hilbert Space**. Amsterdam: North-Holland, 1970. 400p. $20.75.
Contents: Contractions and their dilations. Geometrical and spectral properties of the minimal unitary dilation. Functional calculus. Bounded functions. Functional calculus. Functions bounded or not. Operator valued analytic functions. Characteristic functions and functional models. Regular factorizations of the characteristic function and invariant subspaces. Weak contractions. Problems of similarity, quasi-similarity, and unicellularity.

718. Trèves, F. **Topological Vector Spaces, Distributions and Kernels**. New York: Academic Press, 1967. 565p. $18.50.
Contents: Topological vector spaces. Spaces of functions. Duality: spaces of distributions. Tensor products: kernels. M.R.37 No. 726.

719. Vainberg, M. M. **Variational Methods for the Study of Nonlinear Operators**. Trans. by A. Feinstein. San Francisco: Holden-Day, 1964. 334p.
Contents: Analysis in linear spaces. Potential operators. Operator equations and the extremum of functionals. Conditionally extremal and conditionally critical points of a functional. Proper functions and branch points of nonlinear operators. Operators and functionals of special type. Nonlinear integral equations. Newton's method.

720. Vulikh, B. A. **Introduction to Functional Analysis for Scientists and Technologists.** Elmsford, N.Y.: Pergamon, 1963. 412p.
Contents: Euclidean spaces. Metric spaces. Continuous operators. Normed spaces. Hilbert spaces. The L^2-space. Linear operators. Adjoint and self-adjoint operators. Completely continuous operators. Partially ordered normed spaces.

721. Wilde, Carroll O., editor. **Functional Analysis.** New York: Academic Press, 1970. 162p. $6.50.
Contents: Preface. Several variable spectral theory, Joseph L. Taylor. Representation of certain operators in Hilbert space, G. K. Kalisch. Some inequalities involving the Hardy-Littlewood maximal function in a theory of capacities, C. Preston. A decomposition theorem for automorphisms of Von Neumann algebras, Robert R. Kallman. Recent progress in the structure theory of factors, J. T. Schwartz. An uncountable number of II factors, Shoichiro Sakai. An uncountable family of non-hyperfinite type III factors, Shoichiro Sakai. The isomorphism problem for measure-preserving transformations, Donald Ornstein. On certain analagous difficulties in the investigation of flows in a probability space and of transformations in an infinite measure space, U. Krengel. Spectral properties of measure preserving transformations, R. V. Chacon. Optimal conditioning of operators on Hilbert space, C. McCarthy. On the smoothness of functions satisfying certain integral inequalities, Adriana Garcia.

722. Yosida, Kosaku. **Functional Analysis.** 2nd ed. New York: Springer-Verlag, 1969. 465p.
Contents: Preliminaries. Semi-norms. Applications of the Baire-Hausdorff theorem. The orthogonal projection and F. Riesz's representation theorem. The Hahn-Banach theorems. Strong convergence and weak convergence. Fourier transform and differential equations. Dual operators. Resolvent and spectrum. Analytical theory of semi-groups. Compact operators. Normed rings and spectral representation. Other representation theorems in linear spaces. Ergodic theory and diffusion theory. The integration of the equation of evolution. RECOMMENDED. M.R.39 No. 741.

CHAPTER 23

DIFFERENTIAL AND INTEGRAL EQUATIONS

723. Agmon, Shmuel. **Lectures on Elliptic Boundary Value Problems.** New York: Van Nostrand-Reinhold, 1965. 291p. $4.95.
Contents: Calculus of L_2-derivatives. Elliptic operators. Local existence theory. Local regularity of solutions of elliptic systems. Gärding's inequality. Global existence. Coerciveness. Spectral theory of abstract operators. Eigenvalue problems. Completeness of the eigenfunctions.

724. Agranovich, Z. S., and V. A. Marchenko. **The Inverse Problem of Scattering Theory.** Trans. by B. D. Seckler. New York: Gordon & Breach, 1963. 290p.
Contents: Boundary value problems. Operational calculus. Scattering theory.

725. American Mathematical Society. **Thirteen Papers on Differential Equations.** A.M.S. Translations: Series II, Vol. 89. Providence, R.I.: American Mathematical Society, 1970. 300p. $15.20.
Contents: A theorem on an integral inequality and some of its applications, V. M. Alekseev. Expansion in eigenfunctions of selfadjoint operators corresponding to general elliptic eigenvalue problems with boundary conditions, V. V. Barkovskii. On the reduction of a system of linear equations to diagonal form, B. F. Bylov. The topology of Stokes lines for equations of the second order, M. V. Fedorjuk. On boundary problems for equations of higher orders in infinite domains, G. N. Finozenok. On curves defined by Liénard differential equations, L. V. Hohlova. Analytic properties of multipliers of periodic canonical differential systems of positive type, M. G. Krein and G. Ja. Ljubarskii. A method of reducing boundary problems for a system of differential equations of elliptic type to regular integral equations, Ja. B. Lopatinskii. On a mixed problem for a hyperbolic equation in a normal cylinder, V. R. Nosov. On the application of the sweeping method to the solution of nonlinear elliptic equations, V. A. Oliinik. On the reducibility of a system of ordinary differential equations in a neighborhood of a smooth toroidal manifold, A. M. Samoilenko. General theory of partial differential equations and question in the theory of boundary value problems, M. I. Visik and G. E. Silov. On a nonlocal integral manifold of a nonregularly perturbed differential system, K. V. Zadirada.

726. Ames, William F. **Nonlinear Partial Differential Equations in Engineering.** New York: Academic Press, 1965. 511p. $17.00.
Contents: The origin of nonlinear partial differential equations. Transformation and general solutions. Exact methods of solution. Approximate methods. M.R.35 No. 1235.

727. Arscott, F. M. **Periodic Differential Equations.** New York: Macmillan, 1964. 284p.
Contents: Mathieu's equation. Mathieu functions. Hill's equation. Lame's equation. The spheroidal and ellipsoidal wave equations.

728. Bailey, Paul, L. Shampine, and P. E. Waltman. **Nonlinear Two Point Boundary Value Problems.** New York: Academic Press, 1968. 171p. $9.50. Contents: Relations between the first and second boundary value problems. Picard's iteration. The distance between zeros and the uniqueness interval. Comparison theorems. Principal existence theorems. Further existence and uniqueness results. Numerical solution by initial value methods. Numerical solution by boundary value methods. M.R.37 No. 6524.

729. Barbashin, E. A. **Introduction to the Theory of Stability.** Trans. from the Russian. Groningen, The Netherlands: Wolters-Noorhoff, 1970. 223p. $9.95. Contents: Lyapunov functions. Stability of control systems. Stability of solutions of differential equations. M.R.41 No. 8737. M.R.37 No. 515.

730. Bear, H. S. **Differential Equations.** Reading, Mass.: Addison-Wesley, 1962. 207p. Contents: First order equations. Homogeneous equations. Exact equations. Integrating factors. Series solutions. Linear equations. Method of undermined coefficients. Variation of parameters. Laplace transform. Picard's theorem and systems of equations.

731. Bers, L., F. John, and M. Schechter. **Partial Differential Equations.** New York: John Wiley, 1966 (2nd printing). 343p. $11.00. Contents: Part I: Hyperbolic and parabolic equations. Part II: Elliptic equations. Supplement I: Eigenvalue expansions. Supplement II: Parabolic equations.

732. Bhatia, N. P., and G. P. Szego. **Dynamical Systems: Stability Theory and Applications.** Lecture Notes in Mathematics, No. 35. New York: Springer-Verlag, 1967. 416p. $6.00. Contents: Dynamical systems in metric spaces. Lyapunov methods for ordinary differential equations. Existence theorems. Comparison theorems. Autonomous and non-autonomous systems. M.R.36 No. 2917.

733. Bitsadze, A. V. **Boundary Value Problems for Second Order Elliptic Equations.** Trans. from the Russian. Amsterdam: North-Holland, 1968. 211p. $11.00. Contents: Certain qualitative and constructive properties of the solutions of elliptic equations. The Dirichlet problem for a second order elliptic equation. The Dirichlet problem for elliptic systems. The Poincare problem for second order elliptic systems in two independent variables. Certain classes of multi-dimensional singular integral equations and related boundary value problems. M.R.37 No. 1773.

734. Boyce, W. E., and R. C. Diprima. **Elementary Differential Equations and Boundary Value Problems.** New York: John Wiley, 1969. 533p. $11.95. Contents: First order differential equations. Second order linear equations. Series solutions of second order linear equations. Higher order linear equations. The Laplace transform. Systems of first order equations. Numerical methods. Nonlinear differential equations and stability. Partial differential equations and Fourier series. Boundary value problems. Sturm-Liouville theory.

735. Brand, Louis. **Differential and Difference Equations.** New York: John Wiley, 1966. 698p. $13.95.

Contents: Differential equations of the first order. Types of first order equations. Linear equations of the second order. Linear equations with constant coefficients. Systems of equations. Applications. Laplace transform. Linear difference equations. Linear difference equations with constant coefficients. Solutions in series. Mikusinski's operational calculus. Existence and uniqueness theorems. Interpolation and numerical integration. Numerical solutions. M.R.35 No. 430.

736. Brauer, F., and J. Nohel. **Ordinary Differential Equations: A First Course.** New York: W. A. Benjamin, 1967. 457p.

Contents: Variables separable. Homogeneous equations. Series solutions. Regular singular points. Bessel's functions. Systems of equations. Existence and uniqueness theorems. Laplace transform and numerical methods.

737. Burkill, J. C. **The Theory of Ordinary Differential Equations.** Edinburgh: Oliver & Boyd, 1962. 114p.

Contents: Existence of solutions. Oscillation theorems. Series solutions. Singularities of equations. Contour integral solutions. Legendre functions. Bessel functions and asymptotic series.

738. Carroll, Robert W. **Abstract Methods in Partial Differential Equations.** New York: Harper & Row, 1969. 374p. $14.95.

Contents: Functional analysis and distributions. Sobolev spaces. Elliptic theory. Nonelliptic theory. Evolution equations. Semigroups. Uniqueness and regularity for weak problems. Global analysis.

739. Chaundy, T. W. **Elementary Differential Equations.** Oxford: Clarendon Press, 1969. 414p. $6.40.

Contents: Elementary techniques. Linear equations with constant coefficients. Systems of equations with constant coefficients. Homogeneous linear equations. Series solutions. Hypergeometric functions. Legendre's functions. Bessel's functions. M.R.41 No. 2095.

740. Chester, Clive R. **Techniques in Partial Differential Equations.** New York: McGraw-Hill, 1970. 416p. $12.50.

Contents: Introduction to linear first-order equations. One-dimensional wave equations. Potential theory. The reduced wave equation. The heat equation. Classification of equations. The wave equation in higher dimensions. First-order equations in general. Second-order hyperbolic equations in two independent variables. Linear elliptic equations. More general linear hyperbolic equations. First-order systems. One-dimensional compressible flow. Shocks. Overdetermined systems. Variational methods. Transform methods. Integral equations in boundary-value problems. Techniques in solving linear integral equations. RECOMMENDED.

741. Churchill, Ruel V. **Fourier Series and Boundary Value Problems.** 2nd ed. New York: McGraw-Hill, 1963. 250p. $8.50.

Contents: Partial differential equations in physics. Superposition of solutions. Orthogonal sets of functions. Fourier series. Further properties of Fourier series. Fourier integrals. Boundary-value problems. Bessel functions and applications. Legendre polynomials and applications. Uniqueness of solutions.

742. Cochran, James A. **The Analysis of Linear Integral Equations.** New York: McGraw-Hill, 1972. 370p.
Contents: Fredholm integral equations. Schmidt theory and the resolvent kernel. Volterra equations. Hermitian kernels. Characteristic values. The Rayleigh-Ritz procedure. Intermediate operators and the method of Weinstein and Aronszajn. Series expansions. Nuclear kernels. Difference kernels and the method of Wiener and Hopf.

743. Coddington, Earl A. **An Introduction to Ordinary Differential Equations.** Englewood Cliffs, N.J.: Prentice-Hall, 1961. 292p.
Contents: Linear equations of the first order. Linear equations with constant coefficients. Linear equations with variable coefficients. Linear equations with regular singular points. Existence and uniqueness of solutions to first order equations. Existence and uniqueness of solutions to systems and n-th order equations.

744. Cole, Randal H. **Theory of Ordinary Differential Equations.** New York: Appleton, 1968. 273p.
Contents: Existence and uniqueness theorems. Linear differential equations. Series solutions. The two point boundary problems. Eigenvalue problems and expansion theory. Self-adjoint and non-self-adjoint eigenvalue problems. Sturm-Liouville theory. RECOMMENDED. M.R.39 No. 512.

745. Corduneanu, C. **Principles of Differential and Integral Equations.** Boston: Allyn & Bacon, 1971. 201p.
Contents: Existence theorems for differential equations. Systems of differential equations. Stability theory. Volterra integral equations. Fredholm integral equations.

746. Courant, R., and D. Hilbert. **Methods of Mathematical Physics.** New York: John Wiley, 1953, 1962. Vol. 1, 577p.; $14.00. Vol. 2, 850p.; $19.50.
Contents: VOLUME 1. The algebra of linear transformations and quadratic forms. Series expansions of arbitrary functions. Linear integral equations. The calculus of variations. Vibration and eigenvalue problems of mathematical physics. Application of the calculus of variations to eigenvalue problems. Special functions defined by eigenvalue problems. VOLUME 2. General theory of partial differential equations of first order. Differential equations of higher order. Potential theory and elliptic differential equations. Hyperbolic differential equations in two independent variables. Hyperbolic differential equations in more than two independent variables. HIGHLY RECOMMENDED.

747. Davies, T. V., and E. M. James. **Nonlinear Differential Equations.** Reading, Mass.: Addison-Wesley, 1966. 274p.
Contents: Autonomous differential equations of the second order. Singular

points. Cycles without contact and limit cycles. Special investigations of Liénard and LaSalle. The small-parameter method of Poincare and its extension. Differential equations in which a small parameter is associated as a factor with the differential coefficient of the highest order. Theory of centers and its applications by Bautin in a limit cycle investigation. Stability theory and the use of Lyapunov functions.Differential equations of the second order with a forcing term.

748. Dennemeyer, Rene. **Introduction to Partial Differential Equations and Boundary Value Problems.** New York: McGraw-Hill, 1968. 376p. $13.75.
Contents: Introduction to partial differential equations. First-order equations. Linear second-order equations. Elliptic differential equations. The wave equation. The heat equation.

749. Duff, G. F. C., and D. Naylor. **Differential Equations of Applied Mathematics.** New York: John Wiley, 1966. 423p. $13.95.
Contents: Finite systems. Distributions and waves. Parabolic equations and Fourier integrals. Laplace's equation and complex variables. Equations of motion. General theory of eigenvalues and eigenfunctions. Green's functions. Cylindrical eigenfunctions. Spherical eigenfunctions. Wave propagation in space.

750. El'sgol'ts, L. E. **Introduction to the Theory of Differential Equations with Deviating Arguments.** Trans. by Robert J. McLaughlin. San Francisco: Holden-Day, 1966. 111p.
Contents: Basic concepts and existence theorems. Linear equations. Stability theory. Periodic solutions. Some generalizations and a brief survey of other areas of the theory of differential equations with deviating arguments.

751. Epstein, Bernard. **Partial Differential Equations: An Introduction.** New York: McGraw-Hill, 1962. 273p. $10.50.
Contents: Some preliminary topics. Partial differential equations of first order. The Cauchy problem. The Fredholm alternative in Banach spaces. The Fredholm alternative in Hilbert spaces. Elements of potential theory. The Dirichlet problem. The heat equation. Green's functions and separation of variables. RECOMMENDED.

752. Feshchenko, S. F., N. I. Shkil', and L. D. Nikolenko. **Asymptotic Methods in the Theory of Linear Differential Equations.** Trans. from the Russian by Scripta Technica. Edited by Herbert Eagle. New York: American Elsevier, 1968. 286p. $14.00.
Contents: Construction of asymptotic solutions for second-order linear differential equations with slowly varying coefficients. Construction of an asymptotic solution for a system of second order linear differential equations with slowly varying coefficients. Asymptotic decomposition of a system of linear differential equations. Construction of an asymptotic solution in the case of multiple roots of the characteristic equation. Asymptotic solutions of differential equations in a Banach space. Asymptotic methods of solving linear partial differencial equations. M.R.36 No. 4081; M.R.33 No. 6057.

753. Fort, Tomlinson. **Differential Equations**. New York: Holt, Rinehart, and Winston, 1960. 184p.
Contents: Separation of variables. Exact equations. Integrating factors. Bernoulli equation. Existence theorems. Differential equations of first order and higher degree. Adjoint equations. Power series method. Laplace transform. Systems of equations. Boundary value problems. Partial differential equations.

754. Friedman, Avner. **Generalized Functions and Partial Differential Equations**. Englewood Cliffs, N.J.: Prentice-Hall, 1963. 340p.
Contents: Linear topological spaces. Spaces of generalized functions. Theory of distributions. Convolutions and Fourier transforms of generalized functions. W spaces. Fourier transforms of entire functions. The Cauchy problem for systems of partial differential equations. The Cauchy problem in several time variables. S spaces. Further applications to partial differential equations. Differentiability of solutions of partial differential equations.

755. Friedman, Avner. **Partial Differential Equations**. New York: Holt, Rinehart, and Winston, 1969. 262p.
Contents: Part I: Elliptic equations. Part II: Evolution equations. Part III: Selected topics. RECOMMENDED.

756. Friedman, Avner. **Partial Differential Equations of Parabolic Type**. Englewood Cliffs, N.J.: Prentice-Hall, 1964. 384p.
Contents: General notation. The maximum principle and some applications. The first initial-boundary value problem. Derivation of a priori estimates. The second initial-boundary value problem. Asymptotic behavior of solutions. Semilinear equations. Nonlinear boundary conditions. Free boundary problems. Fundamental solutions for parabolic systems. Boundary value problems for elliptic and parabolic equations of any order.

757. Friedrichs, K. O. **Lectures on Advanced Ordinary Differential Equations**. New York: Gordon & Breach, 1965. 205p.
Contents: Existence and uniqueness. Invariant differential equations. Lie's integrating factor. Topology of integral curves. Periodic solutions of differential equations. Linear differential equations for analytic functions. M.R.37 No. 483.

758. Friedrichs, K. O. **Pseudo-Differential Operators: An Introduction**. 2nd rev. ed. New York: Courant Institute, N.Y.U., 1970. 271p. $6.00.
Contents: Introduction. Basic properties of pseudo-differential operators. Boundary value problems for pseudo-differential equations.

759. Garabedian, P. R. **Partial Differential Equations**. New York: John Wiley, 1964. 672p. $14.95.
Contents: The method of power series. Equations of the first order. Classification of partial differential equations. Cauchy's problem for equations with two independent variables. The fundamental solution. Cauchy's problem in space of higher dimension. The Dirichlet and Neumann problems. Dirichlet's principle. Existence theorems of potential theory. Integral equations. Eigenvalue problems.

Tricomi's problem. Formulation of well-posed problems. Finite differences. Fluid dynamics. Free boundary problems. Partial differential equations in the complex domain.

760. Gilbert, Robert. **Function Theoretic Methods in Partial Differential Equations.** New York: Academic Press, 1969. 311p. $17.50.
Contents: An introduction to the theory of several complex variables. Harmonic functions in (p - 2)-variables. Elliptic differential equations in two variables with analytic coefficients. Singular partial differential equations. Applications of integral operators to scattering problems. M.R.39 No.3127.

761. Golomb, Michael, and Merrill E. Shanks. **Elements of Ordinary Differential Equations.** 2nd ed. New York: McGraw-Hill, 1965. 410p. $10.00.
Contents: Techniques for solving first-order equations. Existence, uniqueness, and geometry of solutions. Applications of first-order equations. Second-order differential equations. Linear differential equations of higher order. Constant coefficients. Laplace transforms and operator methods. Systems of differential equations. Linear equations with variable coefficients. Solution in power series.

762. Green, C. D. **Integral Equations Methods.** London: Thomas Nelson, 1969. 243p.
Contents: Function spaces. Integral operators. Integral equations and their applications. Methods of solutions of integral equations. Integral transformations. M.R.40 No. 658.

763. Greenspan, Donald. **Introduction to Partial Differential Equations.** New York: McGraw-Hill, 1961. 195p. $8.50.
Contents: Basic concepts. Fourier series. Second-order partial differential equations. The wave equation. The potential equation. The heat equation. Approximate solution of partial differential equations. RECOMMENDED.

764. Hajek, Otomar. **Dynamical Systems in the Plane.** New York: Academic Press, 1968. 236p. $9.50.
Contents: Differential equations. Abstract dynamical systems. Inherent topology. Continuous dynamical systems. Extension of dynamical systems. Dynamical systems on special carrier spaces. Transversal theory. Limit sets on 2-manifolds. Flows. M.R.39 No. 1767.

765. Hale, J. K. **Ordinary Differential Equations.** New York: John Wiley, 1969. 332p. $14.95.
Contents: General properties of differential equations. Two dimensional systems. Linear systems and linearization. Perturbations of noncritical linear systems. Simple oscillatory phenomena and the method of averaging. Behavior near a periodic orbit. Integral manifolds of equations with a small parameter. Periodic systems with a small parameter. Alternative problems for the solution of functional equations. The direct method of Lyapunov.

766. Hartman, P. **Ordinary Differential Equations.** New York: John Wiley, 1964. 612p. $21.00.
Contents: Existence theorems. Differential inequalities and uniqueness. Linear differential equations. Dependence on initial conditions and parameters. Total and partial differential equations. The Poincare-Bendixson theory. Plane stationary points. Autonomous systems: invariant manifolds and linearizations. Perturbed linear systems. Linear second order equations. Use of implicit function and fixed point theorems. Dichotomies for solutions of linear equations.

767. Hellwig, G. **Partial Differential Equations: An Introduction.** Trans. by E. Gerlach. New York: Blaisdell, 1964. 263p.
Contents: The wave equation. Domain of dependence. The potential equation. Green's functions. Fundamental solutions. Poisson's formula. The heat equation. Hyperbolic, elliptic and parabolic equations. Theory of characteristics. Existence and uniqueness. The energy-integral method. Tools from functional analysis.

768. Hille, Einar. **Lectures on Ordinary Differential Equations.** Reading, Mass.: Addison-Wesley, 1969. 723p.
Contents: Existence and uniqueness theorems. Variation of data. Linear equations in Banach algebras. Sturm-Liouville theory. Oscillation theory. Non-linear equations. M.R.40 No. 2939.

769. Hörmander, Lars. **Linear Partial Differential Operators.** 3rd rev. ed. New York: Springer-Verlag, 1969. 288p.
Contents: Functional analysis. Distributions. Differential operators with constant coefficients. Differential operators with variable coefficients. The Cauchy problem. Elliptic boundary problems. M.R.40 No. 1687.

770. John, Fritz. **Ordinary Differential Equations.** New York: Courant Institute of Mathematical Sciences, 1965. 233p.
Contents: Local existence theorems. Solutions in the large. Autonomous systems. Linear equations. Linear systems with constant coefficients. Linear systems with periodic coefficients. Linear equations in the complex plane. Two point boundary value problems.

771. John, Fritz. **Partial Differential Equations.** New York: Springer-Verlag, 1971. 221p. $6.50.
Contents: The single first order equation. The Cauchy problem for higher order questions. Second order equations with constant coefficients. The Cauchy problem for linear hyperbolic questions.

772. Kells, Lyman M. **Elementary Differential Equations.** 6th ed. New York: McGraw-Hill, 1965. 420p. $9.00.
Contents: Differential equations of the first order and the first degree. Applications involving differential equations of the first order. First order equations of degree higher than the first. Linear differential equations with constant coefficients. Miscellaneous differential equations of order higher than the first. Applications. Existence theorems and applications. Solution by series. Numerical solutions of differential equations. Partial differential equations.

773. Kiselev, A. I., M. L. Krasnov, and G. I. Makarenko. **Ordinary Differential Equations**. Trans. by E. J. F. Primrose. New York: Ungar, 1967. 231p. Contents: The method of isoclines. Euler's method. Separation of variables. Integrating factors. Homogeneous equations. Bernoulli equations. Reduction of the order of an equation. Linear differential equations with constant and variable coefficients. Variation of parameters. Theory of stability. Routh-Hurwitz criterion. Lyapunov stability. Laplace transforms. M.R.34 No. 6176.

774. Krasnosel'skii, M. A. **The Operator of Translation along the Trajectories of Differential Equations**. Providence, R.I.: American Mathematical Society, 1968. 294p. Contents: The translation operator. The method of guiding functions. Uniqueness and stability of periodic solutions. Translation operator and periodic solutions of differential equations in a Banach space. The translation operator and the method of integral equations in the problem of periodic solutions. M.R.36 No. 6688.

775. Krasnosel'skii, M. A. **Topological Methods in the Theory of Nonlinear Integral Equations**. Elmsford, N.Y.: Pergamon, 1964. 406p. Contents: Nonlinear operations. The rotation of a vector field. Existence theorems of solutions. Problems concerning eigenfunctions. Eigenfunctions of positive operators. Variation methods.

776. Kreider, D. L., R. G. Kuller, and D. R. Ostberg. **Elementary Differential Equations**. Reading, Mass.: Addison-Wesley, 1968. 492p. Contents: Linear algebra. The general theory of linear differential equations. Equations with constant coefficients. The Laplace transform. Matrices and systems of linear equations. Series solutions of linear differential equations. First-order differential equations. Existence and uniqueness theorems. Stability. HIGHLY RECOMMENDED.

777. Ladyzenskaja, O. A., V. A. Solonnikov, and N. N. Ural'ceva. **Linear and Quasilinear Equations of Parabolic Type**. Providence, R.I.: American Mathematical Society, 1968. 648p. Contents: Linear equations with discontinuous coefficients. Linear equations with smooth coefficients. Quasi-linear equations of general form. Quasi-linear equations with principal part in divergence form. Systems of linear and quasi-linear equations. M.R.39 No. 5941.

778. Lakshmikantham, V., and S. Leela. **Differential and Integral Inequalities: Theory and Applications**. New York: Academic Press, 1969. Vol. I, 390p.; $18.50. Vol. II, 300p.; $16.50. Contents: VOLUME I. Ordinary differential equations. Volterra integral equations. VOLUME II. Functional differential equations. Partial differential equations. Differential equations in abstract spaces. Complex differential equations.

779. Lanczos, Cornelius. **Linear Differential Operators**. New York: Van Nostrand-Reinhold, 1961. 564p. $15.00.

Contents: Interpolation. Harmonic analysis. Matrix calculus. The function space. The Green's function. Communication problems. Sturm-Liouville problems. Boundary value problems. Numerical solution of trajectory problems.

780. LaSalle, Joseph, and S. Lefschetz. **Stability by Liapunov's Direct Method: With Applications.** New York: Academic Press, 1961. 134p. $6.50.
Contents: Geometric concepts: vectors and matrices. Differential equations. Application of Liapunov's theory to controls. Extensions of Liapunov's method.

781. Lattès, R., and J-L. Lions. **The Method of Quasi-Reversibility: Applications to Partial Differential Equations.** Trans. from the French and edited by Richard Bellman. New York: American Elsevier, 1969. 408p. $20.00.
Contents: Quasi-reversibility and equations of parabolic type. Non-parabolic equations of evolution. Control in the boundary conditions. Quasi-reversibility and analytic continuation of solutions of elliptic equations. Quasi-reversibility and continuation of solutions of parabolic equations. Some extensions. M.R.39 No. 5067.

782. Lax, P. D., and R. S. Phillips. **Scattering Theory.** New York: Academic Press, 1967. 276p. $12.00.
Contents: Representation theory and the scattering operator. A semigroup of operators related to the scattering matrix. The translation representation for the solution of the wave equation in free space. The solution of the wave equation in an exterior domain. Symmetric hyperbolic systems, the acoustic equation with an indefinite energy form, and the Schrödinger equation. M.R.36 No. 530.

783. Lefschetz, S. **Differential Equations: Geometric Theory.** 2nd ed. New York: John Wiley, 1963. 390p. $13.50.
Contents: Existence theorems. General properties of the solutions. Linear systems. Stability. Differential equations. Periodic systems and their stability. Two-dimensional systems. Simple critical points. The index. Behavior at infinity. Differential equations of the second order. Oscillations in systems of the second order. Methods of approximation.

784. Mackie, A. G. **Boundary Value Problems.** University Mathematical Monographs. New York: Hafner, 1965. 252p. $5.50.
Contents: Initial value problems for ordinary differential equations. Two-point boundary value problems. Boundary value problems in partial differential equations. Fourier integrals. Green's functions in partial differential equations. Riemann's method. M.R.35 No. 2519.

785. Magnus, W., and S. Winkler. **Hill's Equations.** New York: John Wiley, 1966. 127p. $8.95.
Contents: General theory. Basic concepts. Characteristic values and discriminant. Details. Elementary formulas. Oscillatory solutions. Intervals of stability and instability. Discriminant. Coexistence. Examples.

786. Massera, J. L., and J. J. Schäffer. **Linear Differential Equations and Function Spaces.** New York: Academic Press, 1966. 404p. $16.00.
Contents: Geometry of Banach spaces. Function spaces. Linear differential equations. Dichotomies. Admissibility and related concepts. Admissibility and dichotomies. Dependence on A. Equations on R. Lyapunov's method. Equations. with almost periodic A. Equations with periodic A. Higher-order equations.

787. Mikhlin, S. G. **Integral Equations and Their Applications to Certain Problems in Mechanics, Mathematical Physics, and Technology.** Elmsford, N.Y.: Pergamon, 1964. 356p.
Contents: Equations of the Fredholm type. Symmetric equations. Singular equations. Dirichlet's problem. The biharmonic equation. Applications.

788. Mikhlin, S. G., editor. **Linear Equations of Mathematical Physics.** New York: Holt, Rinehart, and Winston, 1967. 318p. $10.95.
Contents: Properties of linear partial differential equations. Hyperbolic equations. Laplace and Poisson equations. Helmholtz equation. Boundary value problems for elliptic equations. Parabolic equations and systems. Degenerate hyperbolic and elliptic equations. Equations of mixed type. Diffraction and propagation of waves.

789. Mikhlin, S. G. **Multi-Dimensional Singular Integrals and Integral Equations.** Elmsford, N.Y.: Pergamon, 1965. 272p.
Contents: Introduction. Simplest properties of multi-dimensional singular integrals. Compounding of singular integrals. Properties of the symbol. Singular integral equations.

790. Miller, Kenneth S. **Linear Differential Equations in the Real Domain.** New York: W. W. Norton, 1963. 193p.
Contents: The Wronskian. Adjoint of an operator. One-sided Green's functions. Algebra of differential operators. Distribution theory. Classical Green's function. Sturm-Liouville theory. Series representations. RECOMMENDED.

791. Mizohata, Sigeru. **Lectures on Cauchy Problem.** Bombay: Tata Institute of Fundamental Research, 1965. 226p.
Contents: The Cauchy problem. The Cauchy-Kowelewsky theorem. The Holmgren uniqueness theorem. Linear first order systems. Well-posedness. Hyperbolic systems. Energy inequalities. Gärding's inequality for elliptic operators. Local and global existence theorems.

792. Moon, Parry, and Domina E. Spencer. **Partial Differential Equations.** Lexington, Mass.: D. C. Heath, 1969. 322p.
Contents: Derivatives. Differential geometry. First-order equations. Equations of higher order. The Laplace equation. The diffusion equation. The wave equation. The vector Helmholtz equation.

793. Nelson, Alfred L., Karl W. Folley, and Max Coral. **Differential Equations.** 3rd ed. Lexington, Mass.: D. C. Heath, 1964. 349p.

Contents: Differential equations of the first order and first degree. Linear differential equations of higher order. Numerical methods. Special differential equations of the second order. Differential equations of the first order and not of the first degree. Solution in series. Systems of partial differential equations. Partial differential equations of the first order. The Laplace transformation. Fourier series.

794. Petrovskii, I. G. **Ordinary Differential Equations.** Trans. from the Russian by R. A. Silverman. Englewood Cliffs, N.J.: Prentice-Hall, 1966. 232p.
Contents: Systems of ordinary differential equations. Linear systems: general theory. Linear systems: the case of constant coefficients. Autonomous systems.

795. Petrovskii, I. G. **Partial Differential Equations.** Philadelphia: W. B. Saunders, 1967. 410p.
Contents: Classification of equations. Hyperbolic, parabolic, and elliptic equations.

796. Plaat, Otto. **Ordinary Differential Equations.** San Francisco: Holden-Day, 1971. 320p.
Contents: First-order equations. Linear systems. Second-order autonomous systems. Existence and uniqueness.

797. Pogorzelski, W. **Integral Equations and Their Applications.** Elmsford, N.Y.: Pergamon, 1966. 714p.
Contents: Volterra's integral equations. Fredholm's integral equation of the first and second kind. Weakly singular Fredholm equations. Fredholm's equations with symmetric kernel.

798. Pontryagin, L. S. **Ordinary Differential Equations.** Trans. by L. Kacinskas and W. B. Counts. Reading, Mass.: Addison-Wesley, 1962. 298p.
Contents: Linear equations with constant coefficients. Linear equations with variable coefficients. Existence theorems and stability theory.

799. Protter, Murray, and Hans F. Weinberger. **Maximum Principles in Differential Equations.** Englewood Cliffs, N.J.: Prentice-Hall, 1967. 261p.
Contents: The one-dimensional maximum principle. Elliptic equations. Parabolic equations. Hyperbolic equations.

800. Rabenstein, Albert L. **Introduction to Ordinary Differential Equations.** 2nd ed. New York: Academic Press, 1972. 431p. $11.25.
Contents: Linear differential equations. Further properties of linear differential equations. Complex variables. Series solutions. Bessel functions. Orthogonal polynomials. Eigenvalue problems. Fourier series. Systems of differential equations. Laplace transforms. Partial differential equations and boundary-value problems. Nonlinear differential equations.

801. Rasulov, M. L. **Methods of Contour Integration.** Trans. from the Russian. Amsterdam: North-Holland, 1967. 453p. $19.95.

Contents: The residue method. Contour-integral method of solving mixed problems for second order equations with discontinuous coefficients.

802. Roach, G. F. **Green's Functions: Introductory Theory with Applications.**
New York: Van Nostrand-Reinhold, 1970. 279p. $16.95.
Contents: The concept of a Green's function. Integral operators. Differential operators. Generalized Fourier series.

803. Sanchez, David A. **Ordinary Differential Equations and Stability Theory: An Introduction.** San Francisco: W. H. Freeman, 1968. 164p.
Contents: Existence and uniqueness theorems. Linear equations. Fundamental solutions. The Wronskian. Autonomous systems and phase space. Stability for nonautonomous equations. RECOMMENDED FOR BEGINNERS. M.R.37 No. 3076.

804. Sansone, G., and R. Conti. **Non-Linear Differential Equations.** Trans. and edited by A. H. Diamond. Elmsford, N.Y.: Pergamon, 1964. 552p.
Contents: General theorems about solutions of differential systems. Particular plane autonomous systems. The singularities of Briot-Bouquet. Plane autonomous systems. Autonomous plane systems with perturbations. On some autonomous systems with one degree of freedom. Non-autonomous systems with one degree of freedom. Linear systems. Stability.

805. Shilov, G. E. **Generalized Functions and Partial Differential Equations.**
Trans. by B. Seckler. New York: Gordon & Breach, 1968. 345p.
Contents: Elementary and special topics in the theory of generalized functions. Fundamental functions of differential operators. Local properties of solutions. Equations in a half-space.

806. Simmons, George F. **Differential Equations: With Applications and Historical Notes.** New York: McGraw-Hill, 1972. 465p.
Contents: First and second order equations. Oscillation theory and boundary value problems. Power series solutions. Special functions. Systems. Nonlinear equations. The calculus of variations. Laplace transforms. Existence and uniqueness theorem. HIGHLY RECOMMENDED.

807. Smith, M. G. **Introduction to the Theory of Partial Differential Equations.**
New York: Van Nostrand-Reinhold, 1967. 224p. $9.50.
Contents: Partial differential equations of mathematical physics. Fundamental existence and uniqueness theorems. First-order equations. Second-order linear equations. Linear partial differential equations of second order in more than two independent variables. Laplace equation. Wave equation. Diffusion or heat conduction equation. Green's functions. Riemann theory of the hyperbolic equation. Generalized functions. Systems of equations of order greater than two. Equations of fluid dynamics.

808. Sobolev, S. L. **Partial Differential Equations of Mathematical Physics.**
Elmsford, N.Y.: Pergamon, 1964. 440p.

Contents: Derivation of the equations. Multiple integrals. The heat equation. Theory of integral equations. Fourier's method. Spherical functions.

809. Spiegel, Murray R. **Applied Differential Equations.** 2nd ed. Englewood Cliffs, N.J.: Prentice-Hall, 1967. 412p.
Contents: Differential equations in general. First-order and simple higher-order ordinary differential equations. Applications of first-order and simple higher-order differential equations. Linear differential equations. Applications of linear differential equations. Simultaneous differential equations and their applications. Solution of differential equations by the Laplace transformation. Solution of differential equations by use of series. The numerical solution of differential equations. Partial differential equations. Solutions of boundary-value problems and Fourier series.

810. Stakgold, Ivar. **Boundary Value Problems of Mathematical Physics.** New York: Macmillan, 1967. Vol. I, 340p.; $14.95. Vol. II, 408p.; $14.95.
Contents: VOLUME I. The Green's functions. Elementary theory of distributions. Linear vector spaces. Linear integral equations. Spectral theory of second-order differential operators. VOLUME II. Distributions in higher dimensions. Test functions. Fundamental solutions. Fourier transforms. Potential theory. Equations of evolution. Causal Green's function for heat conduction and the wave equation. Wiener-Hopf methods. Variational and related methods. HIGHLY RECOMMENDED.

811. Stein, Elias M. **Singular Integrals and Differentiability Properties of Functions.** Princeton, N.J.: Princeton University Press, 1971. 209p.
Contents: Functions of a real variable. Singular integrals. Poisson integrals. Spherical harmonics. Littlewood-Paley theorems. Differentiability properties.

812. Struble, Raimond A. **Nonlinear Differential Equations.** New York: McGraw-Hill, 1962. 267p. $9.00.
Contents: The existence and the uniqueness of the solution of the initial-value problem. Properties of solutions. Properties of linear systems. Stability in nonlinear systems. Two-dimensional systems. Perturbations of periodic solutions. A general asymptotic method.

813. Swanson, E. A. **Comparison and Oscillation Theory of Linear Differential Equations.** New York: Academic Press, 1968. 227p. $13.50.
Contents: Sturm-type theorems for second order ordinary equations. Oscillations and non-oscillation theorems for second order ordinary equations. Fourth order ordinary equations. Third order ordinary equations. nth Order ordinary equations and systems. Partial differential equations.

814. Szego, G. P., and N. P. Bhatia. **Stability Theory of Dynamical Systems.** New York: Springer-Verlag, 1970. 225p.
Contents: Dynamical systems. Stability theory. Higher Prolongations. C^1 and non-continuous Lyapunov functions for ordinary differential equations.

815. Trèves, Francois. **Linear Partial Differential Equations.** New York: Gordon & Breach, 1970. 130p.
Contents: Existence and approximation theorems in functional analysis. L^2-inequalities. Necessary and sufficient conditions for existence of solutions. L^2-estimates and pseudo-convexity.

816. Trèves, Francois. **Linear Partial Differential Equations with Constant Coefficients: Existence, Approximation and Regularity of Solutions.** New York: Gordon & Breach, 1966. 534p.
Contents: Existence and approximation theorems. Fundamental solutions. Convexity with respect to a differential polynomial. Interior regularity. Partial hypoellipticity. Existence and approximations theorems in spaces of analytic functions. M.R.37 No. 557.

817. Trèves, Francois. **Locally Convex Spaces and Linear Partial Differential Equations.** New York: Springer-Verlag, 1967. 120p.
Contents: Spectrum of locally convex spaces. Natural fibration over the spectrum. Epimorphisms of Fréchet spaces. Existence and approximation of solutions to a functional equation. Translation into duality. Applications to partial differential equations. M.R.36 No. 6985.

818. Tricomi, F. G. **Differential Equations.** Trans. by A. McHare. New York: Hafner, 1961. 273p.
Contents: Existence and uniqueness theorems. Characteristic of a first order equation. Boundary problems for linear equations of the second order. The oscillation theorem. The theorem of de la Vallee Poussin. Asymptotic methods and Laguerre and Legendre differential equations and polynomials.

819. Tricomi, F. G. **Integral Equations.** New York: John Wiley, 1967 (4th printing). 238p. $11.95.
Contents: Volterra equations. Fredholm equations. Symmetric kernels and orthogonal systems of functions. Some types of singular or non-linear integral equations. Appendix I: Algebraic systems of linear equations. Appendix II: Hadamard's theorem. RECOMMENDED.

820. Tychonov, A. N., and A. A. Samarski. **Partial Differential Equations of Mathematical Physics.** Trans. by S. Radding and edited by J. J. Brandstatter. San Francisco: Holden-Day, 1964, 1967. Vol. I, 390p. Vol. II, 250p.
Contents: VOLUME I. Classification of partial differential equations of the second order. Hyperbolic differential equations. Parabolic differential equations. Elliptical differential equations. VOLUME II. Spatial wave propagation. Spatial heat propagation. Elliptic differential equations. Appendix: Special functions. M.R.35 No. 6957.

821. Urabe, M. **Nonlinear Autonomous Oscillations: Analytical Theory.** New York: Academic Press, 1967. 330p. $16.00.
Contents: Analysis of vectors and matrices. Basic theorems concerning ordinary differential equations. Linear differential systems. Orbits of autonomous

systems. Moving orthonormal systems along a closed orbit. Stability. Perturbation of autonomous systems. Perturbation of fully oscillatory systems. Perturbation of partially oscillatory systems. Analysis of two-dimensional autonomous systems. Numerical computation of periodic solutions. Center of the autonomous system. M.R.36 No. 2898.

822. Vekua, I. N. **New Methods for Solving Elliptic Equations.** Trans. from the Russian. Amsterdam: North-Holland, 1968. 370p. $16.00.
Contents: General representations of the solutions of second order linear differential equations of the elliptic type in two independent variables. Expansions and approximations of solutions. General representations of solutions of a class of elliptic differential equations. M.R.35 No. 3243.

823. Walter, Wolfgang. **Differential and Integral Inequalities.** Trans. by L. Rosenblatt and L. Shampine. New York: Springer-Verlag, 1970. 352p.
Contents: Volterra integral equations. Ordinary differential equations. Hyperbolic and parabolic differential equations.

824. Weinberger, H. F. **A First Course in Partial Differential Equations with Complex Variables and Transform Methods.** New York: Blaisdell, 1965. 446p.
Contents: The wave equation. Linear second order partial differential equations. Elliptic and parabolic equations. Separation of variables. Fourier series. Sturm-Liouville theory. Analytic functions. Calculus of residues. The Laplace and Fourier transforms.

CHAPTER 24

FOURIER ANALYSIS AND HARMONIC ANALYSIS

825. Alexits, G. **Convergence Problems of Orthogonal Series.** New York: Pergamon, 1962. 350p.
Contents: Fundamental ideas. Examples of series of orthogonal functions. Investigation of the convergence behaviour of orthogonal series by methods belonging to the general theory of series. The Lebesgue functions. Classical convergence problems.

826. Amerio, Luigi, and Giovanni Prouse. **Almost Periodic Functions and Functional Equations.** New York: Van Nostrand-Reinhold, 1971. 184p. $13.95.
Contents: PART I. Theory of almost periodic functions. Almost periodic functions in Banach spaces. Harmonic analysis of almost periodic functions. Weakly almost periodic functions. The integration of almost periodic functions. PART II. Applications to almost periodic functional equations. The wave equation. The Schrödinger type equation. The wave equation with nonlinear dissipative term. Results regarding other functional equations.

827. Bachman, George. **Elements of Abstract Harmonic Analysis.** New York: Academic Press, 1964. 256p. $6.50; $3.45pa.
Contents: The Fourier transform on the real line for functions in L_1. Fourier transform on the real line for functions in L_2. Regular points and spectrum. The Gel'fand theory. Topology. Compactness of the space of maximal ideals over a Banach algebra. The quotient group of a topological group. Right Haar measures and the Haar covering function. The existence of a right invariant Haar integral over any locally compact topological group. The Daniell extension from a topological point of view. Characters and the dual group of a locally compact Abelian topological group. Generalization of the Fourier transform to $L_1(G)$ and $L_2(G)$.

828. Bary, N. K. **A Treatise on Trigonometric Series.** Elmsford, N.Y.: Pergamon, 1965. Vol. I, 553p. Vol. II, 508p.
Contents: VOLUME I. Trigonometric series. Fourier coefficients. Fourier series of continuous functions. Convergence and divergence of a Fourier series in a set. VOLUME II. The summability of Fourier series. Conjugate trigonometric series. Convergence and absolute convergence of Fourier series and trigonometric series. Representations of functions.

829. Bochner, Salomon. **Harmonic Analysis and the Theory of Probability.** Berkeley, Calif.: University of California, 1960. 176p.
Contents: Approximations. Fourier expansions. Closure properties of Fourier transforms. Laplace and Mellin transforms. Stochastic processes and characteristic functionals. Analysis of stochastic processes.

140

830. Corduneanu, C. **Almost Periodic Functions**. New York: John Wiley, 1968. 237p. $14.50.
Contents: Periodic functions. Fourier series. Almost periodic functions. Almost periodic functions depending on parameters. Almost periodic analytic functions. Almost periodic solutions of ordinary differential equations. Almost periodic solutions of partial differential equations. Almost periodic functions with values in Banach spaces. Almost periodic functions on groups.

831. Edwards, R. E. **Fourier Series: A Modern Introduction**. New York: Holt, Rinehart, and Winston, 1967. Vol. I, 208p.; $5.95. Vol. II, 318p.; $9.95.
Contents: VOLUME I. Trigonometric series and Fourier series. Group structure and Fourier series. Convolutions of functions. Homomorphisms of convolution algebras. The Dirichlet and Fejer kernels. Cesaro summability. Fourier series in L^2. Positive definite functions and Bochner's theorem. Pointwise convergence of Fourier series. VOLUME II. Spans of translates. Closed ideals. Closed subalgebras. Banach algebras. Distributions and measures. Interpolation theorems. Changing signs of Fourier coefficients. Lacunary Fourier series. Multipliers.

832. Ehrenpreis, L., editor. **Fourier Analysis in Several Complex Variables**. New York: John Wiley, 1970. 506p. $19.95.
Contents: The geometric structure of local ideals and modules. Integral representation of solutions of homogeneous equations. Elliptic and hyperbolic systems. General theory of Cauchy's problem. Lacunary series. Refined comparison theorems. General theory of quasianalytic functions.

833. Hewitt, Edwin, and K. A. Ross. **Abstract Harmonic Analysis**. New York: Springer-Verlag, 1963, 1970. Vol. I, **Structure of Topological Groups, Integration Theory, Group Representations**. 519p. $19.00. Vol. II, **Structure and Analysis for Compact Groups: Analysis on Locally Compact Abelian Groups**. 771p. $38.50.
Contents: VOLUME I. Elements of the theory of topological groups. Integration on locally compact spaces. Invariant functionals. Convolutions and group representations. Characters and duality of locally compact Abelian groups. RECOMMENDED. M.R.28 No. 158. VOLUME II. Representations and duality of compact groups. Unitary representations. The Peter-Weyl theorem. Fourier transforms. Positive definite functions. Analysis on compact groups. Ideals in convolution algebras. Spectral synthesis. L^2 and L^P transforms. Maximal functions. Tensor products and Von Neumann norms. RECOMMENDED. M.R.41 No. 7378.

834. Hua, L. K. **Harmonic Analysis of Functions of Several Complex Variables in the Classical Domains**. Providence, R.I.: American Mathematical Society, 1963. 164p.
Contents: Algebraic machinery. Evaluation of some integrals. Polar coordinates for matrices. Some general theorems and their applications. Harmonic analysis in the space of symmetric and skew-symmetric matrices. Harmonic analysis on Lie spheres.

835. Katznelson, Yitzhak. **An Introduction to Harmonic Analysis.** New York: John Wiley, 1968. 264p. $13.95.
Contents: Fourier series on T. The convergence of Fourier series. The conjugate function and functions analytic in the unit disc. Interpolation of linear operators and the theorem of Hausdorff-Young. Lacunary series and quasi-analytic classes. Fourier transforms on the line. Fourier analysis on locally compact Abelian groups. Commutative Banach algebras. M.R.40 No. 1734.

836. Ritt, Robert K. **Fourier Series.** New York: McGraw-Hill, 1970. 160p. $7.95.
Contents: Periodic functions. Cesaro sum. The general question. Mean square convergence. The first order linear equation with constant coefficients. The Fourier integral. Generalized Fourier series. The Sturm-Liouville problem. The Bessel functions.

837. Stein, Elias M. **Topics in Harmonic Analysis Related to the Littlewood-Paley Theory.** Princeton, N.J.: Princeton University Press, 1970. 146p. $4.50.
Contents: Lie groups. Littlewood-Paley theory for a compact Lie group. General symmetric diffusion semi-groups. The general Littlewood-Paley theory.

838. Wiener, Norbert. **Generalized Harmonic Analysis and Tauberian Theorems.** Cambridge, Mass.: M.I.T. Press, 1964. 242p. $2.95.
Contents: Precursors of the present theory. Spectra. Extensions of spectrum theory. Examples of functions with spectra. Almost periodic functions.

839. Zygmund, A. **Trigonometric Series.** 2nd rev. ed. London: Cambridge University Press, 1968. Vol. I, 383p. Vol. II, 364p. Bound as one vol., $17.50.
Contents: Trigonometric series. Fourier series. Fourier coefficients. Convergence theorems. Summability of Fourier series. Classes of functions and Fourier series. Absolute convergence of Fourier series. Divergence of Fourier series. Interpolation of linear equations. Convergence and summability almost everywhere. Complex methods in Fourier series. Fourier integrals. Fourier-Stieltjes transforms. Multiple Fourier series. M.R.38 No. 4882.

CHAPTER 25

INTEGRAL TRANSFORMS
AND OPERATIONAL CALCULUS

840. Berg, L. **Introduction to the Operational Calculus.** Trans. from the German. Amsterdam: North-Holland, 1967. 304p. $12.00.
Contents: Algebraic foundations. Functions of a discrete variable. Functions of a continuous variable. Applications. Convergent sequences of operators. The Laplace transformation. Applications. Asymptotic properties. Generalizations. Further operational methods.

841. Bracewell, Ron. **The Fourier Transform and Its Applications.** New York: McGraw-Hill, 1965. 381p. $12.50.
Contents: Convolution. The impulse symbol, S(X). The basic theorems. Doing transforms. The two domains. Electrical waveforms, spectra, and filters. Sampling and series. The Laplace transform. Relatives of the Fourier transform. Antennas. Television image formation. Frequency distributions. Scanning theory and restoration. Heat conduction and diffusion. Pictorial dictionary of Fourier transforms.

842. Dickinson, D. R. **Operators: An Algebraic Synthesis.** New York: Macmillan, 1967. 245p. $7.50.
Contents: Sets. Functions. Groups. Generalized addition and multiplication. Algebra of operators. Sequences. Series. Linear difference equations. Linear differential equations. Infinite series of operators. M.R.35 No. 5869.

843. Donoghue, William F. **Distributions and Fourier Transforms.** New York: Academic Press, 1969. 315p. $16.00.
Contents: Distributions. Differentiation of distributions. Topology of distributions. The support of a distribution. The convolution of distributions. Harmonic and subharmonic distributions. Temperate distributions. Fourier transforms of distributions. Distributions in several variables. Singular integrals. Harmonic analysis. Distributions of positive type. The spectrum of a distribution. Tauberian theorems. Convolution operators. The Bessel potential.

844. Gelfand, I. M., G. E. Shilov, and N. Ya. Vilenkin. **Generalized Functions.** New York: Academic Press. Vol. I, **Properties and Operators.** By I. M. Gelfand and G. E. Shilov. Trans. by Eugene Saletan. 1964. 423p. $13.50. Vol. II, **Function and Generalized Function Spaces.** By I. M. Gelfand and G. E. Shilov. Trans. by M. D. Feinstein, A. Feinstein, and C. P. Peltzer. 1966. 261p. $12.50. Vol. III, **Theory of Differential Equations.** By I. M. Gelfand and G. E. Shilov. Trans. by M. E. Mayer. 1967. 222p. $11.00. Vol. IV, **Applications of Harmonic Analysis.** By I. M. Gelfand and N. Ya. Vilenkin. Trans. by A. Feinstein. 1964. 384p. $13.00. Vol. V, **Integral Geometry and Representation Theory.** By I. M. Gelfand, M. I. Graev, and N. Ya. Vilenkin. Trans. by E. Saletan. 1966. 499p. $17.00.

Contents: VOLUME I. Generalized functions. Fourier transforms of generalized functions. Table of Fourier transforms. VOLUME II. Linear topological spaces. Fundamental and generalized functions. Fourier transforms of fundamental and generalized functions. Spaces of type S. VOLUME III. Spaces of type W. Uniqueness classes for the Cauchy problem. Correctness classes for the Cauchy problem. Generalized eigenfunction expansions. VOLUME IV. The Kernel theorem. Nuclear spaces. Rigged Hilbert space. Positive and positive definite generalized functions. Generalized random processes. Measures on linear topological spaces. VOLUME V. The Radon transform. Integral transforms. Representations of the group of unimodular matrices. Integral geometry in a space of constant curvature. Harmonic analysis on spaces homogeneous with respect to the Lorentz group.

845. Jones, D. S. **Generalized Functions.** New York: McGraw-Hill, 1966. 482p. $10.50.
Contents: Generalized functions. The Fourier transform. Fourier and Hermite series of generalized functions. Convolutions. Generalized functions of several variables. Applications to integral equations. Differential equations. Hilbert transforms and weak functions. M.R.36 No. 623.

846. Liverman, T. P. G. **Generalized Functions and Direct Operational Methods.** Vol. I. Englewood Cliffs, N.J.: Prentice-Hall, 1964. 338p.
Contents: Operational calculus on the space of generalized functions. The operational solution of differential equations. The Laplace transform.

847. Mikusinski, J. **Operational Calculus.** Elmsford, N.Y.: Pergamon, 1960. 495p.
Contents: Operational algebra. Sequences and series of operators. The operational differential calculus. An outline of the general theory of linear differential equations with constant coefficients. Integral operational calculus.

848. Murnaghan, Francis D. **The Laplace Transformation.** Washington: Spartan Books, 1962. 125p. $5.75.
Contents: The Fourier integral theorem. Properties of the Laplace transformation. Applications to problems of vibrating strings and to asymptotic series. Laguerre and Bessel differential equations.

849. Smith, Michael. **Laplace Transform Theory.** New York: Van Nostrand-Reinhold, 1966. 140p. $6.00.
Contents: The D-operator. The Laplace integral. Simple properties of the Laplace transform. The inverse integral. The convolution or Faltung integral. Application to ordinary lineal differential equations. Partial differential equations. Linear integral equations. Linear difference equations. Asymptotic formulae. M.R.34 No. 1799.

850. Zemanian, A. H. **Distribution Theory and Transform Analysis: An Introduction to Generalized Functions, with Applications.** New York: McGraw-Hill, 1965. 371p. $15.50.
Contents: Distributions. The calculus of distributions. Properties of distributions.

Distributions of slow growth. Convolution. Convolution equations. The Fourier transformation. The Laplace transformation. The solution of differential and difference equations by transform analysis. Passive systems. Periodic distributions. RECOMMENDED.

851. Zemanian, A. H. **Generalized Integral Transformations.** New York: John Wiley, 1968. 300p. $16.95.
Contents: Countably multinormed spaces, countable union spaces and their duals. Distributions and generalized functions. The two-sided Laplace transformation. The Mellin transformation. The Hankel transformation. The K transformation. The Weierstrass transformation. The convolution transformation. Transformations arising from orthonormal series expansions.

CHAPTER 26

CALCULUS OF VARIATIONS

852. Arthurs, A. M. **Complementary Variational Principles.** Oxford: Clarendon Press, 1970. 95p.
Contents: Variational principles. Linear and non-linear applications in differential, integral and matrix equations.

853. Becher, Martin. **The Principles and Applications of Variational Methods.** Cambridge, Mass.: M.I.T. Press, 1964. 120p.
Contents: Calculus of variations. Approximate computation. M.I.T. Research Monograph.

854. Caratheodory, C. **Calculus of Variations and Partial Differential Equations of the First Order.** Trans. from the German by R. B. Dean and edited by J. J. Brandstatter. San Francisco: Holden-Day. Part I: 1965, 187p., $10.50. Part II: 1967, 240p. $10.95.
Contents: PART I. Partial differential equations of the first order. PART II. Calculus of variations. M.R.33 No. 597 (Part I). M.R.19 No. 655 (Part II).

855. Clegg, J. C. **Calculus of Variations.** New York: John Wiley, 1968. 190p. $4.00.
Contents: The fundamental problem. The DuBois-Reymond equations and corner conditions. Regular arcs and the general geodesic problem. Variable end-points. Sufficient conditions for a minimum. Isoperimetrical problems. Parametric representation. Miscellaneous topics. M.R.39 No. 2036.

856. Dreyfus, S. E. **Dynamic Programming and the Calculus of Variations.** New York: Academic Press, 1965. 248p. $9.00.
Contents: Discrete dynamic programming. The classical variational theory. The simplest problem. The problem of Mayer. Inequality constraints. Problems with special linear structures. Stochastic and adaptive optimization.

857. Edelen, Dominic G. B. **Nonlocal Variations and Local Invariance of Fields.** New York: American Elsevier, 1969. 197p. $14.50.
Contents: Variations with one dependent function. Variational calculus for several dependent functions. Geometric objects and Lie derivatives. Invariance considerations. M.R.41 No. 2491.

858. Ewing, George M. **Calculus of Variations with Applications.** New York: W. W. Norton, 1969. 343p. $10.00.
Contents: The Euler and Weierstrass conditions. Hamilton's principle. The problems of Bolza and Goddard. The Legendre conditions. Hamilton-Jacobi theory. The necessary tools in real analysis.

859. Gelfand, I. M., and S. V. Fomin. **Calculus of Variations.** Trans. from the Russian by Richard A. Silverman. Englewood Cliffs, N.J.: Prentice-Hall, 1963. 240p.
Contents: Elements of the theory. Further generalizations. The general variation of a functional. The canonical forms of the Euler equations and related topics. The second variation. Sufficient conditions for a weak extremum. Fields. Sufficient conditions for a strong extremum. Variational problems involving multiple integrals. Direct methods in the calculus of variations.

860. Hermann, Robert. **Differential Geometry and the Calculus of Variations.** New York: Academic Press, 1968. 440p. $18.50.
Contents: Differentiable manifolds. Tangent vectors. Differential forms. Exterior derivatives. Lie bracket. Stokes' theorem. The calculus of variations. The Hamilton-Jacobi theory. Applications to special relativity. Riemannian geometry. Affine and projective differential geometry. The Morse index theorem. Complex manifolds. M.R.38 No. 1635.

861. Hestenes, M. R. **Calculus of Variations and Optimal Control Theory.** New York: John Wiley, 1966. 405p. $13.95.
Contents: Auxiliary theorems. Maxima and minima of functions of n-variables. Classical fixed end-point problems. The Jacobi condition, field theory, and sufficiency conditions. A generalized multiplier rule. Alternate elementary variational problems. A general fixed end-point problem. Optimal control problems and the problems of Mayer. Control problems with bounded state variables.

862. McIntyre, J. E. **Guidance, Flight, Mechanics and Trajectory Optimization. Vol. VII: The Pontryagin Maximum Principle.** Washington: National Aeronautics and Space Administration, 1968. 126p. $3.00.
Contents: Survey of the calculus of variations designed for people interested in trajectory optimization. M.R.38 No. 5497.

863. Mikhlin, S. G. **Variational Methods in Mathematical Physics.** Elmsford, N.Y.: Pergamon, 1964. 616p.
Contents: On the operators of mathematical physics. Energy convergence. The energy method. Major applications of the energy method. Problem of Eigenvalues. Generalization of preceding results. Estimates of error of an approximate solution. Numerical examples. The Bubnov-Galerkin method. The method of least squares. The methods of finite differences.

864. Rund, H. **The Hamilton-Jacobi Theory of the Calculus of Variations: Its Role in Mathematics and Physics.** New York: Van Nostrand-Reinhold, 1966. 404p. $13.50.
Contents: Preliminary survey. The simplest problem in the calculus of variations—the non-homogeneous case. The simplest problem in the calculus of variations—the homogeneous case. Multiple integral problems. The problem of Lagrange.

865. Sagan, Hans. **Introduction to the Calculus of Variations.** New York: McGraw-Hill, 1969. 400p. $14.50.
Contents: Extreme values of functionals. The theory of the first variation. Theory of fields and sufficient conditions for a strong relative extremum. The homogeneous problem. The Hamilton-Jacobi theory and the minimum principle of Pontryagin. The problem of Lagrange and the isoperimetric problem. The theory of the second variation.

866. Young, L. C. **Lectures on the Calculus of Variations and Optimal Control Theory.** Philadelphia: W. B. Saunders, 1969. 331p.
Contents: Generalities and typical problems. The method of geodesic coverings. Duality and local embedding. Embedding in the large. Hamiltonians in the large, convexity, inequalities and functional analysis. Existence theory and its consequences. Generalized curves and flows. Some further basic notions of convexity and integration. The variational significance and structure of generalized flows. The nature of control problems. Naive optimal control theory. The application of standard variational methods to optimal control. Generalized optimal control. M.R.41 No. 4337.

CHAPTER 27

NUMERICAL ANALYSIS

867. Ahlberg, J. H., E. N. Nilson, and J. L. Walsh. **The Theory of Splines and Their Applications.** New York: Academic Press, 1967. 284p. $13.50.
Contents: The cubic spline. Intrinsic properties of cubic splines. The polynomial spline. Intrinsic properties of polynomial splines of odd degree. Generalized splines. The doubly cubic spline. Generalized splines in two dimensions. M.R.39 No. 684.

868. American Mathematical Society. **Fourteen Papers on Series and Approximation.** A.M.S. Translations: Series II, Vol. 77. Providence, R.I.: American Mathematical Society, 1968. 266p. $14.10.
Contents: Series with gaps, L. A. Balasov. Mean approximation of periodic functions by Fourier series, V. I. Berdysev. On the absolute convergence of Fourier series with small gaps, R. Bojanic and M. Tomic. A generalization of Jackson's theorem, I. I. Cyganok. On best approximations of classes of periodic functions by means of trigonometric polynomials, A. V. Efimov. On the approximation of functions satisfying a Lipschitz condition by the arithmetic means of their Walsh-Fourier series, M. A. Jastrebova. Estimates for trigonometric integrals and the Bernstein inequality for fractional derivatives, P. I. Lizorkin. On the representation of functions by orthogonal series, R. I. Osipov. Convergence and uniqueness constants for certain interpolation problems, Ju. K. Suetin. Two theorems on the approximation of functions by algebraic polynomials, S. A. Teljakovskii. The best approximation of a function and linear methods for the summation of Fourier series, M. F. Timan. Approximation with respect to various metrics of functions defined on the unit circle by sequences of rational fractions with fixed poles, G. C. Tumarkin. Necessary and sufficient conditions for the possibility of approximating a function on a circumference by rational fractions, expressed in terms directly connected with the distribution of poles of the approximating fractions, G. C. Tumarkin. Estimation of trigonometric sums, I. M. Vinogradov.

869. Ames, William F. **Numerical Methods for Partial Differential Equations.** New York: Barnes & Noble, 1969. 291p. $10.50.
Contents: Classification of equations. Finite difference operators. Stability. Finite difference methods for parabolic, elliptic and hyperbolic equations. Methods for singularities. Shocks. Eigenvalues. Multi-dimensional equations. RECOMMENDED. M.R.41 No. 7862.

870. Babuska, I., M. Prager, and E. Vitasek. **Numerical Processes in Differential Equations.** New York: John Wiley, 1966. 351p. $11.95.
Contents: Stability of numerical processes and some processes of optimization of computations. Initial-value problems for ordinary differential equations. Boundary-value problems for ordinary differential equations. Boundary-value

149

problems for partial differential equations of the elliptic type. Partial differential equations of the parabolic type. M.R.36 No. 6150.

871. Beckett, Royce, and James Hurt. **Numerical Calculations and Algorithms.**
New York: McGraw-Hill, 1967. 298p. $9.95.
Contents: Introduction to computers. The flow chart. Nonlinear algebraic equations. Simultaneous linear equations. Determinants and matrices. Interpolation and numerical integration. Initial-value problems. Finite differences and boundary-value problems. Data approximation.

872. Bellman, Richard. **Methods of Nonlinear Analysis.** New York: Academic Press, 1970. Volume 1. 340p. $16.00.
Contents: First and second order differential equations. Matrix theory. Matrices and linear differential equations. Stability theory and related questions. The Bubnov-Galerkin method. Differential approximation. The Rayleigh-Ritz method. Sturm-Liouville theory. M.R.40 No. 7508.

873. Bellman, Richard, and K. Cooke. **Differential-Difference Equations.**
New York: Academic Press, 1963. 462p. $13.75.
Contents: The Laplace transform. Linear differential equations. First-order linear differential-difference equations of retarded type with constant coefficients. Series expansions of solutions of first-order equations of retarded type. First-order linear equations of neutral and advanced type with constant coefficients. Linear systems of differential-difference equations with constant coefficients. The renewal equation. Systems of renewal equations. Asymptotic behavior of linear differential-difference equations. Stability of solutions of linear differential-difference equations. Stability theory and asymptotic behavior for nonlinear differential-difference equations. Asymptotic location of the zeros of exponential polynomials. On the location of the zeros of exponential polynomials.

874. Bellman, Richard, R. E. Kalaba, and JoAnn Lockett. **Numerical Inversion of the Laplace Transform: Applications to Biology, Economics, Engineering, and Physics.** New York: American Elsevier, 1966. 249p. $8.50.
Contents: Elementary properties of the Laplace transform. Numerical inversion of the Laplace transform. Linear functional equations. Nonlinear equations. Dynamic programming and ill-conditioned systems. M.R.34 No. 5282.

875. Bramble, James H., editor. **Numerical Solution of Partial Differential Equations.** New York: Academic Press, 1966. 373p.
Contents: A finite difference scheme for generalized Neumann problems, K. O. Friedrichs and H. B. Keller. Remarks on the order of convergence in the discrete Dirichlet problem, B. Hubbard. Fluid dynamical calculations, E. Isaacson. Difference approximations for hyperbolic differential equations, H. Kreiss. Discrete methods for non-linear two-point boundary-value problems, M. Lees. Remarks on the numerical computation of solutions of $\Delta \mu = f(P, \mu)$, S. V. Parter. Error bounds based on a priori inequalities, L. E. Payne. On admissibility in representations of functions of several variables as finite sums of functions of

one variable, D. A. Sprecher. Stability of nonlinear discretization algorithms, H. J. Stetter. On maximum-norm stable difference operators, V. Thomée. A posteriori error bounds in iterative matrix inversion, H. F. Weinberger. Finite difference methods for solving systems of partial differential equations, I. Flügge-Lotz. Some numerical results in intermediate problems for eigenvalues, A. Weinstein. Stability of linear and non-linear difference schemes, P. D. Lax. The solutions of multidimensional generalized transport equations and their calculation by difference methods, A. Douglis. Application of integral operators to singular differential equations and to computations of compressible fluid flows, S. Bergman. Numerical solution of the telegraph and related equations, G. Birkhoff and R. E. Lynch. Approximation and estimates for eigenvalues, G. Fichera. Approximate continuation of harmonic and parabolic functions, J. Douglas, Jr. Hermite interpolation-type Ritz methods for two-point boundary value problems, R. S. Varga.

876. Butzer, Paul L., and H. Berens. **Semi-Groups of Operators and Approximation.** New York: Springer-Verlag, 1967. 318p. $14.00.
Contents: Theory of semigroups of bounded linear operators on Banach spaces. The theorems of Feller, Miyadera, Phillips, Hille, and Yosida. Approximation theorems for semigroups of operators. Characterization of the Sobolev and generalized Lipshitz spaces. Theory of interpolation spaces. M.R.37 No. 5588.

877. Cheney, E. Ward. **Introduction to Approximation Theory.** New York: McGraw-Hill, 1966. 259p. $10.95.
Contents: The Tchebycheff solution of inconsistent linear equations. Tchebycheff approximation by polynomials and other linear families. Least squares approximation and related topics. Rational approximation. Some additional topics. M.R.36 No. 5568.

878. Collatz, Lothar. **The Numerical Treatment of Differential Equations.** 3rd ed. Berlin: Springer-Verlag, 1960. 568p.
Contents: Hermite's generalization of Taylor's formula. Green's functions. Initial value problems. The Runga-Kutta method. Boundary value problems. Finite difference methods. Ritz's method. Integral and functional equations.

879. Conte, S. D., and Carl de Boor. **Elementary Numerical Analysis.** 2nd ed. New York: McGraw-Hill, 1972. 396p. $10.95.
Contents: Number systems and errors. The solution of nonlinear equations. Matrices and systems of linear equations. Interpolation and approximation. Differentiation and integration. The solution of differential equations. Boundary-value problems in ordinary differential equations.

880. Copson, E. T. **Asymptotic Expansions.** New York: Cambridge University Press, 1965. 120p. $6.00.
Contents: Integration by parts. The method of stationary phase. The method of Laplace. Watson's lemma. The method of steepest descent. The saddle point method. Airy's integral. Uniform asymptotic expansions.

881. Daniel, James W., and Ramon E. Moore. **Computation and Theory in Ordinary Differential Equations.** San Francisco: W. H. Freeman, 1970. 172p. $7.50.
Contents: Initial value problems. Boundary value problems. Methods of approximate solution: basic numerical methods for initial and boundary value problems. Special numerical methods for initial value problems. Transformation of the equations: a priori global transformations. A priori local transformations. A priori time-dependent transformations. A posteriori transformations.

882. Davis, Philip J., and Philip Rabinowitz. **Numerical Integration.** Waltham, Mass.: Blaisdell, 1967. 230p. $7.50.
Contents: Numerical integration on finite and infinite intervals. Error estimates. Methods for multiple integrals. FORTRAN programs. Tables of integration formulas. M.R.35 No. 2482.

883. Dejon, B., and P. Henrici, editors. **Constructive Aspects of the Fundamental Theorem of Algebra.** New York: John Wiley, 1969. 337p. $9.95.
Contents: A never failing, fast convergent root-finding algorithm, B. Dejon, K. Nickel. Finding a zero by means of successive linear interpolation, T. J. Dekker. Remarks on the paper by Dekker, G. E. Forsythe. What is a satisfactory quadratic equation solver?, G. E. Forsythe. Mathematical and physical polynomials, L. Fox. A constructive form of the second Gauss proof of the fundamental theorem of algebra, R. L. Goodstein. Uniformly convergent algorithms for the simultaneous approximation of all zeros of a polynomial, P. Henrici, I. Gargantini. On the notion of constructivity, H. Hermes. Bigradients, Henkel determinants, and the Pade table, A. S. Householder, G. W. Stewart III. An algorithm for an automatic general polynomial solver, M. A. Jenkins, J. F. Traub. Die numerische bestimmung mehrfacher und nahe benachbarter Polynomnullstellen nach einem vergesserten Bernoulli-Verfahren, I. Kupke. Search procedures for polynomial equation solving, D. H. Lehmer. A method for automatic solution of algebraic equations, A. M. Ostrowski. Iteration functions for solving polynomial matrix equations, M. Pavel-Parvu, A. Korganoff. Zur problematik der Nullstellen-bestimmung bei polynomen, H. Rutishauser. Factorization of polynomials by generalized Newton procedures, J. Schröder. The fundamental theorem of algebra in recursive analysis, E. Specker. M.R.40 No. 6769.

884. Demyanov, V. F., and A. M. Rubinov. **Approximate Methods of Optimization Problems.** Trans. from the Russian by Scripta Technica; edited by George M. Kranc. New York: American Elsevier, 1970. 256p. $18.50.
Contents: Necessary conditions for an extremum. Successive approximation methods. Solution of certain optimal-control problems. Some problems of finite-dimensional spaces.

885. Dorn, William S., and Daniel D. McCracken. **Numerical Methods with FORTRAN IV Case Studies.** New York: John Wiley, 1972. 447p.
Contents: Solution of equations. Errors. Numerical instabilities and their cure. Simultaneous linear algebraic equations. Numerical differentiation and integration. Interpolation. Least squares approximations. Ordinary differential equations. RECOMMENDED.

886. Faddeev, D. K., and V. N. Faddeeva. **Computational Methods of Linear Algebra**. Trans. from the Russian by Robert C. Williams. San Francisco: W. H. Freeman, 1963. 621p. $12.00.
Contents: Basic material from linear algebra. Exact methods for solving systems of linear equations. Iterative methods for solving systems of linear equations. The complete eigenvalue problem. The special eigenvalue problem. The method of minimal iterations and other methods based on the idea of orthogonalization. Gradient iterative methods. Iterative methods for solving the complete eigenvalue problem. Universal algorithms.

887. Falb, Peter L., and J. L. de Jong. **Some Successive Approximation Methods in Control and Oscillation Theory**. New York: Academic Press, 1969. 240p. $13.50.
Contents: Iterative methods. The contraction principle. Newton's method. Operator representations of discrete and continuous two-point boundary value problems. Convergence theorems for the iterative solution. Numerical solutions of some problems. RECOMMENDED. M.R.41 No. 9446.

888. Fike, C. T. **Computer Evaluation of Mathematical Functions**. Englewood Cliffs, N.J.: Prentice-Hall, 1968. 227p. $10.50.
Contents: A study of computer programs for the evaluation of mathematical functions. Newton-Raphson root evaluation. Chebychev approximations. Rational and asymptotic approximations. M.R.38 No. 4003.

889. Forsythe, G. E., and W. R. Wasow. **Finite Difference Methods for Partial Differential Equations**. New York: John Wiley, 1960. 444p.
Contents: Finite difference methods to the wave equation. Finite difference methods for systems of quasilinear hyperbolic equations. Integration along characteristics and Adam's method. Methods for parabolic and elliptic equations. M.R.33 No. 6066.

890. Fox, L. **An Introduction to Numerical Linear Algebra with Exercises**. New York: Oxford University Press, 1965. 327p.
Contents: Solutions of systems of equations by elimination methods. Orthogonalization methods. Iterative methods. Eigenvalue problems. Error analysis.

891. Gastinel, Noel. **Linear Numerical Analysis**. New York: Academic Press, 1970. 341p. $15.00.
Contents: Elementary properties of matrices. Vector and matrix norms. Inversion of matrices—theory. Direct methods for the solution of a system of linear equations. Indirect methods. Invariant subspaces. Some applications of the properties of invariant subspaces. Numerical methods for the calculation of eigenvalues and eigenvectors.

892. Ghizetti, A., and A. Ossini. **Quadrature Formulae**. New York: Academic Press, 1970. 192p. $10.00.
Contents: Additional results on linear differential equations. Elementary quadrature formulae and a general procedure for constructing them. Special functions.

Various examples of elementary quadrature formulae. Generalized quadrature formulae and questions of convergence. Solutions to problems.

893. Greenspan, Donald. **Lectures on the Numerical Solutions of Linear, Singular, and Non-Linear Differential Equations.** Englewood Cliffs, N.J.: Prentice-Hall, 1968. 185p.
Contents: Numerical solution of the interior and the exterior Dirichlet problems. Numerical solutions of linear and singular elliptic equations. Mildly nonlinear elliptic parabolic and hyperbolic problems. Approximate extremization of functionals. Steady state Navier Stokes problems. M.R.38 No. 2958.

894. Greenspan, Donald, editor. **Symposium on Numerical Solution of Non-linear Differential Equations.** New York: John Wiley, 1966. 347p. $9.95.
Contents: Numerical solution of the reactor kinetics equations, Garrett Birkhoff. Some numerical results for the solution of the heat equation backwards in time, J. R. Cannon. The solution of van der Pol's equation in Chebyshev series, C. W. Clenshaw. Monotonicity and related methods in non-linear differential equations problems, L. Collatz. On rigorous error bounds in the numerical solution of ordinary differential equations, Germund G. Dahlquist. The numerical solution of non-linear optimal control problems, Stuart E. Dreyfus. Numerical studies of steady viscous flow about cylinders, H. B. Keller and H. Takami. Difference approximations for the initial-boundary value problem for hyperbolic differential equations, Heinz-Otto Kreiss. A Leray-Schauder principle for A-compact mappings and the numerical solution of non-linear two-points boundary value problems, M. Lees and M. H. Schultz. A general theory of the numerical solution of the equations of hydrodynamics, William F. Noh. Maximal solutions of mildly non-linear elliptic equations, Seymour V. Parter. On some non-well-posed problems for partial differential equations, L. E. Payne. Approximate computational solution of non-linear parabolic partial differential equations by linear programming, J. B. Rosen. Galerkin's procedure for non-linear periodic systems and its extension to multi-point boundary value problems for general non-linear systems, Minoru Urabe. Abstracts of contributed papers.

895. Greville, T. N. E., editor. **Theory and Applications of Spline Functions.** Proceedings of an Advanced Seminar at the Mathematics Research Center, United States Army, University of Wisconsin, Madison, Wisconsin, October 7-9, 1968. New York: Academic Press, 1969. 212p. $4.95.
Contents: Introduction to spline functions, T. N. E. Greville. An introduction to the application of spline functions to initial value problems, Frank R. Loscalzo. Approximation by splines, L. L. Schumaker. Some algorithms for the computation of interpolating and approximating spline functions, L. L. Schumaker. Generalizations of spline functions and applications to nonlinear boundary value and eigenvalue problems, J. W. Jerome and R. S. Varga. Monosplines and quadrature formulae, I. J. Schoenberg.

896. Henrici, Peter. **Discrete Variable Methods in Ordinary Differential Equations.** New York: John Wiley, 1962. 407p. $9.95.
Contents: One-step methods for initial value problems. Multistep methods for initial value problems. Boundary value problems.

897. Henrici, Peter. **Elements of Numerical Analysis.** New York: John Wiley, 1964. 328p. $11.95.
Contents: Numerical analysis. Complex numbers and polynomials. Difference equations. Solution of equations. Iteration. Iteration for systems of equations. Linear difference equations. Bernoulli's method. The quotient-difference algorithm. Interpolation and approximation. The interpolating polynomial. Construction of the interpolating polynomial: methods using ordinates. Construction of the interpolating polynomial: methods using differences. Numerical differentiation. Numerical integration. Numerical solution of differential equations. Computation. Number systems. Propagation of round-off error.

898. Henrici, Peter. **Error Propagation for Difference Methods.** New York: John Wiley, 1963. 73p. $4.95.
Contents: Basic concepts. Stability, consistency, and convergence. Asymptotic behavior of discretization error. Asymptotic behavior of round-off error.

899. Hildebrand, Francis B. **Finite-Difference Equations and Simulations.** Englewood Cliffs, N.J.: Prentice-Hall, 1968. 338p. $12.75.
Contents: Difference equations. Numerical solution of ordinary differential equations. Numerical solution of partial differential equations. Answer to problems. RECOMMENDED. M.R.37 No. 3769.

900. Householder, Alston S. **The Numerical Treatment of a Single Nonlinear Equation.** New York: McGraw-Hill, 1970. 224p. $10.50.
Contents: Localization of roots. Koenig's theorem. The qd and related algorithms. Bernoulli's method and Aitken's algorithm. The ϵ-algorithm. One-point iterations. The method of Laguerre. Graeffe's method.

901. Householder, Alston S. **The Theory of Matrices in Numerical Analysis.** New York: Blaisdell, 1964. 257p.
Contents: Lanczos algorithm for tridiagonalization. Vector norms. Bounds for matrix operators. The Perro-Frobenius theory of non-negative matrices. Localization and exclusion theorems for matrices. Gerschgorin circle theorem. Iterative solution of linear equations. Methods of solving matrix equations and eigenvalue problems. RECOMMENDED.

902. Isaacson, E., and H. B. Keller. **The Analysis of Numerical Methods.** New York: John Wiley, 1966. 541p. $13.95.
Contents: Norms, arithmetic, and well-posed computations. Numerical solution of linear systems and matrix inversion. Iterative solution of non-linear equations. Computation of eigenvalues and eigenvectors. Basic theory of polynomial approximation. Differences, interpolation, polynomials, and approximate differentiation. Numerical integration. Numerical solution of ordinary differential equations. Difference methods for partial differential equations. M.R.34 No. 924.

903. Keller, Herbert B. **Numerical Methods for Two-Point Boundary-Value Problems.** Waltham, Mass.: Blaisdell, 1968. 184p. $7.50.
Contents: Existence and uniqueness theorems for initial and two-point boundary-

value problems. Consistency. Convergence. Stability. Numerical integration. Contraction maps. The Newton-Kantorovich method. Initial value and shooting methods. Finite difference methods. Integral-equation methods. Eigenvalue problems. M.R.37 No. 6038.

904. Khabaza, I. M. **Numerical Analysis.** Elmsford, N.Y.: Pergamon, 1965. 264p.
Contents: Digital computers. Desk machines. Errors in computations. Finite-difference methods. Recurrence relations and algebraic equations. Numerical solution of ordinary differential equations. Matrices. Relaxation methods. Numerical methods for unequal intervals.

905. Kowalik, J., and M. R. Osborne. **Methods for Unconstrained Optimization Problems.** New York: American Elsevier, 1969. 160p. $9.50.
Contents: Direct search methods. Descent methods. Least squares methods. Constrained problems.

906. Künzi, Hans P., H. G. Tzschach, and C. A. Zehnder. **Numerical Methods of Mathematical Optimization: With ALGOL and FORTRAN Programs.** Rev. ed. Trans. by Werner C. Rheinboldt and C. J. Rheinboldt. New York: Academic Press, 1971. 227p. $10.50.
Contents: Linear optimization. Nonlinear optimization. Explanations of the computer programs. ALGOL and FORTRAN programs.

907. Lapidus, Leon, and J. H. Seinfeld. **Numerical Solution of Ordinary Differential Equations.** New York: Academic Press, 1971. 310p. $16.50.
Contents: Fundamental definitions and equations. Runge-Kutta and allied single-step methods. Stability of multistep and Runge-Kutta methods. Predictor-corrector methods. Extrapolation methods. Numerical integration of stiff ordinary differential equations.

908. Lorentz, G. G. **Approximation of Functions.** New York: Holt, Rinehart and Winston, 1966. 188p. $5.00.
Contents: Chebychev approximations. Linear approximations. Degree of approximation by trigonometric polynomials. Degree of approximation by algebraic polynomials. Approximation by linear polynomial operators. Rational approximations. Width and metric entropy. M.R.35 No. 4642.

909. Luenberger, David G. **Optimization by Vector Space Methods.** New York: John Wiley, 1969. 326p.
Contents: Vector spaces. Hilbert spaces. Approximation by the least-square methods. Dual spaces. Calculus of variations. Local optimization. Optimization with constraints. Lagrange multipliers. Iterative methods. M.R.38 No. 6748.

910. Lyusternik, L. A., O. A. Chervonenkis, and A. R. Yanpol'skii. **Handbook for Computing Elementary Functions.** Elmsford, N.Y.: Pergamon, 1965. 251p.
Contents: Rational and power functions. Exponential and logarithmic functions. Trigonometric, hyperbolic, inverse trigonometric and inverse hyperbolic

functions. Algorithms used for computing elementary functions on some Soviet computers.

911. Macon, N. **Numerical Analysis.** New York: John Wiley, 1963. 161p. $6.95.
Contents: Basic concepts. Approximation of functions by polynomials. Iterative methods of solving equations. Matrices and systems of linear equations. Computational methods with matrices. The characteristic values and characteristic vectors of a matrix. Interpolation. Differentiation and integration. Remainder terms for the integration formulas. Ordinary differential equations. Systems of first-order equations. Difference equations.

912. Mikhlin, S. G., and K. L. Smolitskiy. **Approximate Methods for Solution of Differential and Integral Equations.** Trans. from the Russian by Scripta Technica. Edited by Robert Kalaba. New York: American Elsevier, 1967. 323p. $14.00.
Contents: Approximate solution of the Cauchy problem for ordinary differential equations. Grid methods. Variational methods. Approximate solution of integral equations. M.R.36 No. 1108.

913. Mitchell, A. R. **Computational Methods in Partial Differential Equations.** New York: John Wiley, 1969. 255p. $11.00.
Contents: Basic linear algebra. Parabolic equations. Elliptic equations. Hyperbolic systems. Hyperbolic equations of the second order. Applications in fluid mechanics and elasticity.

914. Moore, Ramon E. **Interval Analysis.** Englewood Cliffs, N.J.: Prentice-Hall, 1966. 145p. $9.00.
Contents: Introduction. Interval numbers. Interval arithmetic. A metric topology for intervals. Matrix computations with intervals. Values and ranges of values of real functions. Interval contractions and root-finding. Interval integrals. Integral equations. The initial-value problem in ordinary differential equations. The machine generation of Taylor coefficients. Numerical results with the Kth order method. Coordinate transformations for the initial-value problem. M.R.37 No. 7069.

915. Moursund, David G., and Charles S. Duris. **Elementary Theory and Application of Numerical Analysis.** New York: McGraw-Hill, 1967. 297p. $9.50.
Contents: Solution of equations by fixed-point iteration. Matrix computations and solution of linear equations. Iterative solution of systems of equations. Polynomials, Taylor's series, and interpolation theory. Errors and floating-point arithmetic. Numerical differentiation and integration. Introduction to the numerical solution of ordinary differential equations. Numerical solution of ordinary differential equations.

916. Ortega, James M. **Numerical Analysis: A Second Course.** New York: Academic Press, 1972. 201p.
Contents: Linear algebra. Mathematical stability and ill conditioning. Discretization error. Convergence of iterative methods. Rounding error.

917. Ortega, James M., and W. C. Rheinboldt. **Iterative Solution of Non-linear Equations in Several Variables.** New York: Academic Press, 1970. 572p. $24.00.
Contents: Linear algebra. Analysis. Gradient mappings and minimization. Contractions and the continuation property. The degree of a mapping. General iterative methods. Minimization methods. Rates of convergence—general, one-step stationary methods, multi-step methods and additional one-step methods. Contractions and non-linear majorants. Convergence under partial ordering. Convergence of minimization methods.

918. Ostrowski, A. M. **Solution of Equations and Systems of Equations.** 2nd ed. New York: Academic Press, 1966. 338p. $11.95.
Contents: Divided differences. Method of false position. Iteration. Newton-Raphson method. The square root iteration. A general theorem on zeros of interpolating polynomials. Norms of vectors and matrices. Method of steepest descent. M.R.35 No. 7575.

919. Polozhii, G. N. **The Method of Summary Representation for Numerical Solution of Problems of Mathematical Physics.** Elmsford, N.Y.: Pergamon, 1965. 283p.
Contents: Ordinary finite-difference equations. Solution of particular boundary-value problems. Formulae of summary representation for finite difference equations. Differential equation for the transverse vibrations of beams.

920. Rall, L. B. **Computational Solution of Nonlinear Operator Equations.** New York: John Wiley, 1969. 225p. $14.95.
Contents: Linear spaces, operators, and equations. The contraction mapping principle. Differentiation of operators. Newton's method and its application. M.R.39 No. 2289.

921. Ralston, Anthony. **A First Course in Numerical Analysis.** New York: McGraw-Hill, 1965. 578p. $13.50.
Contents: Polynomial approximation. Interpolation. Numerical differentiation. Numerical quadrature and summation. The numerical solution of ordinary differential equations. Functional approximation—least square techniques. Functional approximation—minimum-maximum error techniques. The solution of nonlinear equations. The solution of simultaneous linear equations. The calculation of eigenvalues and eigenvectors of matrices.

922. Richtmyer, R. D., and K. W. Morton. **Difference Methods for Initial-Value Problems.** 2nd ed. New York: John Wiley, 1967. 405p. $15.95.
Contents: Linear difference equations. Pure initial-value problems with constant coefficients. Linear problems with variable coefficients. Non-linear problems. Mixed initial-boundary-value problems. Multi-level difference equations. Diffusion and heat flow. The transport equation. Sound waves. Elastic vibrations. Fluid dynamics in one space variable. Multi-dimensional fluid dynamics. M.R.36 No. 3515.

923. Saulyev, V. K. **Integration of Equations of Parabolic Type by the Method of Nets.** Elmsford, N.Y.: Pergamon, 1964. 364p.
Contents: PART 1. Construction of difference equations, their stability and their accuracy. PART 2. The solution of difference equations.

924 Schoenberg, J. J., editor. **Approximations with Special Emphasis on Spline Functions.** New York: Academic Press, 1969. 488p. $10.00.
Contents: Splines in the complex plane, J. H. Ahlberg. Prolongement d'une fonction en une fonction différentiable, Christian Coatmelec. Spline interpolation near discontinuities, Michael Golomb. Generalized spline interpolation and nonlinear programming, Klaus Ritter. Splines via optimal control, O. L. Mangasarian and L. L. Schumaker. On the approximation by y-polynomials, Carl de Boor. Piecewise bicubic interpolation and approximation in polygons, Garrett Birkhoff. Distributive lattices and the approximation of multivariate functions, William J. Gordon. Multivariate spline functions and elliptic problems, Martin H. Schultz. On the degree of convergence of nonlinear spline approximation, John R. Rice. Error bounds for spline interpolation, Richard S. Varga. Multipoint expansions of finite differences, A. Meir and A. Sharma. One-sided L_1-approximation by splines of an arbitrary degree, Zvi Zeigler. Construction of spline functions in a convex set, P. J. Laurent. Best quadrature formulas and interpolation by splines satisfying boundary conditions, Samuel Karlin. The fundamental theorem of algebra for monosplines satisfying certain boundary conditions and applications to optimal quadrature formulas, Samuel Karlin.

925. Singer, James. **Elements of Numerical Analysis.** New York: Academic Press, 1964. 395p. $11.00.
Contents: Numbers and errors. The approximating polynomial: approximation at a point. The approximating polynomial: approximation in an interval. The numerical solution of algebraic and transcendental equations in one unknown: geometric methods. The numerical solution of algebraic and transcendental equations in one unknown: arithmetic methods. The numerical solution of simultaneous algebraic and transcendental equations. Numerical differentiation and integration. The numerical solution of ordinary differential equations. Curve fitting.

926. Smith, G. D. **Numerical Solution of Partial Differential Equations.** London: Oxford University Press, 1965. 179p.
Contents: Finite difference formulae. Parabolic equations. Convergence. Stability. Systematic iterative methods. Hyperbolic equations. Characteristics and elliptic equations.

927. Snyder, Martin Avery. **Chebyshev Methods in Numerical Approximation.** Englewood Cliffs, N.J.: Prentice-Hall, 1966. 114p. $7.50.
Contents: The Chebyshev polynomials. The use of Chebyshev polynomials in polynomial approximations. Rational approximation and continued fractions. Rational approximation with Chebyshev polynomials. Appendix: Linear difference equations with constant coefficients. The Bessel functions. M.R. 34 No. 5257.

928. Stanton, Ralph G. **Numerical Methods for Science and Engineering.** Englewood Cliffs, N.J.: Prentice-Hall, 1961. 266p.
Contents: Difference methods. Interpolation. Computation with series and integrals. Numerical solution of differential equations. Linear systems and matrices. Difference equations. Principles of automatic computation.

929. Stiefel, Eduard L. **An Introduction to Numerical Mathematics.** Trans. by W. C. Rheinboldt. New York: Academic Press, 1963. 286p.
Contents: Linear and non-linear algebra. Linear programming. Least-squares approximation. Eigenvalue problems. Differential equations. Approximations.

930. Stroud, A. H., and Don Secrest. **Gaussian Quadrature Formulas.** Englewood Cliffs, N.J.: Prentice-Hall, 1966. 374p.
Contents: Properties of Gaussian quadrature formulas. Computation of formulas. Various uses of the tabulated formulas. Error estimates. Survey of other tables. Tables of orthogonal polynomials. Table of factorials. M.R.34 No. 2185.

931. Talbot, A., editor. **Approximation Theory.** New York: Academic Press, 1970. 356p. $11.00.
Contents: Approximation by polynomials. Error estimates for best polynomial approximations. Linear approximation. Characterization of best spline approximations with free knots. Non-linear approximation. Tchebyscheff-approximation with sums of exponentials. Approximation in abstract linear spaces. M.R.41 No. 8879.

932. Tompkins, Charles B., and Walter L. Wilson. **Elementary Numerical Analysis.** Englewood Cliffs, N.J.: Prentice-Hall, 1969. 396p.
Contents: Taylor's formula. Truncation error. Iteration process. Newton's method. Systems of linear equations. Eigenvalues and eigenvectors. Finite differences. Interpolation. Least square estimates. Numerical differentiation. Numerical integration. Difference equations. Numerical solution of differential equations. M.R.40 No. 3670.

933. Traub, J. F. **Iterative Methods for the Solution of Equations.** Englewood Cliffs, N.J.: Prentice-Hall, 1964. 310p.
Contents: General theorems on iteration functions. The mathematics of difference relations. Interpolatory iteration functions. One-point iteration functions.

934. Varga, Richard S. **Matrix Iterative Analysis.** Englewood Cliffs, N.J.: Prentice-Hall, 1962. 322p.
Contents: Matrix properties. Non-negative matrices. Iterative methods. Elliptic difference equations. Alternating-direction implicit iterative methods. Matrix methods for parabolic partial differential equations.

935. Von Rosenberg, Dale U. **Methods for the Numerical Solution of Partial Differential Equations.** New York: American Elsevier, 1969. 140p. $9.50.
Contents: Linear ordinary differential equations. Linear parabolic equations in one space dimension. Linear hyperbolic equation in two independent variables.

Alternate forms of coefficient matrices and solution algorithms. Non-linear parabolic equations with one space dimension. Non-linear hyperbolic equations. Treatment of non-linear boundary conditions. Elliptic equations and parabolic equations with more than one space variable. Hyperbolic equations in two space variables.

936. Wasow, W. **Asymptotic Expansions for Ordinary Differential Equations.** New York: John Wiley, 1966. 362p. $14.95.
Contents: Some basic properties of linear differential equations in the complex domain. Regular singular points. Asymptotic power series. Irregular singular points. Generalizations by means of Jordan's canonical form. Some special asymptotic methods. Asymptotic expansions with respect to a parameter. Turning point problems. Nonlinear equations. Singular perturbations. Integration of differential equations by factorial series.

937. Watson, W. A., T. Philipson, and P. J. Oates. **Numerical Analysis: The Mathematics of Computing.** New York: American Elsevier, 1969, 1970. Vol. 1, 236p., $4.50. Vol. 2, 176p., $5.50.
Contents: VOLUME 1. Introduction and the use of hand-calculating machines. Programming calculations. Curve sketching. Iterative methods. Differences. The solution of linear simultaneous equations. Roots of polynomial equations. Linear interpolation. Numerical integration. VOLUME 2. Interpolation formulae. Inverse interpolation. Lagrange interpolation. Numerical integration. Numerical differentiation. An introduction to the numerical solution of differential equations. Curve fitting by the method of least squares. Summation of series with slow convergence.

938. Weeg, Gerard P., and G. B. Reed. **Introduction to Numerical Analysis.** Waltham, Mass.: Blaisdell, 1966. 184p.
Contents: Errors in numerical computations. Polynomial approximations. Numerical integration. Ordinary differential equations. Equations and matrices. Least-squares approximation. Gaussian quadrature.

939. Wendroff, Burton. **First Principles of Numerical Analysis: An Undergraduate Text.** Reading, Mass.: Addison-Wesley, 1969. 118p.
Contents: Digital computation. Applications of Taylor's theorem. The theory of interpolation. Applications of interpolation. Gaussian elimination. RECOMMENDED.

940. Wendroff, Burton. **Theoretical Numerical Analysis.** New York: Academic Press, 1967 (2nd printing). 239p. $11.95.
Contents: Interpolation and quadrature. Approximation. Ordinary differential equations. Solution of equations. Partial differential equations. HIGHLY RECOMMENDED. M.R.33 No. 5080.

941. Westlake, J. R. **A Handbook of Numerical Matrix Inversion and Solution of Linear Equations.** New York: John Wiley, 1968. 171p. $10.95.
Contents: Direct methods. Iterative methods. Ill-conditioning—measures of

condition. Error measures. Scaling. Work required—number of arithmetic operations. Comments and comparisons. Appendix A, Glossary of matrix terminology. Appendix B, Theorems on matrix algebra. Appendix C, Test matrices. M.R.36 No. 4794.

942. Wilcox, C. H., editor. **Asymptotic Solutions of Differential Equations and Their Applications.** New York: John Wiley, 1964. 294p. $8.50.
Contents: Asymptotic expansions for ordinary differential equations: trends and problems, Wolfgang Wasow. Solvable related equations pertaining to turning point problems, Hugh L. Turrittin. Asymptotic methods for the solution of dispersive hyperbolic equations, Robert M. Lewis. Asymptotic solutions and indefinite boundary value problems, Robert W. McKelvey. Some examples of asymptotic problems in mathematical physics, C. C. Lin. On the problem of turning points for systems of linear ordinary differential equations of higher orders, Yasutaka Sibuya. Error bounds for asymptotic expansions, Frank W. J. Olver. Asymptotic solutions of elastic shell problems, Robert A. Clark. The integral equations of asymptotic theory, Arthur Erdelyi. Application of Langer's theory of turning points to diffraction problems, Nicholas D. Kazarinoff.

943. Wilkins, B. R. **Analogue and Iterative Methods in Computation, Simulation, and Control.** New York: Barnes & Noble, 1970. 273p.
Contents: Elements of analogue computation. Input/output equipment. Linear and non-linear systems. Applications.

944. Wilkinson, James H. **The Algebraic Eigenvalue Problem.** New York: Oxford University Press, 1965. 662p.
Contents: Elementary hermitians. Perturbation theory. Error analysis. Linear equations. Hermitian matrices. Jacobi methods. Calculation of eigenvectors and eigensystems of matrices. The power, deflation, and inverse iteration methods. HIGHLY RECOMMENDED.

945. Wilkinson, James H. **Rounding Errors in Algebraic Processes.** Englewood Cliffs, N.J.: Prentice-Hall, 1964. 161p.
Contents: The fundamental arithmetic operations. Computations involving polynomials. Matrix computations.

CHAPTER 28

COMPUTER SCIENCE

946. Andrews, Harry C., editor. **Computer Techniques in Image Processing.** New York: Academic Press, 1970. 187p. $10.50.
Contents: Optical data processing. Digital optical processing. Two-dimensional matched filtering. Orthogonal transformations, by Harry Andrews and Kenneth Caspari. Image transforms, by Harry Andrews and William Pratt. Image coding, by William Pratt and Harry Andrews.

947. Arbib, Michael A. **Brains, Machines, and Mathematics.** New York: McGraw-Hill, 1964. 163p. $6.95; $1.95pa.
Contents: Neural nets, finite automata, and Turing machines. Structure and randomness. The correction and errors in communication and computation. Cybernetics. Gödel's incompleteness theorem. The brain-machine controversy. M.R. 40 No. 2458.

948. Arbib, Michael A. **Theories of Abstract Automata.** Englewood Cliffs, N.J.: Prentice-Hall, 1969. 412p.
Contents: An overview of automata theory. Algebraic background. Finite automata. Turing machines and effective computations. Post systems and context free languages. Partial recursive functions. Complexity of computation. Algebraic decomposition theory. Stochastic automata. Machines which compute and construct.

949. Artiaga, Lucio. **Algorithms and Their Computer Solutions.** Columbus, Ohio: Charles E. Merrill, 1972. 384p. $12.95.
Contents: FORTRAN. Writing a complete program. Numerical algorithms. Polynomials. Solutions of equations and plotting. Systems of numeration. Sorting. Applications. Simulation.

950. Banerji, R. B. **Theory of Problem Solving: An Approach to Artificial Intelligence.** New York: American Elsevier, 1969. 203p. $14.50.
Contents: Structure of problems. Structure of games. Description languages. Learning. Problems and patterns. Sundry possibilities of solution based on problem structure. Case histories—interpretations in terms of basic theory. M.R. 40 No. 8305.

951. Barron, D. W. **Assemblers and Loaders.** New York: American Elsevier, 1969. 88p. $4.50.
Contents: Why and how. User's view. Symbol tables. Two-pass assemblers. Loaders and linkage editors. One-pass assemblers. Macro-assemblers. Meta-assemblers.

952. Bartee, Thomas C. **Digital Computer Fundamentals.** 2nd ed. New York: McGraw-Hill, 1966. 480p. $6.95.
Contents: Computer operations. Programming. Number systems. Basic logic circuits. Logical design. The arithmetic element. The memory element. Input-output devices. Computer organization and control.

953. Bellman, Richard, K. L. Cooke, and J. A. Lockett. **Algorithms, Graphs and Computers.** New York: Academic Press, 1970. 246p. $9.75.
Contents: Commuting and computing. The method of successive approximations. From Chicago to the Grand Canyon by car and computer: difficulties associated with large maps. Proof of the validity of the method. Juggling jugs. The Sawyer graph and the billiard ball computer. Cannibals and missionaries. The "travelling salesman" and other scheduling problems. M.R.41 No. 4779.

954. Bernard, Solomon Martin. **System/360 COBOL.** Englewood Cliffs, N.J.: Prentice-Hall, 1968. 312p.
Contents: Fundamentals. Program environment. Data description. Program logic. Other programming considerations. Standards.

955. Braun, Edward L. **Digital Computer Design.** New York: Academic Press, 1963. 606p. $18.00.
Contents: The nature of automatic computation. Boolean algebra. Circuit descriptions of switching and storage elements. Large capacity storage systems. Arithmetic operations. System design of GP computers. Digital differential analyzer. The detection and correction of errors. Appendix: input-output equipment.

956. Chapin, Ned. **Programming Computers for Business Applications.** New York: McGraw-Hill, 1961. 275p. $8.75.
Contents: Programmers and programming. Automatic computers. Preparation for programming. Translation programming. Development programming. Programming to save storage. Programming for accuracy. Programming for speed. Subroutine and library programs.

957. Colman, Harry L., and Clarence P. Smallwood. **Computer Language: An Auto Instructional Introduction to FORTRAN.** New York: McGraw-Hill, 1962. 196p. $6.50; $4.95 text edition.
Contents: Directions for readers. Introduction. Program structure. Variables and constants. Input statements. Arithmetic expressions. Arithmetic statements. Control statements. Output statements. Answers for exercises. FORTRAN coding sheets.

958. Cutler, Donald. **Introduction to Computer Programming.** Englewood Cliffs, N.J.: Prentice-Hall, 1964. 216p.
Contents: Beginnings. Number systems. Data representation and organization. Flow diagramming. The Ex-1 computer. Ex-1 arithmetic operations. Programming techniques. Input-output for the Ex-1. Advanced fixed point techniques. A typical modern giant computer. Trivial—a higher level language. Programming systems.

959. Davis, Gordon B. **An Introduction to Electronic Computers**. New York: McGraw-Hill, 1965. 541p. $9.75.
Contents: The historical development of automatic electronic computers. The uses for computers. Computer arithmetic and storage. The internal operation of a computer. The programming of a computer. Basic computer instructions. Program modification. Input and output. Operating the computer. Detecting and controlling errors. Introduction to problem-oriented languages. FORTRAN—basic elements. FORTRAN—additional features. COBOL—overview and procedure division. COBOL—data, environment, and identification divisions. Evaluating computers. Current and prospective developments in computer hardware, computer software, and computer systems.

960. DeAngelo, Salvatore, and Paul Jorgenson. **Mathematics for Data Processing**. New York: McGraw-Hill, 1970. 300p. $9.95.
Contents: Review of basic algebra. Functions, equations, and graphs. Systems of linear equations. Matrices and matrix methods. Sequences, series, and number bases. Elementary logic and set theory. Boolean algebra. Algorithms and iterative techniques—algorithms, sequences and flowcharts. Computer computation and error analysis. Characteristics of programming languages. FORTRAN. COBOL.

961. Dietmeyer, Donald L. **Logic Design of Digital Systems**. Boston: Allyn & Bacon, 1971. 800p.
Contents: Representation and manipulation of information. Combinatorial and sequential logic. Algorithmic synthesis of combinatorial logic. Asynchronous sequential logic. Digital information processing systems. Switching circuit theory. Design of logic network systems.

962. Dorn, William S., Gary G. Bitter, and David L. Hector. **Computer Applications for the Calculus**. Boston: Prindle, Weber & Schmidt, 1972. 264p.
Contents: Functions. Limits. The derivative. Root finding. Integration. Differential equations. Sequences and series. Linear algebra. Miscellaneous problems.

963. Dorn, William S., and H. J. Greenberg. **Mathematics and Computing: With FORTRAN Programming**. New York: John Wiley, 1967. 595p.
Contents: Linear equations and inequalities. Computer arithmetic and FORTRAN programming. Infinite algorithms. Iteration methods. Algorithms for areas. Mathematical logic. Probability.

964. Emery, Glyn. **Electronic Data Processing**. New York: American Elsevier, 1969. 253p. $8.75.
Contents: What is data processing? Some fundamental ideas. Instructions. Modifying and counting. Subroutines. How programs are assembled. How programs are run and checked. The system centre. Floating point. Programming at a higher level. Input and output. Files and file devices. Multiprogramming. The architecture of a range.

965. Farina, Mario V. **COBOL Simplified**. Englewood Cliffs, N.J.: Prentice-Hall, 1968. 528p.

Contents: Idea behind COBOL. Sample COBOL program. COBOL program organization. Simple industrial program. Editing program.

966. Farina, Mario V. **Computers: A Self-Teaching Introduction.** Englewood Cliffs, N.J.: Prentice-Hall, 1969. 225p.
Contents: History of computers. We design a computer. The rocky road to a computer program. What is a compiler. Data cards. Learning a simple programming language. Flow charting the language of logic. Debugging. The nature of computer applications. Files on magnetic tape. File maintenance. A computer center. Representative types of computers. A computer study in depth. Multiprogramming. Software. Teletype time sharing. Automation.

967. Farina, Mario V. **Elementary BASIC with Applications.** Englewood Cliffs, N.J.: Prentice-Hall, 1970. 309p.
Contents: Teletype time sharing. Loops. Data reading. Labels and headings printing. Flow charting. Sorting. Plotting. Finding areas under curves.

968. Farina, Mario V. **FORTRAN IV Self-Taught.** Englewood Cliffs, N.J.: Prentice-Hall, 1966. 448p.
Contents: Telling the computer about numbers. Some basic concepts. Getting answers. Continuations, comments, flow charts, writing your first program. Telling the computer what to do with numbers. Arithmetic. Hierarchy. Mathematical subroutines. Constant and variable types. Three useful tables. Making decisions. Logical values, logical if. Your second program. Loops. Setting up arrays. Subscribing. Multiple subscripts. Subscript forms. Exponential notation. Read statement. Additional read forms. Write statement and format generator. Arithmetic statement functions. Function sub-programs. Subroutine sub-programs. Communications. Built-in functions. Built-in subroutines. Octal and Hollerith formats. Data statement. Namelist statement. Debugging aids. Control cards. Update-alter.

969. Fenves, Steven J. **Computer Methods in Civil Engineering.** Englewood Cliffs, N.J.: Prentice-Hall, 1967. 242p.
Contents: Introduction to computers and programming. Programming fundamentals. Marching methods: single-subscripted variables. Marching methods: multiply subscripted variables. Successive approximations. Successive approximations combined with iteration. Logical methods. Organization of programs. Programming of engineering applications. A survey of programming languages.

970. Finerman, Aaron, editor. **University Education in Computing Science.** New York: Academic Press, 1968. 327p. $12.00.
Contents: University education in computing science—introduction, Aaron Finerman. Keynote address, John R. Pierce. Computers and education, Anthony G. Oettinger. Graduate computer science program at American universities, Frank S. Beckman. The dilemma of computer sciences, Lotfi A. Zadeh. Computer science is neither mathematics nor electrical engineering, Alan J. Perlis. The science and engineering of information, Vladimir Slamecka. How many computers per university?, Calvin C. Gotlieb. Industry's view of computing science,

Eric A. Weiss. Planning a profession, Stanley Gill. The Master's program in computing science—a report of the workshop, Robert L. Ashenhurst. The doctoral program in computing science—a report of the workshop, Thomas E. Hull. The position of computing science in the university structure—a report of the workshop, William F. Atchison. The computing center and the academic program— a report of the workshop, Kenneth King. University education in computing science—summary, Aaron Finerman. Appendix: Computers in higher education.

971. Flores, Ivan. **Computer Logic: The Functional Design of Digital Computers.** Englewood Cliffs, N.J.: Prentice-Hall, 1960. 458p.
Contents: First principles and definitions. The computer and the problem. The flow and control of information. Coding. Machine arithmetic. Number systems and counting. Machine languages. Logic. Logical construction. Functional units. The logic of arithmetic. Memory devices and their logic. The control unit. Input and output equipment. A problem.

972. Flores, Ivan. **The Logic of Computer Arithmetic.** Englewood Cliffs, N.J.: Prentice-Hall, 1963. 493p.
Contents: Signed binary arithmetic. Multiplication and division in different binary representations. Fast adders. The carry lookahead adder. The conditional sum adder. A simple parallel arithmetic unit. Multiplication. Ternary multiplication. Multiplication by shifting over 1's and 0's. Compound multiplication. Comparative summary. Division I, II, and III. Floating-point numbers, addition and subtraction. Floating-point multiplication and division. Special arithmetic and compound arithmetic. Modulus arithmetic.

973. Foster, J. M. **Automatic Syntactic Analysis.** New York: American Elsevier, 1970. 96p. $4.00.
Contents: Context-free grammars. Parsing. Universal parsing methods. Special parsing methods. Transformations on grammars. Using grammatical analyses for compilation. Appendix 1: Elementary list processing. Appendix 2: An algorithm for top-to-bottom parsing.

974. Foster, J. M. **List Processing.** New York: American Elsevier, 1967. 60p. $4.50.
Contents: The representation of lists. Operations on lists. More advanced features. An example of list processing. Garbage collection. Some typical list languages. The future of list processing.

975. Galler, Bernard A. **The Language of Computers.** New York: McGraw-Hill, 1962. 254p. $9.50.
Contents: The change problem. Expressions. Conditional statements and iteration statements. The social security problem. The secret-code problem. Monte Carlo methods. A sorting problem. The correlation coefficient. A program to produce programs. Simultaneous linear equations. The MAD language.

976. Gavrilov, M. A., and A. D. Zakrevskii, editors. **LYaPAS: A Programming Language for Logic and Coding Algorithms.** Trans. from the Russian by Morton Nadler. New York: Academic Press, 1969. 475p. $24.50.

Contents: Preface, M. A. Gavrilov and A. D. Zakrevskii. Description of LYaPAS, A. D. Zakrevskii. Representation of input information to the LYaPAS compiler, M. Ya. Tovshtein. The programming system PS-LYaPAS, A. D. Zakrevskii, M. Ya. Tovshtein, and N. R. Toropov. Translator for high speed computers, M. Ya. Tovshtein. The LYaPAS compiler, A. D. Zakrevskii. Detection of syntactic errors in L-programs, N. A. Usacheva. Checking out L-programs on Ural-1, A. D. Zakrevskii. Translator for Ural-1, N. R. Toropov. Programming Boolean computations, A. D. Zakrevskii. Optimal coverage of sets, A. D. Zakrevskii. The solution of systems of logical equations, A. D. Zakrevskii and A. Yu. Kalmykova. Testing for identities in Boolean algebra, A. D. Zakrevskii. Determination of the connectivity of a graph, A. D. Zakrevskii. Algorithms for the minimization of Boolean functions, V. G. Novoselov. The decomposition problem for Boolean functions, I. L. Fadeev. The construction of particular minimal normal forms of Boolean functions, V. P. Videnko. An algorithm for obtaining factored forms of Boolean functions, V. P. Videnko and V. Sh. Okudzhava. Approximate method for obtaining minimal factored forms of a certain class, V. I. Ostrovskii. Realization of Boolean functions by threshold elements, E. A. Butakov, S. V. Bykova, and V. A. Vorob'ev. An algorithm for the synthesis of majority-element logical circuits, V. L. Pavlov. Simulation of switching circuits, A. A. Utkin. Algorithms for state minimization of a discrete automaton, Yu. V. Pottosin. Sak-LYaPAS—a system of coding theory algorithms in LYaPAS, G. P. Agibalov.

977. Genuys, F., editor. **Programming Languages.** New York: Academic Press, 1968. 395p. $15.00.
Contents: Abstract algorithms and diagram closure, C. Elgot. Co-operating sequential processes, E. W. Dijkstra. Compiler writing techniques, L. Bolliet. Record handling, C. A. R. Hoare. Discrete event simulation languages, O. J. Dahl.

978. Germain, Clarence B. **Programming the IBM 1620.** 2nd ed. Englewood Cliffs, N.J.: Prentice-Hall, 1965. 191p.
Contents: Computing fundamentals. 1620 instructions. Operation of the 1620. Programming. Introduction to FORTRAN. The symbolic programming system. Address modification. Advanced operating techniques. Disk storage. FORTRAN subprograms.

979. Germain, Clarence B. **Programming the IBM 360.** Englewood Cliffs, N.J.: Prentice-Hall, 1967. 366p.
Contents: The punched card. Mathematical notation. Nature of computers. Instructions. Input/output programming. Input/output devices. Programming considerations. Introduction to PL/I. Writing PL/I programs. Advanced PL/I topics. Introduction to FORTRAN. Additional FORTRAN statements. Advanced FORTRAN topics. Introduction to COBOL. Assembly language. DOS macro statements. OS macro statements.

980. Gildersleeve, Thomas R. **Computer Data Processing and Programming.** Englewood Cliffs, N.J.: Prentice-Hall, 1970. 176p.
Contents: Introduction to computer data processing. Representation of infor-

mation. Introduction to programming. Introduction to the RPG. Printer output vertical format. Printer output horizontal format. Extension and summarization. Multiple card types. Multiple inputs. Tables. Random access. Levels of control. Updated tables.

981. Gillman, Leonard, and A. L. Rose. **APL/360: An Interactive Approach.** New York: John Wiley, 1971. 335p.
Contents: Scalar dyadic functions. Relational functions. Logical functions. Function editing. Mixed functions. Branching. Inner product. APL functions. Commands and features. APL language. RECOMMENDED FOR BEGINNERS.

982. Golden, James T. **FORTRAN IV: Programming and Computing.** Englewood Cliffs, N.J.: Prentice-Hall, 1965. 270p.
Contents: Introduction to FORTRAN IV: subscripted variables. Input-output operations. Subroutines. Complex numbers, Boolean algebra. Simulation. Practices and pitfalls in computing. Problems.

983. Golden, James T., and R. Leichus. **IBM 360: Programming and Computing.** Englewood Cliffs, N.J.: Prentice-Hall, 1967. 342p.
Contents: Introduction to computing. Decimal programming. Binary programming. Indexing. Logical operations. Floating-point operations. Macros. Subroutines. System/360 input/output operations. 1/0 software. Operation systems. Disk operating system.

984. Griswold, Ralph E., James F. Poage, and Ivan Polonsky. **The SNOBOL 4 Programming Language.** Englewood Cliffs, N.J.: Prentice-Hall, 1968. 221p.
Contents: Introduction to SNOBOL 4 programming language. Pattern matching. Primitive functions. Programmer-defined functions. Arrays, data types, and keywords. Details of evaluation. Tracing. Input and output. Structure of a SNOBOL 4 run. Programming details.

985. Hanson, Peggy. **Keypunching.** Englewood Cliffs, N.J.: Prentice-Hall, 1967. 148p.
Contents: Introduction to keypunching. Some keypunch terms you should know. The punch card. Learning to read a punch card. Taking a card count. Operating features of the 24 and 26 keypunches. Operating suggestions. The program drum. Finger dexterity. The program card. Normal program codes guide sheet. Planning and punching a program card. Preparing layout and program cards. Invoices. The 56 card verifier. Payrolls. Cost reports. Time cards. Payroll masters. Spread cards. Spread cards: 999 punching. Adding machine tapes. Alternate programming. Normal and alternate program—codes guide sheet. Alternate program—invoices. Alternate program—payroll. Alternate program—time cards. Alternate program—adding machine tape. Alternate program—name and address mailing list. Punching computer programs. The 29 card punch and the 59 card verifier.

986. Harrison, Michael A. **Lectures on Linear Sequential Machines.** New York: Academic Press, 1970. 210p. $10.50.

Contents: Special topics in linear algebra. Basic notions of sequential machine theory. Basic properties of linear sequential machines. Relations and decision problems for LSM's. Decomposition of LSM's.

987. Hart, J. F., E. W. Cheney, C. Lawson, C. Mesztenyi, J. R. Rice, H. Thacher, Jr., C. Witzgall, and H. Maehly. **Computer Approximations.** New York: John Wiley, 1968. 343p. $17.95.
Contents: The design of function subroutines. General methods of computing functions. Least maximum approximations. The choice and application of approximations.

988. Harvill, John B. **Basic FORTRAN Programming.** 2nd ed. Englewood Cliffs, N.J.: Prentice-Hall, 1968. 268p.
Contents: Fundamental concepts. The programming language. Control statements and flow charts. Sub-programs. Memory arrays. Additional capabilities. Appendices: 1. Explanation of the basic FORTRAN format statements. 2. Coding conventions for "look-alike" characters. 3. Instructions for punching basic FORTRAN input data. 4. Explanation of the memory chart conventions.

989. Hassitt, Anthony. **Computer Programming and Computer Systems.** New York: Academic Press, 1967. 374p. $12.50.
Contents: Computer systems. Machine language and assembly language. Central processing units. Compiler language. The computer system. FORTRAN subroutines. Character manipulation. Efficiency. The dynamic use of memory. Programming language.

990. Higman, Bryan. **A Comparative Study of Programming Languages.** New York: American Elsevier, 1967. 172p. $8.50.
Contents: The nature of language in general. Recursion. Polish notations. Theory of names. Systems aspects. Formal language structure. Microgenerator. From machine code to FORTRAN. COBOL. ALGOL. List processing languages. C.P.L. and the I.B.M. share issue. Input and output. M.R.40 No. 5168.

991. Hopgood, F. R. A. **Compiling Techniques.** New York: American Elsevier, 1969. 136p. $6.00.
Contents: Data structure. Data structure mapping. Tables. Language description. Lexical analysis. Syntax analysis. Code generation for arithmetic expressions. Storage allocation. Compiler-compilers.

992. Hull, T. E. **Introduction to Computing.** Englewood Cliffs, N.J.: Prentice-Hall, 1966. 212p.
Contents: Algorithms. Stored-program computers. Basic programming techniques. Compilers and monitors. FORTRAN programs. FORTRAN constants and variables. FORTRAN expressions and assignment statements. FORTRAN control statements. FORTRAN input and output statements. FORTRAN subprograms. Program planning and debugging. Numerical methods. Non-numerical applications. Simulation. Algorithms, automata and languages. Suggestions for the exercises of chapters.

993. Husson, Samir S. **Microprogramming: Principles and Practices.** Englewood Cliffs, N.J.: Prentice-Hall, 1970. 512p.
Contents: Principles of microprogramming. Microprogram control. Microprogramming applications. Writable control storage. ROS memory technology. IBM System/360 microprogramming. IBM System/360 Model 40. IBM System/360 Model 50. RCA Spectra 70/45. Microprogramming. The Honeywell H4200.

994. Hyndman, D. E. **Analog and Hybrid Computing.** Elmsford, N.Y.: Pergamon, 1970. 218p.
Contents: Introduction to the basic units of electronic analog computers. Solution of non-linear differential equations, with variable coefficients. Iterative operation and hybrid computing.

995. Iliffe, J. K. **Basic Machine Principles.** New York: American Elsevier, 1968. 96p. $5.25.
Contents: General principles. Some related systems. Basic machine. Basic language. Techniques.

996. Inman, Kenneth L. **Fundamentals of Electronic Data Processing: A Programmed Text.** Englewood Cliffs, N.J.: Prentice-Hall, 1965. 158p.
Contents: Introduction to electronic data processing. Computer classification and electronic data processing systems. Numbering systems and data representation. Computer word. Flow charting. Programming languages. Programming techniques.

997. Jones, Robert Lloyd. **Fundamental COBOL for IBM System /360.** Englewood Cliffs, N.J.: Prentice-Hall, 1969. 245p.
Contents: COBOL programming language. The program coding sheet. Entry formats. Program structure. Identification division. Environment division. Data division. Procedure division. COBOL vocabulary.

998. Kain, Richard Y. **Automata Theory: Machines and Languages.** New York: McGraw-Hill, 1972. 301p. $12.50.
Contents: Mathematical linguistics. Finite-state machines. Turing machines. Linear-bounded automata. Pushdown automata. Operations on languages. Solvable and unsolvable linguistic questions.

999. Kapur, G. K. **IBM/360 Assembler Language Programming.** New York: John Wiley, 1971. 560p.
Contents: Computing. Data formats. Storage allocation. Logic of assembler language. Binary arithmetic. Looping indexing. Subroutines.

1000. Katzan, Harry, Jr. **APL Programming and Computer Techniques.** New York: Van Nostrand-Reinhold, 1970. 329p.
Contents: The structure of computers. Elements of APL programming. Arrays. Computer programming systems. Programming languages.

1001. Katzan, Harry, Jr. **APL User's Guide.** New York: Van Nostrand-Reinhold, 1971. 126p.
Contents: The APL terminal system. Arithmetic operations. Terminal operations. APL programming. Scalar and vector functions. Input/output. Matrices. RECOMMENDED.

1002. Levison, Michael, and W. Alan Sentance. **Introduction to Computer Science.** New York: Gordon & Breach, 1970. 160p.
Contents: Digital computers and their applications. Processes. Automatic programming languages. Subroutines. Input/output and memory devices. Errors and precautions. Machine language I—arithmetic and control instructions. Machine language II—input/output. Program input. Interpreters and compilers. Further programming techniques. An introduction to ALGOL. Writing a compiler.

1003. Lott, Richard W. **Basic Data Processing.** Englewood Cliffs, N.J.: Prentice-Hall, 1967. 228p.
Contents: Data processing. System analysis and design. Punched card processing. A conventional punched card application. Computer concepts. Programming fundamentals. The computer installation cycle. Magnetic tape processing. Random access processing. Data transmission: concepts and applications. Other media and methods. Data processing costs. Data control. Decision tables.

1004. Louden, Robert K. **Programming the IBM 1130 and 1800.** Englewood Cliffs, N.J.: Prentice-Hall, 1967. 443p.
Contents: Programming concepts. Bits, codes and computing equipment. FORTRAN. Running, debugging and documenting a program. More FORTRAN. Functions subroutines and subprogram. Disk storage and monitor operation. Floating-point manipulation in FORTRAN. Integer manipulation in FORTRAN. Character manipulation in FORTRAN. Symbolic assembly language programming. More symbolic assembly language programming. Assembly subprograms and the subroutine library. The 1800 time sharing executive programs (TSX). More TSX programming.

1005. McCormick, John M., and Mario G. Salvadori. **Numerical Methods in FORTRAN.** Englewood Cliffs, N.J.: Prentice-Hall, 1964. 324p.
Contents: Computers and programming. Approximate computations. Differentiation, integration, interpolation and extrapolation. Solution of algebraic and transcendental equations. Simultaneous linear algebraic equations. Ordinary boundary-value problems. Ordinary initial-value problems. Two-dimensional problems. FORTRAN programs. Beginner's hints. Flow charts.

1006. McKeeman, W. M., and others. **A Compiler Generator.** Englewood Cliffs, N.J.: Prentice-Hall, 1970. 527p.
Contents: Language. Translators. Algorithms. The language XPL. BNF programming. XCOM self-compiling compiler. SKELETON proto-compilers.

1007. Minsky, Marvin. **Computation: Finite and Infinite Machines.** Englewood Cliffs, N.J.: Prentice-Hall, 1967. 317p.

172

Contents: Physical machines and their abstract counterparts. Finite-state machines. Neural networks. Automata made up of parts. The memories of events in finite-state machines. Computability, effective procedures, and algorithms—infinite machines, Turing machines. Universal Turing machines. Limitations of effective computability: some problems not solvable by instruction-obeying machines. The computable real numbers. The relations between Turing machines and recursive functions. Models similar to digital computers. The symbol manipulation systems of Emil Post. Post's normal form theorem. Very simple bases for computability.

1008. Moore, Ramon E. **Interval Analysis.** Englewood Cliffs, N.J.: Prentice-Hall, 1966. 145p.
Contents: Interval numbers. Interval arithmetic. A metric topology for intervals. Matrix computations with intervals. Values and ranges of values of real functions. Interval contractions and root finding. Interval integrals. Interval equations. The initial value problem in ordinary differential equations. The machine generation of Taylor co-efficients. Numerical results with the Kth-order method. Coordinate transformations for the initial value problem. M.R.37 No. 7069. M.R.41 No. 4856.

1009. Nenadal, Z., and B. Mirtes. **Analogue and Hybrid Computers.** Trans. from the Czech by R. J. M. Grew. New York: American Elsevier, 1969. 628p. $16.00.
Contents: Operational amplifiers. Passive and a.c. computing elements. Electromechanical, servomechanical and mechanical computing elements. Electronic nonlinear circuits, digital and hybrid elements, analogue computer programming, auxiliary circuits and apparatus. Analogue and hybrid computers.

1010. Oberle, Aloyse P. **Programming the I.B.M. System/360 Model 20 with R.P.6.** Englewood Cliffs, N.J.: Prentice-Hall, 1970. 166p.
Contents: RP6 system designs. Operational concepts. Card input. Calculations on data. Tape and disk files.

1011. Pugh, Alexander, L., III. **DYNAMO II User's Manual.** Cambridge, Mass.: M.I.T. Press, 1971. 73p.
Contents: DYNAMO II for the IBM S/360.

1012. Randell, B., and L. J. Russell. **ALGOL 60 Implementation: The Translation and Use of ALGOL 60 Programs on a Computer.** New York: Academic Press, 1964. 418p. $14.00.
Contents: ALGOL compilers. ALGOL translation techniques. Translation of arithmetic expressions. The whetstone compiler. The object program. Assignment statements. Blocks and procedures. Arrays. Labels and switches. Parameters for statements. Code procedures. The translator: introduction. Translator stack: name list. Translation techniques. Translator routines.

1013. Rosenfeld, Azriel. **Picture Processing by Computer.** New York: Academic Press, 1969. 196p. $11.50.
Contents: Pictures and picture processing. Picture coding. Approximation of

pictures. Position-invariant operations on pictures. Picture properties and pictorial pattern recognition. Picture segmentation. Geometrical properties of picture subsets. Picture description and picture languages.

1014. Rubenstein, Moshe F. **Matrix Computer Analysis of Structures.** Englewood Cliffs, N.J.: Prentice-Hall, 1966. 402p.
Contents: Computers—fundamental concepts. Structures—fundamental concepts. Characteristics of structures—stiffness and flexibility. Determinants and matrices. Solution of linear equations. Energy concepts in structures. Transformation of information in structures. The flexibility method. The stiffness method. Analysis by substructures and by recursion. Analysis by iteration. Analysis of plates and shells.

1015. Sammet, Jean E. **Programming Languages: History and Fundamentals.** Englewood Cliffs, N.J.: Prentice-Hall, 1969. 785p.
Contents: General introduction. Functional characteristics of programming languages. Technical characteristics of programming languages. Languages for numerical scientific problems. Languages for business data processing problems. String and list processing languages. Formal algebraic manipulation languages. Multipurpose languages. Specialized languages. Significant unimplemented concepts. Future long range developments. Appendices. Bibliography arrangements and author list. Language summary.

1016. Saxon, James A. **Programming the RCA 301: A Self-Instructional Programmed Manual.** Englewood Cliffs, N.J.: Prentice-Hall, 1965. 247p.
Contents: RCA 301. Flow charting. Basic instructions. Decision, looping. End of file, end of tape procedures. Printing. Symbolic programming. Address modification. Arithmetic. Variable data and switches. Editing. Printouts. Batching.

1017. Saxon, James A., and Richard W. Senseman. **Programming and Wiring the UNIVAC 1004 Card Processor: A Self-Instructional Programmed Manual.** Englewood Cliffs, N.J.: Prentice-Hall, 1964. 254p.
Contents: UNIVAC 1004 basics. Basic processes. Wiring basics. Character identification and insertion. Input/output. Selector techniques. Testing. Multiply-divide standard report programs. Advanced techniques.

1018. Shorter, Edward. **The Historian and the Computer: A Practical Guide.** Englewood Cliffs, N.J.: Prentice-Hall, 1971. 149p.
Contents: Some historic background. Codebook designing. Data processing. Programming. Data analyzing. RECOMMENDED FOR BEGINNERS.

1019. Skelton, J. E. **An Introduction to the BASIC Language.** New York: Holt, Rinehart and Winston, 1971. 158p.
Contents: The problem-solving process. Computation: the LET statement. Input/output: the INPUT and PRINT statements. Control statements: the GO TO, IF, and END statements. Lists and tables: the DIM statement. Computing the value of polynomials: a second example. Loops: the FOR and NEXT statements. The READ and the DATA statements. Functions and subroutines: DEF and GOSUB

statements. Some more programming techniques. An interesting problem. A summary of BASIC. HIGHLY RECOMMENDED.

1020. Southworth, Raymond W., and Samuel L. DeLeeuw. **Digital Computation and Numerical Methods.** New York: McGraw-Hill, 1965. 544p. $10.50.
Contents: Flow charting. The FORTRAN language. Number systems and machine language. Rounding and truncation errors. Roots of equations. Simultaneous linear equations. Interpolation. Numerical differentiation and integration. Taylor's series. Numerical solution of ordinary differential equations. Empirical formulas and approximation.

1021. Stein, Marvin L., and W. D. Munro. **Computer Programming.** New York: Academic Press, 1964. 459p. $13.25.
Contents: Number systems. Machine organization. Elementary coding. Fixed and floating point arithmetic. Scaling. Non-arithmetic operations subroutines. Input-output. Assembly of complete programs. FORTRAN. Mixed language programs.

1022. Stein, Marvin L., and W. D. Munro. **A FORTRAN Introduction to Programming and Computers.** New York: Academic Press, 1966. 122p. $5.25pa.
Contents: Elements of FORTRAN. Input and output. Program control and organization. Extended modes of arithmetic. Program checking and execution. Miscellany. Some nonarithmetic aspects of FORTRAN. Introduction to modern digital computing systems.

1023. Steinhart, R. F., and S. V. Pollack. **Programming the IBM System/360.** New York: Holt, Rinehart, and Winston, 1970. 576p.
Contents: Introduction to System/360 organization and data handling. Assembler language. Binary arithmetic. Branching and decimal operations. Floating point arithmetic. FORTRAN programming. COBOL programming. PL/I input/output programming. ALGOL language. HIGHLY RECOMMENDED.

1024. Tedeschi, F. P., and J. A. Scigliano. **Digital Computers and Logic Circuits.** Beverly Hills, Calif.: Glencoe Press, 1971. 226p.
Contents: Computer arithmetic. Boolean algebra. Integrated circuit devices. Computer design. Processing systems.

1025. Walnut, Francis K. **Introduction to Computer Programming and Coding.** Englewood Cliffs, N.J.: Prentice-Hall, 1968. 429p.
Contents: Introduction to computers and programming. Language of computers. Input-output. Structure of a computing system. Instructions. Programming procedures. Precoding requirements. Basic coding techniques. Arithmetic and data manipulation coding. Processing of the radix point (scaling). Logical operations. Decision coding. Input-output coding. Address modification. Loops and loop control. Program switches. Subroutines. Table processing techniques. Table look-up procedures. Sorting. Chained list processing. Magnetic tape coding techniques. Coding for random access devices.

1026. Ward, B. **Computer Technician's Handbook**. Blue Ridge Summit, Pa.: TAB, 1971. 480p.
Contents: Computers. Number and coding systems. Elements of Boolean algebra. Circuits. Decoding. Shift registers and memories. Maintenance and diagnostic information.

1027. Weiss, Eric A. **Programming the IBM 1620: The Hands-on Approach**. New York: McGraw-Hill, 1965. 316p. $8.50.
Contents: Instructions, storage, and first steps. Dump, branch, transmit digit, fields, and flags. Comparing and indicators. How to go about creating a program. Arithmetic and negative numbers. Immediate commands, flag manipulation, No Op, and decimal point. Branching and making a program. Flow charts, typewriter control, and a loop program. Alphameric characters. Punched cards. Symbolic programming system. SPS illustrative program. Special SPS commands. Floating point arithmetic. FORTRAN, variables, constants, statements, expressions. FORTRAN, parentheses, order of operations. FORMAT, preparing and running of FORTRAN program. FORTRAN, IF, lists, subscripts, and DO loops. FORTRAN, functions, subroutines.

CHAPTER 29

PROBABILITY AND STATISTICS

1028. Alder, Henry L., and Edward B. Roessler. **Introduction to Probability and Statistics.** San Francisco: W. H. Freeman, 1968. 333p. $7.50.
Contents: Analysis of data. Elementary probability, permutations, and combinations. The binomial distribution. The normal distribution. Random sampling. Testing hypotheses, significance levels, confidence limits. Student's t-distribution. Nonparametric statistics. Regression and correlation. Chi-square distribution. Index numbers. Time series. The F-distribution. The analysis of variance.

1029. Amstadter, Bertram L. **Reliability Mathematics: Fundamentals, Practices, Procedures.** New York: McGraw-Hill, 1971. 408p. $17.50.
Contents: Elements of statistics. Distributions and reliability. Testing of hypothesis. Mathematical models. Prediction. Reliability and reliability growth. Methods of assessment. Tables. ARBA 1972.

1030. Barber, Michael, and B. W. Ninham. **Random and Restricted Walks: Theory and Applications.** New York: Gordon & Breach, 1970. 176p.
Contents: Random walk. Lattice walk. Correlated walks. Self-avoiding walks. Diffusion. Brownian motion. Polymers. Solid state applications.

1031. Barlow, R. E., and F. Proschan. **Mathematical Theory of Reliability.** New York: John Wiley, 1965. 256p. $11.95.
Contents: Definitions of reliability. Failure distributions. Renewal theory. Replacement policies. Stochastic models for complex systems. Markov chains and semi-Markov processes. Redundancy optimization. Qualitative relationships for multicomponent structure.

1032. Barr, Donald R. **Finite Statistics.** Boston: Allyn & Bacon, 1969. 193p.
Contents: Finite probability models. Sampling distributions. Estimation. Testing hypotheses. Confidence intervals.

1033. Bechhofer, Robert E., J. Kiefer, and M. Sobel. **Sequential Identification and Ranking Procedures with Special Reference to Koopman-Darmois Populations.** Chicago: University of Chicago Press, 1968. 420p. $17.50.
Contents: PART I. General theory. PART II. Application to ranking Koopman-Darmois parameters with special reference to normal means and unsolved problems. PART III. Monte Carlo sampling results for the problem of ranking means of normal populations with a common known variance. M.R.39 No. 6445.

1034. Bharucha-Reid, A. T., editor. **Probabilistic Methods in Applied Mathematics.** New York: Academic Press, 1968. Vol. I, 291p.; Vol. II, 220p.
Contents: VOLUME I. Random eigenvalue problems, W. E. Boyce. Wave propagation in random media, U. Frish. Branching processes in neutron transport

theory, T. W. Mullikin. VOLUME II. Random algebraic equations, A. T.
Bharucha-Reid. Axiomatic quantum mechanics and generalized probability
theory, S. Gudder. Random differential equations in control theory, W. M.
Wonham. M.R.41 No. 4701a, b.

1035. Blackwell, David. **Introduction to Statistics.** New York: McGraw-Hill,
1969. 150p. $5.50.
Contents: Probability. Variables. Densities. Mean. Variance. Work of a predictor.
Correlation. Multiple and partial correlation. Independence. The binomial dis-
tribution. The normal approximation. Inference. Inference about proportions:
I, Inference about proportions. II, Independent proportions. Chi-square.

1036. Blanc-Lapierre, A., and R. Fortet. **Theory of Random Functions.** Vol. I.
Trans. from the French by J. Gani. New York: Gordon & Breach, 1967. 443p.
$29.50.
Contents: Axioms and basic concepts and theorems of probability theory.
Random functions. Stochastic processes. Markov processes and permanent,
continuous and discontinuous Markov chains. M.R.36 No. 7188.

1037. Blumenthal, R. M., and R. K. Getoor. **Markov Processes and Potential
Theory.** New York: Academic Press, 1968. 313p. $15.00.
Contents: Markov processes. Excessive functions. Multiplicative functionals
and subprocesses. Additive functionals and their potentials. Further properties
of continuous additive functionals. Dual processes and potential theory.
M.R.41 No. 9348.

1038. Breiman, L. **Probability and Stochastic Processes.** Boston: Houghton,
1969. 324p.
Contents: The basic probability model. Some classical models. Random variables. Independent random variables. Conditional probability. Markov chains.
Continuous time Markov processes. Vector independence and the multivariate normal distribution. Stationary time series. M.R. 40 No. 8149.

1039. Bulmer, M. G. **Principles of Statistics.** Cambridge, Mass.: M.I.T. Press,
1965. 214p. $7.50.
Contents: Elementary introduction to probability. Random variables. Probability distributions. Expectations. Moment generating functions. Binomial distributions. Poisson distributions. Exponential distributions. Normal distributions.
Chi-squared distributions. t-distributions. F-distributions. Statistical inference.
Testing of hypothesis. Point estimation. Regression and correlation.

1040. Carlson, Roger. **Introduction to Probability and Statistics.** San Fran-
cisco: Holden-Day, 1970. 550p.
Contents: Probability theory. Random variables and probability distributions.
Limit theorems in probability theory. Theory of sampling. Statistical inference.

1041. Chakravarti, I. M., R. G. Laha, and J. Roy. **Handbook of Methods of
Applied Statistics. Vol. I: Techniques of Computation, Descriptive Methods,**

and Statistical Inference. **Vol. II: Planning of Surveys and Experiments.** New York: John Wiley, 1967. Vol. I: 460p.; $12.95. Vol. II: 160p.; $9.00.
Contents: VOLUME I. Techniques of computation. Descriptive methods. Statistical inference. VOLUME II. Sample survey methods. Experimented designs. REFERENCE BOOK FOR PRACTICING STATISTICIANS.

1042. Chover, J., editor. **Markov Processes and Potential Theory.** New York: John Wiley, 1967. 235p. $8.95.
Contents: Construction of Markov processes in axiomatic potential theory, Heinz Bauer. Hunt processes and standard processes, R. M. Blumenthal and R. G. Getoor. Markov processes with infinities, Kai Lai Chung. Semiclassical potential theory, Z. Ciesielski. Excessive functions for a class of continuous time Markof chains, G. E. Denzel, J. C. Kemeny, and J. L. Snell. On random time substitutions and the Feller property, John Lamperti. On the multiplicative decomposition of positive supermartingales, P. A. Meyer. Polar sets for processes with stationary independent increments, Steven Orey. Random walks on the line, Donald S. Ornstein. Potentials associated with recurrent stable processes, Sidney C. Port. Application of unsmoothing and Fourier analysis to random walks, Charles J. Stone. Penetration times and passage times, D. Stroock. Limit theorem for a class of branching processes, Shinzo Watanabe. M.R. 37 No. 3647.

1043. Chung, Kai Lai. **A Course in Probability Theory.** New York: Harcourt, 1968. 331p. $12.00.
Contents: Sums of random variables. Random walks. Law of large numbers. Central limit theorem. Iterated logarithmic law. Markov property and martingales. M.R.37 No. 4842.

1044. Cohen, J. W. **The Single Server Queue.** Amsterdam: North-Holland, 1969. 671p. $35.00.
Contents: Markov chains with a discrete time parameter. Markov chains with continuous time parameter. Birth and death processes. Derived Markov chains. Renewal theory and regenerative processes. Priority disciplines for single server queue. Uniformly bounded actual waiting time. The finite dam. Uniformly bounded virtual waiting time. Limit theorems for single server queues.

1045. Conover, W. J. **Practical Nonparametric Statistics.** New York: John Wiley, 1971. 462p.
Contents: Elements of probability theory. The binomial distribution. Statistical inference. Contingency tables. The Kolmogorov-Smirnov-type statistics.

1046. Cramer, H., and M. R. Leadbetter. **Stationary and Related Stochastic Processes: Sample Function Properties and Their Applications.** New York: John Wiley, 1967. 348p. $12.50.
Contents: Sample functions of stationary stochastic processes. Sample continuity and differentiability. Processes with finite second moments. Processes with orthogonal increments. Weak and strict stationarity. Stationary and non-stationary normal processes. Limit theorems for random variables. M.R.38 No. 4751.

1047. Davenport, Wilbur B., Jr. **Probability and Random Processes**. New York: McGraw-Hill, 1970. 542p.
Contents: Probability. Sample points. Sample spaces. Random variables. Random process. Functions of a random variable. Average. Sampling techniques. Prediction. Estimation. The Poisson and Gaussian processes.

1048. David, H. A. **Order Statistics**. New York: John Wiley, 1970. 272p.
Contents: Distributions. Bounds for moments of order statistics. Estimation. Testing of hypothesis.

1049. DeGroot, Morris H. **Optimal Statistical Decisions**. New York: McGraw-Hill, 1970. 480p. $14.50.
Contents: Experiments, sample spaces, and probability. Random variables, random vectors, and distribution functions. Some special univariate distributions. Some special multivariate distirbutions. Subjective probability. Utility. Decision problems. Conjugate prior distributions. Limiting posterior distributions. Estimation, testing hypotheses, and linear statistical models. Sequential sampling. Optimal stopping. Sequential choice of experiments.

1050. Derman, Cyrus. **Finite State Markovian Decision Processes**. New York: Academic Press, 1970. 159p. $10.00.
Contents: Finite horizon expected cost minimization. Some existence theorems. Computational methods for the discounted cost problem. Computational procedures. State-action frequencies and problems with constraints. Optimal stopping of a Markov chain.

1051. Dixon, Wilfrid J., and Frank J. Massey, Jr. **Introduction to Statistical Analysis**. 3rd ed. New York: McGraw-Hill, 1969. 608p. $11.50.
Contents: Distributions. Introduction to measures of central value and dispersion. Population and sample. The normal distribution. Statistical inference: estimation and tests. Inference: single population. Inference: two populations. Efficiency and various statistics. Analysis of variance. Regression and correlation. Analysis of covariance. Enumeration statistics. Probability of accepting a false hypothesis. More analysis of variance. Questions of normality. Nonparametric statistics. Sequential analysis. Sensitivity experiments. Probability.

1052. Drake, Alvin W. **Fundamentals of Applied Probability Theory**. New York: McGraw-Hill, 1967. 283p. $9.75.
Contents: Events, sample spaces and probability. Random variables. Transforms. Sums of independent random variables. Probabilistic processes. Discrete-state Markov processes. Limit theorems.

1053. Draper, N. R., and H. Smith. **Applied Regression Analysis**. New York: John Wiley, 1966. 407p. $11.75.
Contents: Fitting a straight line by least squares. The matrix approach to linear regression. The examination of residuals. Two independent variables. More complicated models. Selecting the "best" regression equations. A specific problem. Multiple regression and mathematical model building. Multiple regres-

sion applied to analysis of variance problems. An introduction to nonlinear estimation. M.R.35 No. 2415.

1054. Dynkin, E. B. **The Theory of Markov Processes.** Elmsford, N.Y.: Pergamon, 1961. 219p.
Contents: Markov processes. Subprocesses. Construction of Markov processes with given transition functions. Strictly Markov processes. Conditions of boundedness and continuity of Markov processes.

1055. Dynkin, E. B., and A. A. Yushkevich. **Markov Processes: Theorems and Problems.** Trans. from the Russian by James S. Wood. New York: Plenum, 1969. 237p. $15.00.
Contents: Metric random walk. The Wiener process. The problem of optimal choice. Boundary conditions. The birth and death process. M.R.39 No. 3585a.

1056. Eisen, Martin. **Introduction to Mathematical Probability Theory.** Englewood Cliffs, N.J.: Prentice-Hall, 1969. 535p.
Contents: Finite probability spaces. Random variables and combinations of events. Dependence and independence. Some elementary limit theorems. Infinite probability spaces. Theory of measure. Integration. Probability and measure. Distribution and moments. Characteristic functions. Independence. Series of independent random variables. Limit theorems for sums of independent random variables.

1057. Feller, W. **An Introduction to Probability Theory and Its Applications.** New York: John Wiley. Vol. I: 1968 (3rd ed.), 509p., $13.95. Vol. II: 1971 (2nd ed.), 669p., $15.95.
Contents: VOLUME I. The sample space. Combinatorial analysis. Conditional probability. Stochastic independence. The binomial and Poisson distributions. Normal approximation to the binomial distribution. Random variables. Laws of large numbers. Generating functions. Branching processes. Renewal theory. Random walk and ruin problems. Markov chains. VOLUME II. The exponential and the uniform densities. Randomization. Normal densities and processes. Probability measures and spaces. Laws of large numbers. The basic limit theorems. Markov processes and semi-groups. Renewal theory. Random walks. Tauberian theorems. Resolvents. Harmonic analysis. HIGHLY RECOMMENDED.

1058. Ferguson, T. S. **Mathematical Statistics: A Decision Theoretic Approach.** New York: Academic Press, 1967. 396p. $15.25.
Contents: Game theory and decision theory. The main theorems of decision theory. Distributions and sufficient statistics. Invariant statistical decision problems. Testing hypotheses. Multiple decision problems. Sequential decision problems.

1059. Fraser, D. A. S. **The Structure of Inference.** New York: John Wiley, 1968. 344p. $16.00.
Contents: Measurement models. Structural model. Structural distribution. Linear models. Error distributions. Approximations. M.R. 38 No. 3946.

1060. Freedman, David A. **Approximating Countable Markov Chains.** San Francisco: Holden-Day, 1971. 304p.
Contents: Restricting the range: applications. Constructing the general Markov chain. Introduction to discrete time. Introduction to continuous time. The general case.

1061. Freedman, David A. **Brownian Motion and Diffusion.** San Francisco: Holden-Day, 1971. 248p.
Contents: Brownian motion. Diffusion.

1062. Freedman, David A. **Markov Chains.** San Francisco: Holden-Day, 1971. 382p.
Contents: Discrete and continuous time. Ratio limit theorems. Invariance principles.

1063. Fryer, H. C. **Concepts and Methods of Experimental Statistics.** Boston: Allyn & Bacon, 1966. 602p. $12.50.
Contents: Sampling binomial and multinomial populations. Sampling one normal population. Sampling two normal populations. Sampling populations with unspecified distributions. Linear regression and correlation. Sampling more than two normal populations simultaneously. Introductory analysis of variance. Models and expected mean squares for the analysis of variance for completely randomized and randomized complete block designs. More complex analyses of variance. Simple analysis of covariance. Multiple linear regression and correlation analysis. Curvilinear regression analysis. The Poisson and negative binomial distributions. Introductory discriminatory analysis. RECOMMENDED. M.R.36 No. 6034.

1064. Ghosal, A. **Some Aspects of Queueing and Storage Systems.** New York: Springer-Verlag, 1970. 93p. $2.80.
Contents: Queueing and storage problems. Probability distributions. First passage problems and duality relations. Cybernetic queueing and storage systems. Optimal capacity of a storage system. M.R.41 No. 7785.

1065. Girault, M. **Stochastic Process.** New York: Springer-Verlag, 1966. 126p. $7.00.
Contents: Introduction. Poisson processes. Numerical processes with independent random increments. Markov processes. Laplace processes and second-order processes. Some Markov processes on continuous time. M.R.35 No. 2331.

1066. Gnedenko, B. V., Yu. K. Belyayev, and A. D. Solovyev. **Mathematical Methods of Reliability Theory.** Trans. by Scripta Technica, Inc., and edited by R. E. Barlow. New York: Academic Press, 1969. 506p. $24.50.
Contents: Fundamentals of probability theory and mathematical statistics. Characteristics of reliability. Evaluation of reliability factors from experimental data. Testing of reliability hypotheses. Standby redundancy without renewal. Standby redundancy with renewal. Statistical methods of quality control and reliability of mass production.

1067. Gnedenko, B. V., and A. Ya. Khinchin. **An Elementary Introduction to the Theory of Probability.** Trans. from the Russian by W. R. Stahl and edited by J. B. Roberts. San Francisco: W. H. Freeman, 1961. 139p. $1.75pa.
Contents: The probability of events. Rules for adding probabilities. Conditional probabilities and the rule for multiplication. Consequences of addition and multiplication rules. Bernoulli's method. Bernoulli's theorem. Random variables and the distribution law. Mean values. Mean values of sums and products. Scattering and mean (standard) deviation. The law of large numbers. The normal laws.

1068. Good, I. J. **The Estimation of Probabilities: An Essay on Modern Bayesian Methods.** Cambridge, Mass.: M.I.T. Press, 1965. 109p.
Contents: Theories of probabilities. Bayesian methods. Estimation of probabilities from binomial and multinomial sampling. Using beta and Dirichlet distributions. Tests for independence in multidimensional contingency tables.

1069. Grenander, U. L. F. **Probabilities on Algebraic Structures.** New York: John Wiley, 1963. 218p. $12.00.
Contents: Probabilities and probability measures on compact groups. Lie groups. Locally compact groups. Banach spaces. Topological algebras.

1070. Guenther, William C. **Concepts of Probability.** New York: McGraw-Hill, 1968. 384p. $8.50.
Contents: Definitions and interpretation of probability. The calculation of probabilities. Random variables plus probability distributions and expectation. Some important probability distributions. Random variables and distributions in two and more dimensions. Continuous random variables and large sample approximations. Statistical applications of probability: estimation and testing hypotheses. Sampling. Markov chains.

1071. Guenther, William C. **Concepts of Statistical Inference.** New York: McGraw-Hill, 1965. 353p. $8.50.
Contents: Probability. Probability models. Parameters, statistics, and sampling distributions. Testing statistical hypotheses. Some statistical inference for means and variances. Some statistical inference for parameters of discrete random variables. Approximate Chi-square tests. Analysis of variance. Regression and correlation.

1072. Gupta, Shanti S., and James Yackel, editors. **Statistical Decision Theory and Related Topics.** New York: Academic Press, 1971. 394p. $11.50.
Contents: Molecular studies of evolution: a source of novel statistical problems, Jerzy Neyman. Asymptotically efficient estimation of non-parametric regression coefficients, L. Weiss and J. Wolfowitz. Optimal allocation of observations when comparing several treatments with a control (III): globally best one-sided intervals for unequal variances, Robert Bechhofer and Bruce Turnbull. On some contributions to multiple decision theory, Shanti S. Gupta and Klaus Nagel. A decision-theoretic approach to the problem of testing a null hypothesis, Herman Rubin. The role of symmetry and approximation in exact design optimality, J. Kiefer. Symmetric binomial group-testing with three outcomes, M. Sobel,

S. Kumar, and S. Blumenthal. Detection of outliers, A. P. Dempster and Bernard Rosner. Empirical Bayes slippage tests, J. Van Ryzin. Analogues of linear combinations of order statistics in the linear model, P. J. Bickel. A theorem on exponentially bounded stopping time of invariant SPRT's with applications, R. A. Wijsman. Some aspects of search strategies for Wiener processes, E. M. Klimko and James Yackel. Optimal pari-mutual wagering, James N. Arvesen and Bernard Rosner. Nonparametric procedures for selecting fixed-size subsets, M. M. Desu and Milton Sobel. On a subset selection procedure for the most probable event in a multinomial distribution, S. Panchapakesan. On approximating constants required to implement a selection procedure based on ranks, Gary C. McDonald. Selection procedures with respect to measures of association, Z. Govindarajulu and Anil P. Gore. Sample size for selection, Edward J. Dudewicz and Nicholas A. Zaino, Jr. Optimal confidence intervals for the largest location parameter, Edward J. Dudewicz and Yung Liang Tong. Non-optimality of likelihood ratio tests for sequential detection of signals in Gaussian noise, Bennett Eisenberg.

1073. Hadley, George. **Elementary Statistics.** San Francisco: Holden-Day, 1969. 467p.
Contents: Probability theory. Random variables. Estimation. Hypothesis testing. Modern decision theory. Analysis of variance. Regression and correlation analysis. Time series.

1074. Hajek, Jaroslav. **Nonparametric Statistics.** San Francisco: Holden-Day, 1969. 192p.
Contents: Uniform distribution over the space of permutations. Hypothesis of randomness. Testing randomness. The hypothesis of symmetry and of random blocks. The hypothesis of independence. M.R.39 No. 7771.

1075. Hajek, Jaroslav, and Z. Sidak. **Theory of Rank Tests.** New York: Academic Press, 1967. 297p. $8.00.
Contents: Theory of rank tests. Selected rank tests. Exact distributions of test statistics under the hypotheses and their computation. Limiting distributions of test statistics under the hypotheses. Limiting distributions of test statistics under the alternatives. Asymptotic optimality and efficiency of rank tests. M.R. 37 No. 4925.

1076. Hannan, E. J. **Multiple Time Series.** New York: John Wiley, 1970. 536p.
Contents: Prediction theory. Spectral theory of vector processes. Laws of large numbers. The central limit theorem. Regression. Inference about spectra.

1077. Harris, Bernard, editor. **Spectral Analysis of Time Series.** New York: John Wiley, 1967. 319p. $9.95.
Contents: Introduction to the theory of spectral analysis of time series, B. Harris. An introduction to the calculations of numerical spectrum analysis, J. W. Tukey. Quartic statistics in spectral analysis, S. K. Zaremba. Some problems in the application of the cross-spectral method, H. Akaike. Meteorological applications of cross-spectrum analysis, H. A. Panofsky. Estimation of coherency,

L. J. Tick. Asymptotic theory of estimates of k-th order spectra, D. R. Brillinger and M. Rosenblatt. Computation interpretation of k-th order spectra, D. R. Brillinger and M. Rosenblatt. Time series analysis for models of signal plus white noise, E. Parzen. Prediction for non-stationary stochastic processes, M. D. Godfrey. Models for forecasting seasonal and non-seasonal time series, G. E. P. Box, G. M. Jenkins, and D. W. Bacon. M.R.35 No. 1168.

1078. Harris, Bernard. **Theory of Probability.** Reading, Mass.: Addison-Wesley, 1966. 294p. $9.75.
Contents: Elementary probability theory. Random variables. Probability distributions. Expectation. Limit theorems. Random experiments. M.R.34 No. 6811.

1079. Harris, Theodore E. **The Theory of Branching Processes.** Englewood Cliffs, N.J.: Prentice-Hall, 1964. 230p.
Contents: The Galton-Watson branching process. Processes with a finite number of types. The general branching process. Neutron branching processes. Markov branching processes. Age-dependent branching processes. Branching processes in the theory of cosmic rays.

1080. Hodges, J. L., Jr., and E. L. Lehmann. **Basic Concepts of Probability and Statistics.** 2nd ed. San Francisco: Holden-Day, 1970. 456p.
Contents: Probability models. Sampling. Product models. Conditional probability. Random variables. Special distributions. Multivariate distributions. Introduction to statistics. Estimation. Estimation of measurement and sampling models. Optimum methods of estimation. Tests of significance. Tests for comparative experiments. The concept of power. M.R.40 No. 6602.

1081. Hogg, R. V., and A. T. Craig. **Introduction to Mathematical Statistics.** 3rd rev. ed. New York: Macmillan, 1970. 415p.
Contents: Distributions of random variables. Conditional probability and stochastic independence. Distributions of functions of random variables. Interval estimation. Order statistics. Limiting distributions. Sufficient statistics. Point estimation. Statistical hypotheses. The analysis of variance. Tables: the Poisson distribution, the chi-square distribution, the normal distribution, the t-distribution, and the F-distribution. RECOMMENDED.

1082. Ivakhnenko, A. G., and V. G. Lapa. **Cybernetics and Forecasting Techniques.** Trans. from the Russian by Scripta Technica, edited by Robert N. McDonough. New York: American Elsevier, 1967. 196p. $13.75.
Contents: Prediction of deterministic processes. Interpolation and extrapolation. Prediction of stationary random processes. Summary of probability theory and the theory of random functions. Predicting nonstationary random processes. Cognitive systems used as predicting filters and regulators.

1083. Jenkins, Gwilym M., and Donald G. Watts. **Spectral Analysis and Its Applications.** San Francisco: Holden-Day, 1968. 525p. $16.25.
Contents: Aims and means in time series analysis. Fourier analysis. Probability theory. Introduction to statistical inference. Introduction to time series. Uni-

variate spectra. Examples of univariate spectral analysis. The cross-correlation function and cross spectrum. Estimation of cross-power spectra. Estimation of frequency-response functions. Multivariate spectral analysis. M.R.37 No. 6000.

1084. John, Peter W. M. **Statistical Design and Analysis of Experiments.** New York: Macmillan, 1971. 356p.
Contents: Linear models. Experiments with several factors. Factors with two levels. Factors with three levels. Existence and construction of balanced incomplete designs and partially balanced designs. Latin square designs. Applications in engineering experiments.

1085. Johnson, N. L., and F. C. Leone. **Statistics and Experimental Design in Engineering and the Physical Sciences.** New York: John Wiley, 1964. 2v.
Contents: VOLUME I. Sampling. Probability and distribution theory. VOLUME II. Analysis of variance. Sequential analysis. Multivariate observations.

1086. Kahane, Jean-Pierre. **Some Random Series of Functions.** Lexington, Mass.: D. C. Heath, 1968. 184p. $7.95.
Contents: Random series in Banach spaces and Hilbert spaces. Random Taylor series. Random Fourier series. Covering the circle of circumference one by random intervals. M.R.40 No. 8095.

1087. Kalinin, V. M., and O. V. Shalaevskii. **Investigations in Classical Problems of Probability Theory and Mathematical Statistics.** Trans. from the Russian. New York: Plenum Press, 1971. 141p.
Contents: Probability distributions and their limit properties. Hotelling's T^2 test and R^2 test and their minimax characters.

1088. Kappos, Demetrios. **Probability Algebras and Stochastic Spaces.** New York: Academic Press, 1969. 267p. $12.50.
Contents: Probability algebras. Extension of probability algebras. Cartesian product of probability algebras. Stochastic spaces. Expectation of random variables. Moment spaces Lq. Generalized random variables (random variables having values in any space). Complements.

1089. Karlin, Samuel. **A First Course in Stochastic Processes.** New York: Academic Press, 1966. 502p. $14.50.
Contents: Elements of stochastic processes. Markov chains. The basic limit theorem of Markov chains and applications. Algebraic methods in Markov chains. Ratio theorems of transition probabilities and applications. Sums of independent random variables as a Markov chain. Classical examples of continuous time Markov chains. Continuous time Markov chains. Order statistics. Poisson processes and applications. Brownian motion. Branching processes. Compounding stochastic processes. Deterministic and stochastic genetic and ecological processes. Queueing processes.

1090. Karlin, Samuel, and W. J. Studden. **Tchebycheff Systems: With Applications in Analysis and Statistics.** New York: John Wiley, 1966. 586p. $19.50.

Contents: Tchebycheff systems on a closed interval: definitions, examples, and preliminaries. Moment spaces induced by T-systems and their duals. The Markov-Krein theory and ramifications. Complements and the classical moment spaces. Tchebycheff systems and moment spaces on the interval. Moment spaces of periodic functions and T-systems on . . . moment spaces for Tchebycheff systems defined on discrete sets. Moment spaces generated by restricted measures. Minimax approximation, the Markoff-Bernstein inequality, and related matters. Some problems of best interpolating systems, maximization of moment determinants, and applications to optimum experimental designs. Generalized convex functions induced by Et-Systems. Generalized Tchebycheff inequalities. Multi-variate Tchebycheff inequalities. Tchebycheff type inequalities for sums of random variables and non-linear problems. M.R.34 No. 4757.

1091. Kemeny, John G., J. L. Snell, and A. W. Knopp. **Denumerable Markov Chains.** Princeton, N.J.: D. Van Nostrand, 1966. 439p. $12.50.
Contents: Denumerable matrices. Measure theory. Probability theory. Random variables. Martingale theory. Markov chains. Transient chains. Recurrent chains. Convergence theorems. Potential theory. Brownian motion. Random walks. The potential theory of transient and recurrent chains. HIGHLY RECOM-MENDED. M.R. 34 No. 6858.

1092. Kendall, M. G., and A. Stuart. **The Advanced Theory of Statistics.** New York: Hafner. Vol. I: **Distribution Theory.** 3rd ed., 1969. 439p. $13.95. M.R.23 No. A2247. Vol. II: **Statistical Inference and Statistical Relationship.** 2nd ed. 1967. 690p. $22.95. M.R.39 No. 4969. Vol. III: **Design and Analysis, and Time-Series.** 2nd ed. 1968. 557p. $21.95. (Includes tables and bibliography) M.R. 37 No. 999.

1093. Kingman, J. F. C., and S. J. Taylor. **Introduction to Measure and Probability.** New York: Cambridge University Press, 1966. 401p. $12.50.
Contents: Set theory. Point set topology. Measure theory. Lebesgue measure. Integration. Product measures. Measure in functions spaces. Convergence in function spaces. Complete orthonormal sets in Hilbert spaces. Differentiation and relation to Lebesgue measure. Haar measure on locally compact groups. Probability. Random variables. Distributions. Characteristic functions. Sums of independent variables. Joint distributions. Stochastic processes.

1094. Krishnaiah, Paruchuri, editor. **Multivariate Analysis II: Proceedings of the Second International Symposium on Multivariate Analysis held at Wright State University, Dayton, Ohio, June, 1968.** New York: Academic Press, 1969. 596p. $28.50.
Contents: Collection of papers on nonparametric methods. Multivariate analysis of variance and related topics. Distribution theory. Characteristic functions and characterization problems. Time series and stochastic processes. Decision procedures. Econometrics: principal components, reliability, and applications. M.R.40 No. 5069.

1095. Krutchkoff, Richard G. **Probability and Statistical Inference.** New York: Gordon & Breach, 1970. 306p.
Contents: Elements of probability. Random variables and their distributions. The Gamma and Beta distributions and distributions related to them. Multivariate distributions. Moment generating functions. Distribution of functions of random variables. An introduction to statistical inference. Sample moments. Properties of estimators. Uniformly minimum variance. Unbiased estimators. Hypothesis testing. Confidence intervals.

1096. Kyburg, Henry E., Jr. **Probability Theory.** Englewood Cliffs, N.J.: Prentice-Hall, 1969. 294p.
Contents: Sets and their properties. Calculus of measure. Frequency functions and distribution functions. Distributions of several random quantities. Limit theorems and the normal distribution. Other important distributions. Conditional distributions and correlation. Probability and measure. Applications of probability. Estimation. Interval estimation. Inductive behavior.

1097. Lamperti, John. **Probability: A Survey of the Mathematical Theory.** New York: W. A. Benjamin, 1966. 150p. $8.00.
Contents: Probability. Independence. Laws of large numbers. Central limit theorem. Stable laws. Brownian paths. The invariance principle. The Dirichlet problem. The distribution of eigenvalues. M.R.34 No. 6812.

1098. Lancaster, H. O. **The Chi-Squared Distribution.** New York: John Wiley, 1969. 356p.
Contents: Historical survey. Properties of the chi-squared distribution. Discrete and continuous variables. The binomial, hypergeometric, and the Poisson distributions. Relationship of the chi-squared distribution to the multinomial. HIGHLY RECOMMENDED. M.R.40 No. 6667.

1099. Lange, H. F. **Correlation Techniques.** New York: Van Nostrand-Reinhold, 1967. 464p. $13.50.
Contetns: The concept of correlation. The correlation function in correlation analysis. Instrument engineering in correlation analysis. The correlation function as a parameter in telecommunications. The correlation function as a means of investigating communications transmission systems. The correlation detector as a signal receiver. Recent trends in correlation techniques.

1100. Loève, Michel. **Probability Theory.** 3rd ed. New York: Van Nostrand-Reinhold, 1963. 704p. $16.75.
Contents: Probability theory. Independence and the Bernoulli case. Dependence and chains. Notions of measure theory: sets. Classes and functions. Measurable functions and integration. General concepts and tools of probability theory: probability concepts. Distribution functions and characteristic functions. Independence: sums of independent random variables. Central limit problems. Dependence: conditioning. From independence to dependence. Ergodic theorems. Second order properties. Elements of random analysis. Foundations. Martingales and decomposability. Markov processes.

1101. Lukacs, E. **Probability and Mathematical Statistics: An Introduction.**
New York: Academic Press, 1972. 242p.
Contents: The probability space. Elementary properties of probability spaces.
Random variables and their probability distributions. Typical values. Limit
theorems. Some important distributions. Sampling. Estimation. Testing hypo-
theses.

1102. Lukacs, E. **Stochastic Convergence.** Lexington, Mass.: D. C. Heath, 1968.
142p. $6.95.
Contents: The basic definitions and theorems of probability. Metric spaces of
random variables. The Levy metric. The KyFan metric. Kolmogorov strong law
of large numbers. The law of iterated logarithm. Stochastic integrals and the
Wiener process.

1103. Maksoudian, Y. Leon. **Probability and Statistics with Applications.**
Scranton, Penn.: Intext Educational Publishers, 1969. 416p. $9.95.
Contents: Univariate probability functions. The moments of probability
functions. Joint probability functions. Sampling distributions. Statistical decision-
making. Curve fitting. Design of experiments and analysis of variance. Latin
and Graeco-Latin squares. Factorial designs.

1104. Martin, J. J. **Bayesian Decision Problems and Markov Chains.** New York:
John Wiley, 1967. 202p. $10.95.
Contents: Markov sequential decision processes. Regularity theorems. The
Whittle distribution. M.R.36 No. 4761.

1105. Massey, L. Daniel. **Probability and Statistics.** New York: McGraw-Hill,
1971. 180p.
Contents: Probability theory. Discrete and continuous random variables.
Markov process. Queueing. Sampling. Point estimation. Statistical inference.
Testing of hypothesis. Confidence intervals.

1106. McKean, H. P., Jr. **Stochastic Integrals.** New York: Academic Press,
1969. 140p. $9.00.
Contents: Brownian motion. Stochastic integrals and differentials. Stochastic
integral equations. HIGHLY RECOMMENDED. M.R.40 No. 947.

1107. Miller, K. S. **Multidimensional Gaussian Distributions.** New York: John
Wiley, 1964. 129p. $9.95.
Contents: Quadratic forms. Multidimensional distribution. Miscellaneous
results. Some applications to Gaussian noise.

1108. Miller, Rupert G., Jr. **Simultaneous Statistical Inference.** New York:
McGraw-Hill, 1966. 272p. $11.50.
Contents: Normal univariate. Techniques. Regression techniques. Nonparametric
techniques. Multivariate techniques.

1109. Milton, Roy C., and J. Nelder, editors. **Statistical Computation: Proceedings of a Conference held at the University of Wisconsin, Madison, Wisconsin, April 1969.** New York: Academic Press, 1969. 462p. $11.00.
Contents: Collection of papers on specifications for statistical data structures. Statistical systems and languages. Statistical data screening with computers. Teaching of statistics with computers. Current techniques in numerical analysis related to statistical computation.

1110. Mine, Hisashi, and Shunji Osaki. **Markovian Decision Processes.** New York: American Elsevier, 1970. 142p. $12.50.
Contents: Markov chains. Semi-Markov processes. Markovian decision processes. Markovian decision processes with discounting. Markovian decision processes with no discounting. Markovian decision processes with absorbing states. Semi-Markovian decision processes. Sequential decision processes. Applications of Markovian decision processes.

1111. Mode, Charles J. **Multitype Branching Process: Theory and Applications.** New York: American Elsevier, 1970. 330p.
Contents: Galton-Watson processes. Multitype age-dependent branching processes. Markov branching processes with discrete and continuous time parameter.

1112. Mode, Elmer B. **Elements of Probability and Statistics.** Englewood Cliffs, N.J.: Prentice-Hall, 1966. 356p. $8.00.
Contents: Sets. Probability. Probability theorems. Some miscellaneous problems. Introduction to statistics. Frequency distributions and probability functions. The discrete case. The binomial probability function. Frequency distributions and probability density functions. The normal probability density function. The multinomial and chi-square distributions. Distributions related to the binomial and normal distributions. Inferences from sample means. The bivariate binomial probability function. Discrete bivariate probability functions. Estimation. Line fitting: regression. Correlations. Markov chains. HIGHLY RECOMMENDED FOR BEGINNERS. M.R.35 No. 3768.

1113. Mood, Alexander M., and Franklin A. Graybill. **Introduction to the Theory of Statistics.** 2nd ed. New York: McGraw-Hill, 1963. 443p. $10.00.
Contents: Probability. Discrete random variables. Continuous random variables. Expected values and moments. Special continuous distributions. Sampling. Point estimation. The multivariate normal distribution. Sampling distributions. Interval estimation. Tests of hypotheses. Regression and linear hypotheses. Experimental design models. Sequential tests of hypotheses. Nonparametric methods.

1114. Morrison, Donald F. **Multivariate Statistical Methods.** New York: McGraw-Hill, 1967. 338p. $9.95.
Contents: Basic concepts. Matrix algebra. Multinomial distributions. Estimation. Test procedures. Maximum likelihood and likelihood-ratio methods. Multivariate analysis of variance. Canonical analysis. Factor analysis. Principal components. M.R.35 No. 3811.

1115. Mulholland, H., and C. R. Jones. **Fundamentals of Statistics.** New York: Plenum, 1969. 288p. $5.95.
Contents: The collection, organization and representation of numerical data. Elementary probability. The binomial and Poisson distributions. Measures of central tendency. Measures of dispersion. Continuous distributions. The normal distribution. Significance testing and confidence intervals. Quality control. Chi-squared distribution. The F-distribution (variance ratio). Bivariate distribution. Mathematical expectation, variance and covariance. Weighted averages, death rates and time series.

1116. Papoulis, Athanasios. **Probability, Random Variables, and Stochastic Processes.** New York: McGraw-Hill, 1965. 583p. $12.75.
Contents: Probability theory. Functions of random variables. Sequences of random variables. Correlation and power spectrum of stationary processes. Linear mean-square estimation. Non-stationary processes. Transients in linear systems with stochastic inputs. Harmonic analysis of stochastic processes. Stationary and nonstationary normal processes. Brownian movement and Markoff processes. Poisson process and shot noise.

1117. Parthasarthy, K. R. **Probability Measures on Metric Spaces.** New York: Academic Press, 1967. 276p. $12.00.
Contents: Borel sets of complete separable metric spaces. Weak topology for probability measures. Regularity of measures. Perfectness of measures. Convergence of sample distributions. Metrizability and compactness in probability measure spaces. Probability on metric groups. Decomposability. Idempotence. Sums of infinitesimal random variables. Limit theorem for probability measures on Hilbert spaces. M.R.37 No. 2271.

1118. Parzen, Emanuel. **Stochastic Processes.** San Francisco: Holden-Day, 1962. 335p.
Contents: Random variables. Stochastic processes. Conditional probability and expectation. Normal processes and covariance stationary processes. Counting processes and Poisson processes. Renewal counting processes. Markov chains: discrete parameter. Markov chains: continuous parameter.

1119. Parzen, Emanuel. **Time Series Analysis Papers.** San Francisco: Holden-Day, 1967. 580p. $8.50.
Contents: 21 papers on Time Series Analysis, one written by G. M. Jenkins and the rest by E. Parzen. M.R.36 No. 6091.

1120. Peng, K. C. **The Design and Analysis of Scientific Experiments: An Introduction with Some Emphasis on Computation.** Reading, Mass.: Addison-Wesley, 1967. 252p. $12.50.
Contents: Two-way arrangements. Three-way and multi-way arrangements. Methods of partitioning sum of squares. Nested experiments. Fixed, random, and mixed models. Randomized blocks. Latin squares. Split-plot designs. Fractional factorial designs and confounding. Response surface designs. Special topics. Analysis of covariance. Non-factorial experiments. A computer program for the analysis

of Latin square and Graeco-Latin square experiments. A computer program for the analysis of fractional experiments with factors at two levels. M.R.35 No. 5093.

1121. Pfeiffer, Paul E. **Concepts of Probability Theory**. New York: McGraw-Hill, 1965. 399p. $11.00.
Contents: A mathematical model for probability. Random variables and probability distributions. Sums and integrals. Mathematical expectation. Sequences and sums of random variables. Random processes.

1122. Pinsker, M. S. **Information and Information Stability of Random Variables and Processes**. Trans. by A. Feinstein. San Francisco: Holden-Day, 1964. 256p.
Contents: Information and information stability of random variables. Information rate and information stability of random processes. Information, information rate, and information stability of Gaussian random variables and processes.

1123. Prabhu, N. U. **Queues and Inventories: A Study of Their Basic Stochastic Process**. New York: John Wiley, 1965. 275p. $12.95.
Contents: Queueing systems. Single server systems. Markov chains. Combinatorial methods in single server systems. Inventories. Theory of collective risk. Moran's theory of storage systems. Continuous time storage processes. M.R.35 No. 2374.

1124. Prohorov, Yu. V., and Yu. A. Rozanov. **Probability Theory**. Trans. by K. Krickeberg and H. Urmitzer. New York: Springer-Verlag, 1969. 401p. $18.70.
Contents: Basic concepts of elementary probability theory. Spaces and measures. Limit theorems in probability theory. Markov processes. Stationary processes. M.R.40 No. 4981.

1125. Pugachev, V. S. **Theory of Random Functions**. Elmsford, N.Y.: Pergamon, 1965. 850p.
Contents: The probability of an event and its properties. Random and random vector variables. Functions with random arguments. The law of large numbers. Characteristics of systems. Accuracy of linear systems. Tables of formulae and tables of functions.

1126. Raj, Des. **Sampling Theory**. New York: McGraw-Hill, 1968. 302p. $11.50.
Contents: Variance of a product. Conditional covariance. Tchebycheff inequality. Planning of sample surveys. Unequal probability sampling. Unbiased estimators. Construction and choice of the number of strata. Ratio and regression estimation. Multistage sampling. Double sampling. Nonsampling errors and stability of variance estimators. M.R.37 No. 6002.

1127. Raj, Des. **The Design of Sample Surveys**. New York: McGraw-Hill, 1972. 390p. $12.95.
Contents: Principles of sampling. Techniques of sampling. Planning and execution of surveys. Data collection in selected fields. RECOMMENDED.

1128. Rényi, A. **Probability Theory.** Amsterdam: North-Holland, 1971.
666p. $21.50.
Contents: Algebras of events. Probability. Discrete random variables. General
theory of random variables. More about random variables. Characteristic
functions. Laws of large numbers. The limit theorem of probability theory.
Appendix: Introducton to information theory. HIGHLY RECOMMENDED.

1129. Revesz, Pál. **The Laws of Large Numbers.** New York: Academic Press,
1968. 176p. $9.50.
Contents: Independent random variables. Orthogonal random variables.
Stationary sequences. Subsequences of sequences of random variables. Symmetri-
cally dependent random variables. Markov chains. Weakly dependent random
variables. Independent random variables in an abstract space. Sum of a random
number of independent random variables. M.R.39 No. 6391.

1130. Rickmers, Albert D., and Hollis N. Todd. **Statistics: An Introduction.**
New York: McGraw-Hill, 1967. 576p. $7.95.
Contents: Variability. The normal curve. Test of hypothesis of the mean. Interval
estimates for the mean. Statistical inference about variances. Sample size for
variables. Additional uses of the chi-square distribution. Analysis of variance.
Components of variance. Crossed and nested experiments. Studying individual
effects. Regression analysis. Planning experiments for statistical analysis. Based
experiment designs. Factorial experiments. Fractional factorials. Determination
of optimum conditions. Nonparametric statistics. Probability. Acceptance
sampling.

1131. Ross, S. M. **Applied Probability Models with Optimization Applications.**
San Francisco: Holden-Day, 1970. 198p. $12.95.
Contents: Poisson processes. Markov chains. Renewal theory. Markov renewal
processes. Markov decision processes. Inventory theory. Brownian motion.
Continuous-time optimization problems.

1132. Rozanov, Yu. A. **Stationary Random Processes.** Trans. by A. Feinstein.
San Francisco: Holden-Day, 1967. 216p.
Contents: Harmonic analysis of stationary random processes. Linear forecasting
of stationary discrete-parameter processes. Linear forecasting of continuous-
parameter stationary processes. Random processes stationary in the strict sense.

1133. Rumshiskii, L. Z. **Elements of Probability Theory.** Elmsford, N.Y.:
Pergamon, 1966. 172p.
Contents: Events and probabilities. Random variables and probability distribu-
tions. The law of large numbers. Limit theorems and estimates of the mean.
Linear correlation.

1134. Saaty, Thomas L. **Elements of Queueing Theory, with Applications.**
New York: McGraw-Hill, 1961. 440p. $13.00.
Contents: Queueing models. Probability. Markoff chains and processes. Ergodic
properties and queues. The birth-death process in queueing theory. Poisson and

non-Poisson queues. Poisson and arbitrary input. Arbitrary service. The waiting time for single and multiple channels. General independent input. Exponential or Erlangian service. Applications and renewal theory.

1135. Sage, Andrew P., and J. L. Melsa. **Estimation Theory with Applications to Communications and Control.** New York: McGraw-Hill, 1971. 529p.
Contents: Elements of probability theory. Random variables. Stochastic processes. Stochastic differential equations. Decision theory. Estimation theory and the optimum linear filter.

1136. Sawaragi, Y., Y. Sunahara, and T. Nakamizo. **Statistical Decision Theory in Adaptive Control Systems.** New York: Academic Press, 1967. 211p. $12.00.
Contents: Random processes. Statistical decision theory. Statistical decision concept in control processes. Non-sequential decision approaches in adaptive control systems. Sequential decision approaches in adaptive control systems. Adaptive adjustments of parameters of non-linear control systems.

1137. Seber, G. A. F. **The Linear Hypothesis: A General Theory.** New York: Hafner, 1966. 115p. $4.15.
Contents: Vector spaces. Range and nullspace of matrices. Distributions. The linear hypothesis. Least squares estimation. The Gauss-Markov theorem. Hypothesis testing. The likelihood-ratio test. The F-test. The nesting procedure. Modified hypothesis. Missing observations. Non-linear regression.

1138. Smillie, K. W. **An Introduction to Regression and Correlation.** New York: Academic Press, 1966. 168p. $7.00.
Contents: Simple linear regression. Correlation between two variables. The general linear regression model. Non-linear regression. Some miscellaneous problems. The programming of regression calculations.

1139. Spitzer, Frank. **Principles of Random Walk.** New York: Van Nostrand-Reinhold, 1964. 406p. $13.50.
Contents: The classification of random walk. Harmonic analysis. Two-dimensional recurrent random walk. Random walk on a half-line. Random walk on an interval. Transient random walk. Recurrent random walk.

1140. Srinivasan, S. K. **Stochastic Theory and Cascade Processes.** New York: American Elsevier, 1969. 232p. $12.50.
Contents: Branching phenomena. Point processes—general approach. Point processes—product densities. Multiple product densities and sequent correlations. Electromagnetic cascades—mathematical techniques. Electromagnetic cascades—analytic and computational methods. Extensive air showers. Polarisation in cascades. Population growth. M.R.41 No. 6324.

1141. Stratonovich, R. L. **Conditional Markov Processes and Their Application to the Theory of Optimal Control.** Trans. from the Russian by Scripta Technica. Edited by Richard Bellman. New York: American Elsevier, 1968. 367p. $14.75.
Contents: Markov processes. The main results of the theory of conditional Mar-

kov processes. Application of the theory of conditional Markov processes to the theory of optimal control. M.R.36 No. 4912.

1142. Thomas, John B. **An Introduction to Applied Probability and Random Processes.** New York: John Wiley, 1971. 338p.
Contents: Probability theory. Random variables. Density functions. Expectations. Moments. Binomial, Poisson, and normal distributions. Random processes and applications.

1143. Thomasian, Aram J. **The Structure of Probability Theory with Applications.** New York: McGraw-Hill, 1969. 832p. $14.95.
Contents: Discrete probability densities. Independent events. Independent experiments. The mean and variance. Random variables. The law of large numbers. The central limit theorem. Conditional expectations and branching processes. Probability spaces. Continuous probability densities. Distribution functions. Multivariate normal densities and characteristic functions. Stochastic processes. The Poisson process and some of its generalizations. The Wiener process.

1144. Tucker, Howard G. **A Graduate Course in Probability and Mathematical Statistics.** New York: Academic Press, 1967. 273p. $12.00.
Contents: Convergence theorems. Sums of independent random variables. The central limit theorem. Stochastic process. Poisson process and Brownian motion. M.R.36 No. 4593.

1145. Tucker, Howard G. **An Introduction to Probability and Mathematical Statistics.** New York: Academic Press, 1962. 228p. $11.00.
Contents: Events and probabilities. Dependent and independent events. Random variables and probability distributions. Expectation and limit theorems. Point estimation. The multivariate normal density. Testing statistical hypotheses: simple hypothesis vs. simple alternative. Testing simple and composite hypotheses. Confidence intervals.

1146. Wadsworth, George A., and Joseph G. Bryan. **Introduction to Probability and Random Variables.** New York: McGraw-Hill, 1960. 292p. $9.50.
Contents: Probability concepts. Discrete random variables. Continuous random variables. Joint distributions. Derived distributions. Mathematical expectation.

1147. Wasan, M. T. **Parametric Estimation.** New York: McGraw-Hill, 1970. 240p. $12.50.
Contents: Least-squares estimation method. Optimal spacing of a regressor variable. Completeness. Sufficiency. Convex functions. Unbiased estimation. A lower bound on the variance of an estimation. Maximum-likelihood estimation. Admissible and minimax estimation. The empirical Bayes method of estimation.

1148. Wasan, M. T. **Stochastic Approximation.** New York: Cambridge University Press, 1969. 216p. $9.50.
Contents: Robbins-Monro's method. Kiefer-Wolfowitz method. Multivariate

stochastic approximation methods. Asymptotic normality. The approximations for continuous random processes. Up-and-down method. M.R.40 No. 975.

1149. Whittle, P. **Prediction and Regulation by Linear Least-Square Methods.** New York: Van Nostrand-Reinhold, 1963. 147p. $3.75.
Contents: Stationary processes. The prediction problem. Least-square approximation. Projection on finite, semi-infinite, and infinite samples. Deviations from stationarity. Multivariate processes. Regulation.

1150. Wiener, Norbert, A. Siegel, B. Rankin, and W. T. Martin. **Differential Space, Quantum Systems, and Prediction.** Cambridge, Mass.: M.I.T. Press, 1966. 176p. $7.50.
Contents: Brownian motion. Differential space. Integration in differential space. Matrix factorization. The differential space theory of quantum systems. Dichotomic algorithm. M.R.35 No. 7394.

1151. Wolf, Frank L. **Elements of Probability and Statistics.** New York: McGraw-Hill, 1962. 322p. $7.95.
Contents: Empirical frequency distributions. Sets and set operations. Measures of dispersion. Probability. Distributions. Normal distributions. Chi-square distributions. Distributions. Student's distributions. Bivariate distributions.

1152. Zehna, Peter W. **Finite Probability.** Boston: Allyn & Bacon, 1969. 193p.
Contents: Sample spaces and events. Probability spaces. Conditional probability and independence. Compound experiments. Random variables and probability distributions. Mathematical expectation.

1153. Zyskind, G., O. Kempthorne, F. B. Martin, E. J. Carney, and E. N. West. **Linear Models and Analysis of Variance Research Procedures.** Wright-Patterson Air Force Base, Ohio: Aerospace Research Laboratories, Office of Aerospace Research, U.S. Air Force, 1968. 180p. $3.00.
Contents: Collection of papers: On canonical forms. Non-negative covariance matrices. Best and simple. Least squares. Linear estimators in linear models. A general Gauss-Markov theorem for linear models with arbitrary non-negative covariance structure. Combination of information from uncorrelated linear models by simple weighing. Relationship of generalized polykays to unrestricted sums for balanced complete finite populations. A Monte Carlo study of the Behrens-Fisher problem. Grouping of observations and the Wilcoxon signed-rank test. M.R.38 No. 6690.

CHAPTER 30

BIBLIOGRAPHIES

1154. American Mathematical Society. **Author Index of Mathematical Reviews, 1960-1964.** Providence, R.I.: American Mathematical Society, 1966. Part I: A-K. 688p. Part II: L-Z. 672p.
Contents: Author index to *Mathematical Reviews* from 1960 to 1964. For an index to *Mathematical Reviews* before 1960, see *20-Volume Author Index of Mathematical Papers, 1940-1959.* In preparation: *Author Index of Mathematical Reviews, 1965-1969.*

1155. American Mathematical Society. **Author Index to Soviet Mathematics— Doklady, 1960-1969.** Providence, R.I.: American Mathematical Society, 1971. 136p. $8.00.
Contents: Author index to the articles that appeared in *Soviet Mathematics— Doklady.* Each entry includes author, title of article, page numbers of the Russian original and the English translation, and *Mathematical Reviews* numbers.

1156. American Mathematical Society. **Contemporary Chinese Research Mathematics.** Providence, R.I.: American Mathematical Society, 1961 (Vol. I), 1964 (Vol. II). 83p. (Vol. I). 19p. (Vol. II). $1.00ea.
Contents: VOLUME I. A bibliography of Chinese mathematical literature from 1949 to 1960. Compiled by Chia Kuei Tsao. VOLUME II. A report by S. H. Gould on Chinese-English mathematical dictionaries.

1157. American Mathematical Society. **Index to Translations Selected by the American Mathematical Society.** Providence, R.I.: American Mathematical Society, 1966. 90p. $6.60.
Contents: This is an author index to A.M.S. Translations—Series I, A.M.S. Translations—Series II (Vols. 1-50), and Selected Translations in Mathematical Statistics and Probability (Vols. 1-5). Each entry includes author, title of article, volume number, and *Mathematical Reviews* number.

1158. American Mathematical Society. **20-Volume Author Index of Mathematical Papers, 1940-1959.** Rev. ed. Providence, R.I.: American Mathematical Society, 1966. Part I: A-K. 1092p. Part II: L-Z. 1115p. $35.50.
Contents: This is an author index to *Mathematical Reviews* from 1940 to 1959. See *Author Index of Mathematical Reviews, 1960-1964.*

1159. Kendall, M. G., and D. G. Alison. **Bibliography of Statistical Literature 1940-1949.** Edinburgh: Oliver & Boyd, 1965. 190p. $10.50.

1160. Kendall, M. G., and D. G. Alison. **Bibliography of Statistical Literature Pre-1940, with Supplements to the Volumes for 1940-1949 and 1950-1958.** Edinburgh: Oliver & Boyd, 1968. 356p. $23.50.

1161. Lancaster, H. O. **Bibliography of Statistical Bibliographies.** Edinburgh: Oliver & Boyd, 1968. 103p. $11.75.
Contents: Personal bibliographies. Subject bibliographies. National bibliographies. Subject index. Author index.

1162. Lebedev, A. V., and R. M. Fedorova. **A Guide to Mathematical Tables.** Elmsford, N.Y.: Pergamon, 1960. 632p. Supplement No. 1, by N. M. Burunova. 1960. 228p.
Contents: A guide to mathematical tables published in books or periodicals up to 1958.

1163. Robin, Richard S. **Annotated Catalogue of the Papers of Charles S. Peirce.** Amherst: University of Massachusetts Press, 1967. 268p. $15.00.
Contents: Manuscripts and correspondence of Charles S. Peirce.

1164. Schutte, K. **Index of Mathematical Tables from All Branches of Sciences./Index mathematischer Tafelwerke und Tabellen aus allen Begieten der Naturwissenschaften.** 2nd rev. ed. Munich-Vienna: R. Oldenbourg, 1966. 239p. M.R.35 No. 2453.

1165. Steenrod, N. E., compiler. **Reviews of Papers in Algebraic and Differential Topology, Topological Groups, and Homological Algebra. Parts I, II.** Providence, R.I.: American Mathematical Society, 1968. Part I: 862p. Part II: pp. 863-1448. $26.00set.
Contents: Reviews of all papers that appeared in *Mathematical Reviews* from 1940 to 1967 under the above headings.

CHAPTER 31

DICTIONARIES

1166. Breuer, H. **Dictionary for Computer Languages**. New York: Academic Press, 1966. 332p. $12.50.
Contents: Methods of translating a program. ALGOL 60–FORTRAN II and IV. FORTRAN II and IV–ALGOL 60. Lists of computers considered and their properties.

1167. DeFrancis, John, compiler. **A Chinese-English Glossary of the Mathematical Sciences**. Providence, R.I.: American Mathematical Society, 1964. 275p. $3.00.

1168. Feys, R., and F. B. Fitch, editors. **Dictionary of Symbols of Mathematical Logic**. Amsterdam: North-Holland, 1969. 171p. $9.50.

1169. Freiberger, W. F., et al., editors. **International Dictionary of Applied Mathematics**. New York: Van Nostrand-Reinhold, 1960. 1173p. $32.50.
Contents: 8,000 terms. Russian, German, French, Spanish.

1170. Freund, John E., and Frank J. Williams. **Dictionary/Outline of Basic Statistics**. New York: McGraw-Hill, 1966. 224p. $5.95; $2.95pa.
Contents: PART I: Dictionary of basic statistics. PART II: Statistical formulas. References.

1171. Gould, S. H., and P. E. Obreanu. **Romanian-English Dictionary and Grammar for the Mathematical Sciences**. Providence, R.I.: American Mathematical Society, 1967. 51p. $7.10.
Contents: Romanian grammar and Romanian-English dictionary (1,700 entries).

1172. Hyman, C. **German-English Mathematics Dictionary**. New York: Interlanguage Dictionaries Publishing Corp., 1960. 131p. $8.00.
Contents: 8,500 entries.

1173. James, Glenn, and Robert C. James. **Mathematics Dictionary: Regular and Multilingual Editions**. 3rd ed. New York: Van Nostrand-Reinhold, 1968. 448p. $13.50. Multilingual edition: 517p. $17.50.
Contents: Regular edition: 8,000 entries. Multilingual edition: Russian, German, French, and Spanish.

1174. Kluznin, L. A., V. A. Mihailov, and L. P. Niznik, compilers. **German-Russian Mathematical Dictionary**. 2nd rev. ed. Moscow: Izdat. Nauka, 1968. 387p. M.R.39 No. 7.

1175. Kurtz, Albert, and H. A. Edgerton. **Statistical Dictionary of Terms and Symbols.** New York: Hafner Publ. Co., 1967. 191p. $7.50.

1176. Lohwater, A. J., compiler. **Russian-English Dictionary of the Mathematical Sciences.** With the collaboration of S. H. Gould. Providence, R.I.: American Mathematical Society, 1960. 267p. $7.70.
Contents: Grammar of the Russian language and over 15,000 terms.

1177. Millington, T. Alaric, and William Millington. **Dictionary of Mathematics.** New York: Barnes & Noble, 1971. 259p. $2.00.

1178. Zimmerman, M. **Russian-English Translators Dictionary.** New York: Plenum, 1967. 294p. $12.00.

CHAPTER 32

TABLES

1179. Abramov, A. A. **Tables of** $\ln \Gamma[z]$ **for Complex Argument.** Elmsford, N.Y.: Pergamon, 1960. 336p. $17.50.
Contents: Six-decimal tables.

1180. Aizenshtadt, V. S., V. I. Krylov, and A. S. Metel'skii. **Tables of Laguerre Polynomials and Functions.** New York: Pergamon, 1966. 160p.
Contents: Theory and tables.

1181. Arscott, F. M., and I. M. Khabaza. **Tables of Lame Polynomials.** New York: Pergamon, 1962. 560p.
Contents: Summary of the theory of Lame polynomials, and tables.

1182. Bailey, Leslie F. **Tables of Folded** $-\dfrac{\sin x}{x}$ **Interpolation Coefficients.** Washington: U.S. Government Printing Office, 1966. 161p. $2.75.

1183. Bark, L. S., and P. I. Kuznetsov. **Tables of Lommel's Functions of Two Pure Imaginary Variables.** Elmsford, N.Y.: Pergamon, 1965. 292p.
Contents: Seven-figure tables.

1184. Belousov, S. L. **Tables of Normalized Associated Legendre Polynomials.** Elmsford, N.Y.: Pergamon, 1962. 380p.
Contents: Six-decimal tables.

1185. Belyakov, V. M., R. I. Kravtsova, and M. G. Rappoport. **Tables of Elliptic Integrals, Part I.** Elmsford, N.Y.: Pergamon, 1965. 646p.

1186. Berlyand, O. S., R. I. Gavrilova, and A. P. Prudnikov. **Tables of Integral Error Functions and Hermite Polynomials.** Elmsford, N.Y.: Pergamon, 1962. 164p.

1187. Burington, Richard Stevens. **Handbook of Mathematical Tables and Formulas.** 4th ed. New York: McGraw-Hill, 1965. 423p. $3.95.
Contents: PART I. Formulas, definitions and theorems from elementary mathematics. Tables of derivatives, integrals, and series. PART II. Tables. Logarithms. Trigonometric functions. Exponential functions. Powers. Roots. Reciprocals. Statistics and probability.

1188. Burstein, Herman. **Attribute Sampling: Tables and Explanations.** New York: McGraw-Hill, 1971. 464p. $18.50.
Contents: Tables for the determination of confidence limits and sample size. The tables and formulas are based on the binomial distribution. Some theory, explanations and numerous examples on the use of tables. HIGHLY RECOMMENDED. ARBA 72.

1189. Byrd, P. F., and M. D. Friedman. **Handbook of Elliptic Integrals for Engineers and Scientists.** 2nd rev. ed. New York: Springer-Verlag, 1971. 358p. Contents: Definitions and fundamental relations. Reduction of algebraic, trigonometric, and hyperbolic integrands to Jacobian elliptic functions. Tables of integrals of Jacobian elliptic functions. Elliptic integrals. Expansion in series.

1190. Dekanosidze, E. N. **Tables of Lommel's Functions of Two Variables.** Elmsford, N.Y.: Pergamon, 1960. 500p. Contents: Six-decimal tables.

1191. Faddeyev, D. K. **Tables of the Principal Unitary Representations of Fedorov Groups.** Elmsford, N.Y.: Pergamon, 1964. 180p.

1192. Faddeyeva, V. N., and N. M. Terent'ev. **Tables of Values of the Function**

$$w(z) = e^{-z^2} \left(1 + \frac{2i}{\sqrt{\pi}} \int_0^z e^{t^2} dt\right).$$ Elmsford, N.Y.: Pergamon, 1961. 288p.
Contents: Six-decimal tables.

1193. Fettis, H. E., and J. C. Caslin. **Tables of Modified Bessel Functions.** Wright-Patterson Air Force Base, Ohio: Aerospace Research Laboratories, Office of Aerospace Research, U.S. Air Force, 1969. 232p. $3.00.

1194. Gradshteyn, I. S., and I. M. Ryzhik. **Tables of Integrals, Series and Products.** 4th ed. Translation editor, A. Jeffrey. New York: Academic Press, 1966. 1086p. $10.50.
Contents: Logarithms. Trigonometric functions. Exponential functions. Integrals of functions. The special functions. M.R.33 No. 5952.

1195. Groenewoud, C., D. C. Hoaglin, and J. A. Vitalis. **Bivariate Normal Offset Circle Probability Tables with Offset Ellipse Transformation.** Prepared with the cooperation of the Environmental Science Services Administration, U.S. Department of Commerce. Buffalo, N.Y.: Cornell Aeronautical Laboratory, 1967. Vol. I: pp. 1-594. Vol. II: pp. 595-1320. $12.50set.

1196. Harter, H. L., and D. B. Owen, editors. **Selected Tables in Mathematical Statistics, Vol. I.** Sponsored by the Institute of Mathematical Statistics. Chicago: Markham, 1970. 405p. $5.80.

1197. Ince, E. L. **Mathematical Tables. Vol. IV: Cycles of Reduced Ideals in Quadratic Fields.** Prepared for the British Association Committee for the Calculation of Mathematical Tables. London: Royal Society at the Cambridge University Press, 1966. 80p.

1198. Ingelstam, E., and S. Sjoberg, compilers. **Elphyma Tables: Tables, Formulas, Nomograms within Mathematics—Physics—Electricity.** New York: John Wiley, 1964. 99p. $4.95.
Contents: Logarithms. Reciprocals. Square roots. Cube roots. Squares. Surfaces. Volumes. Centers of gravity. Exponential and hyperbolic functions.

1199. Karpov, K. A. **Tables of Lagrange Interpolation Coefficients. Supplement to Tables of the Function** $w(z) = e^{-z^2} \int_0^z e^{t^2} dt$ **in the Complex Domain.** Elmsford, N.Y.: Pergamon, 1965. 84p.

1200. Karpov, K. A. **Tables of the Function** $F(z) = \int z_0 e^{x^2} dx$ **in the Complex Domain.** Elmsford, N.Y.: Pergamon, 1965. 524p.
Contents: Five-decimal tables.

1201. Karpov, K. A. **Tables of the Function** $w(z) = e^{-z^2} \int_0^z e^{x^2} dx$ **in the Complex Domain.** Elmsford, N.Y.: Pergamon, 1965. 544p.
Contents: Five-decimal tables.

1202. Khrenov, L. S. **Six-Figure Tables of Trigonometric Functions.** Elmsford, N.Y.: Pergamon, 1965. 380p.

1203. Kireyeva, I. Ye., and K. A. Karpov. **Tables of Weber Functions, Vol. 1.** Elmsford, N.Y.: Pergamon, 1961. 364p.
Contents: Five- and six-decimal tables.

1204. Kogbetliantz, Ervand G., and Alice Krikorian. **Handbook of First Complex Prime Numbers.** New York: Gordon & Breach, 1971. Vol. I: 256p. Vol. II: 772p.
Contents: The complex prime numbers of modulus smaller than $10^{3.5}$, and all possible decompositions of real primes of the form 4n-1 as sums of two squares.

1205. Luk'yanov, A. V., I. B. Teplov, and M. K. Akimova. **Tables of Coulomb Wave Functions (Whittaker Functions).** Elmsford, N.Y.: Pergamon, 1965. 240p.
Contents: Five-figure tables.

1206. Lyusternik, L. A. **Ten-Decimal Tables of the Logarithms of Complex Numbers and for the Transformation from Cartesian to Polar Coordinates. Tables of the Functions in** ℓnx, arctan x, ½ln$(1 + x^2)$,$\sqrt{1 + x^2}$. Elmsford, N.Y.: Pergamon, 1965. 124p.

1207. Meredith, William B. **Basic Mathematical and Statistical Tables for Psychology and Education.** New York: McGraw-Hill, 1967. 333p. $3.95.
Contents: Tables with explanations.

1208. Nosova, L. N. **Tables of Thomson Functions and Their First Derivatives.** Elmsford, N.Y.: Pergamon, 1961. 426p.
Contents: Seven-figure tables.

1209. Nosova, L. N., and S. A. Tumarkin. **Tables of Generalized Airy Functions for the Asymptotic Solution of the Differential Equations** $e(py')' + (q + er)y = f$. Elmsford, N.Y.: Pergamon, 1965. 124p.

1210. Owen, D. B. **Handbook of Statistical Tables.** Elmsford, N.Y.: Pergamon, 1962. 592p.

1211. Pagurova, V. I. **Tables of the Exponential Integral**

$$E_n(x) = \int_\pi^\infty e^{-xu}\, u^{-n} du$$

Elmsford, N.Y.: Pergamon, 1961. 170p.
Contents: Seven-decimal tables.

1212. Pearson, K., editor. **Tables of the Incomplete Beta-function.** 2nd ed., rev. by E. S. Pearson and N. L. Johnson. Published for the Biometrika Trustees. London: Cambridge University Press, 1968. 505p. $15.50.

1213. Peters, J. **Eight-place Tables of Trigonometric Functions for Every Second of Arc: With an Appendix on the Computation to Twenty Places.** New York: Chelsea, 1968. 954p. $25.00.

1214. Salzer, H. E., and N. Levine. **Tables of Sines and Cosines to Ten-Decimal Places at Thousandths of a Degree.** Elmsford, N.Y.: Pergamon, 1962. 916p.

1215. Schofield, C. W. **Technical Tables for Schools and Colleges.** Elmsford, N.Y.: Pergamon, 1966. 80p.
Contents: Logarithmic tables. Exponential functions. Hyperbolic functions. Arithmetical tables. Financial tables. Trigonometrical tables. Mensuration. Metric. Calculus.

1216. Sheppard, W. F. **The Probability Integral.** Compiled and edited by the Committee for the Caluculation of Mathematical Tables. British Association for the Advancement of Science, Mathematical Tables, Vol. VII. London: Cambridge University Press, 1966. 34p. $5.50.

1217. Smirnov, A. D. **Tables of Airy Functions and Special Confluent Hypergeometric Functions, for Asymptotic Solutions of Differential Equations of the Second Order.** Elmsford, N.Y.: Pergamon, 1960. 268p.

1218. Smirnov, N. V. **Tables for the Distribution and Density Functions of the t-distribution (Students' distribution).** Elmsford, N.Y.: Pergamon, 1961. 130p.

1219. Smirnov, N. V. **Tables of the Normal Probability Integral, the Normal Density and Its Normalized Derivatives.** Elmsford, N.Y.: Pergamon, 1965. 144p.

1220. **Tables of the Clebsch-Gordan Coefficients.** Compiled by the Institute of Atomic Energy, Academia Sinica, Peking, Science Press, 1965. 564p. M.R.35 No.3973.

1221. **Tables of the Individual and Cumulative Terms of the Poisson Distribution.** New York: Van Nostrand-Reinhold, 1962. 224p. $8.00.
Contents: Computation to eight decimal places.

1222. **Tables Relating to Mathieu Functions: Characteristic Values, Coefficients and Joining Factors.** 2nd ed. Washington: U.S. Government Printing Office, 1967. 311p. $3.25.
Contents: Tables. Bibliography on texts and tables on Mathieu functions.

1223. Von Vega, Baron. **Logarithmic Tables of Numbers and Trigonometrical Functions.** 84th ed. New York: Van Nostrand-Reinhold, 1960. 603p. $4.95.
Contents: Table of common logarithms of the natural numbers from 1 to 100,000. Logarithms of the sines and tangents from second to second. Logarithms of the trigonometrical functions from ten to ten seconds. Table for the conversion of sidereal time into mean time. Table for the conversion of mean time into sidereal time. Tables of refraction. Constants.

1224. Warmus, M. W. **Tables of Elementary Functions: Also Supplementary Table of Proportional Parts.** Elmsford, N.Y.: Pergamon, 1961. 576p.
Contents: Six-figure tables.

1225. Wheelon, Albert D. **Tables of Summable Series and Integrals Involving Bessel Functions.** San Francisco: Holden-Day, 1968. 131p.
Contents: PART I. A short table of summable series. PART II. A table of integrals involving Bessel functions. M.R.37 No. 1060.

1226. Zhurina, M. I., and L. N. Karmazina. **Tables and Formulae for the Spherical Functions.** Elmsford, N.Y.: Pergamon, 1966. 100p.
Contents: Spherical functions: theory and tables.

1227. Zhurina, M. I., and L. N. Karmazina. **Tables of the Legendre Functions:** $P_{-\frac{1}{2} + it}(x)$. Elmsford, N.Y.: Pergamon, 1965. Vol. 1: 324p. Vol. 2: 422p.
Contents: Seven-decimal tables.

CHAPTER 33

MISCELLANEOUS

1228. Abadie, J., editor. **Integer and Nonlinear Programming.** Amsterdam: North-Holland, 1970. 560p. $28.50.
Contents: Convergence theory in nonlinear programming. Rank one methods for unconstrained optimization. A class of methods of nonlinear programming with termination and convergence properties. Minimization of a convex function by relaxation. Computational methods for least squares. Linear least squares and quadratic programming. Ritter's cutting plane method for nonconvex quadratic programming. The asymptotic integer algorithm. Minimax and duality for linear and nonlinear mixed integer programming. Branch and bound methods for integer and nonconvex programming. Dynamic programming of some integer and nonlinear programming problems.

1229. Abadie, J., editor. **Non-Linear Programming.** Amsterdam: North-Holland, 1967. 316p. $13.00.
Contents: Non-linear programming and duality, S. Vajda. On the Kuhn-Tucker theorem, J. Abadie. On a pair of dual non-linear programs, H. W. Kuhn. Positive (semi)-definite programming, G. B. Dantzig and R. W. Cottle. Some applications of the conjugate function theory to duality, A. Whinston. Methods of non-linear programming, P. Wolfe. Numerical methods, E. M. L. Beale. Resolution of mathematical programming with non-linear constraints by the methods of centres, P. Huard. Some general methods in integer programming, M. L. Balinski. A study of some inequalities for nonlinear stochastic programming, B. T. Lieu. One-sided constraints in hydrodynamics, J. J. Moreau. Application of generalized linear programming to control theory, D. B. Dantzig. Optimal control and convex programming, J. B. Rosen. The product form of the simplex method, P. Wolfe. M.R.35 No. 6454.

1230. Abraham, Ralph. **Foundations of Mechanics: A Mathematical Exposition of Classical Mechanics with an Introduction to the Qualitative Theory of Dynamical Systems and Applications to the Three-Body Problem.** New York: W. A. Benjamin, 1967. 296p. $14.75.
Contents: Differential manifolds. Vector bundles. Vector fields. Differential geometry. Hamiltonian and Lagrangian systems. Legendre transformation. Applications to the three-body problem.

1231. Aczel, J. **On Applications and Theory of Functional Equations.** New York: Academic Press, 1969. 64p. $3.75pa.
Contents: On applications and theory of functional equations. International meeting on functional equations—what are they anyway. M.R.39 No. 1838.

1232. Adams, Roy N., and Eugene D. Denman. **Wave Propagation and Turbulent Media.** New York: American Elsevier, 1967. 134p. $7.50.

Contents: Wave propagation in an inhomogeneous medium. Derivation of the recurrence equations using signal flow graphs. Solution to the Epstein variation. Turbulent medium. Results and accuracy.

1233. American Mathematical Society. **Eleven Papers on Differential Equations, Functional Analysis and Measure Theory.** A.M.S. Translations–Series II, Vol. 51. Providence, R.I.: American Mathematical Society, 1966. 332p. $17.30.

Contents: On the existence of eigenvectors for positive noncompletely continuous linear operators, I. A. Bahtin. On measures invariant under a group of transformations, S. V. Fomin. On the factorization of operators in Hilbert space, I. C. Gohberg and M. G. Krein. On Sommerfeld-type conditions for a certain class of partial differential equations, V. V. Grusin. On the solubility and stability of the Dirichlet problem, M. V. Keldys. Structure of the set of solutions of an equation of parabolic type, M. A. Krasnosel'skii and P. E. Sobolevskii. On the stability of solutions of differential-operator equations with time lag under perturbations bounded in the mean, L. H. Liberman. On a generalization of the concept of spectral measure, V. E. Ljance. A variation problem, S. M. Nikol'skii. A global existence theorem for a nonlinear system and the defect index of a linear operator, A. Povzner. Proof of Mahler's conjecture on the measure of the set of S-numbers, V. G. Sprindzuk.

1234. American Mathematical Society. **Twelve Papers on Topology, Algebra and Number Theory.** A.M.S. Translations–Series II, Vol. 52. Providence, R.I.: American Mathematical Society, 1966. 276p. $14.50.
Contents: Radicals of associative rings, I and II, V. A. Andrunakievic. Pairs of quadratic forms over a complete field with discrete norm and with finite residue class field, V. B. Dem'janov. Rings of quotients of associative rings, V. P. Elizarov. Categories of finite-dimensional spaces, I. M. Gel'fand and G. E. Shilov. Some relations between eigenvalues and matrix elements of linear operators, I. C. Gohberg and A. S. Markus. On a criterion of completeness, Ju. A. Kaz'min. On imbedding certain monotone images of E^n in E^n and E^{n+1}, L. V. Keldys. Cohomological dimensions of compacta, V. Kuz'minov. A new method in probabilistic number theory, E. V. Novoselov. Some questions on the general theory of representations of groups, B. I. Plotkin. The equivalence of cohomology theories, M. M. Postnikov.

1235. American Mathematical Soceity. **Eleven Papers on Analysis.** A.M.S. Translations–Series II, Vol. 53. Providence, R.I.: American Mathematical Society, 1966. 286p. $15.00.
Contents: Some questions of approximation and representation of functions, V. I. Arnol'd. The spectrum of singular boundary problems, M. S. Birman. Some integral inequalities and the solvability of degenerate quasilinear elliptic systems of differential equations, Ju. A. Dubinskii. On the approximation of continuous functions in closed domains with corners and on a problem of S. M. Nikol'skii, I and II, V. K. Dzjadyk. Some boundary-value problems in infinite regions, D. M. Eidus. Asymptotic behavior of the spectral function of the Schrödinger operator near a plane part of the boundary, M. G. Gasymov

207

and B. M. Levitan. On the space of bounded regular functions, V. P. Havin. On the spectra of some non-self-adjoint operators, R. M. Martirosjan. On the structure of Mikusinski operators in a pseudonormed space, I. I. Rjabcev. Local properties of Mikusinski operators, I. I. Rjabcev.

1236. American Mathematical Society. **Fifteen Papers on Analysis.** A.M.S. Translations—Series II, Vol 54. Providence, R.I.: American Mathematical Society, 1966. 281p. $14.80.
Contents: Existence conditions for wave operators, M. S. Birman. Criteria for completeness of the system of root vectors of a contraction, I. C. Gohberg and M. G. Krein. The global behavior of solutions of high-order differential equations, A. M. Krasnodembskii. Some nonlinear problems with many solutions, M. A. Krasnosel'skii and V. Ja. Stecenko. Separators of nonlinear functionals and their linear extensions, D. P. Mil'man. A remark on the evaluation of the trace of a nonnegative completely continuous operator, L. V. Ovsjannikov. Conditions for correct solvability in the large of a certain class of equations with constant coefficeints, V. P. Palamodov. On the singularity of fundamental solutions of hypoelliptic equations, V. P. Palamodov. A study of the "triangular form" of nonselfadjoint operators, L. A. Sahnovic. Korovkin systems in spaces of continuous functions, Ju. A. Saskin. On the solution of equations in functional derivatives, A. N. Serstnev. Some properties of CP-integrals, V. A. Skvorcov. On the boundedness of nonlinear functionals, S. B. Steckin. On the A-integration of a function which is conjugate to a summable function, I. A. Vinogradova. Harmonic analysis of functions on semisimple Lie groups, I, D. P. Zelobenko.

1237. American Mathematical Society. **Eight Papers on Differential Equations and Functional Analysis.** A.M.S. Translations—Series II, Vol. 56. Providence, R.I.: American Mathematical Society, 1966. 295p. $15.50.
Contents: On the Lee model, F. A. Berezin. Some bounds for the solutions of elliptic equations, V. I. Judovic. Calculation of the index of a fixed point of a vector field, M. A. Krasnosel'skii and P. P. Zabreiko. Boundary-value problems for linear and quasi-linear equations and systems of parabolic type, III, O. A. Ladyzenskaja and N. N. Ural'ceva. The canonical regularization of functions with nonintegrable singularities, V. P. Palamodov. On general boundary problems for systems which are elliptic in the sense of A. Douglis and L. Nirenberg, I, V. A. Solonnikov. The Martin boundary for a linear elliptic second-order operator, M. G. Sur. Some correct problems on the whole plane for hypoelliptic equations, B. R. Vainberg.

1238. American Mathematical Society. **Nine Papers on Foundations, Measure Theory and Analysis.** A.M.S. Translations—Series II, Vol. 57. Providence, R.I.: American Mathematical Society, 1966. 304p. $15.90.
Contents: Finite measures invariant under flows, S. Fomin. A two-dimensional problem of unsteady flow of an ideal incompressible fluid across a given domain, V. I. Judovic. Convergence of orthogonal series with respect to subsequences of partial sums, M. I. Livsic. Approximation of polynomials of continuous functions on Jordan arcs in the space K^n of n complex variables,

L. A. Markusevic. Application of Walsh polynomials in approximate calculations, B. T. Poljak and Ju. A. Sreider. Weak isomorphism of transformations with invariant measure, Ja. G. Sinai. Generalized Sobolev spaces and their application to boundary problems for partial differential equations, L. N. Slobodeckii. The independence of the continuum hypothesis, Petr Vopeka. Some properties of constructive real numbers and constructive functions, I. D. Zaslavskii.

1239. American Mathematical Society. **Twelve Papers on Algebra, Number Theory and Topology.** A.M.S. Translations—Series II, Vol. 58. Providence, R.I.: American Mathematical Society, 1966. 260p. $13.70.
Contents: The second obstruction for the imbedding problem of algebraic number fields, S. P. Demuskin and I. R. Safarevic. On the measure of transcendence, N. I. Fel'dman. Approximation of the logarithms of algebraic numbers by algebraic numbers, N. I. Fel'dman. Summation of mappings and the concept of center in categories, A. H. Livsic. Varieties in categories, A. H. Livsic, M. S. Calenko, and E. G. Sul'geifer. Category-theoretic foundations of the duality of radicality and semisimplicity, A. H. Livsic. On the class number of fields with complex multiplication, A. P. Novikov. Analogue of the Plancherel formula for the real unimodular group of the nth order, B. D. Romm. Lattice isomorphisms of commutative aperiodic semigroups with cancellation, L. N. Sevrin. Some questions in the theory of bicompactifications, E. G. Skljarenko. On the number of integral points in a given domain, I. M. Vinogradov. On the number of integral points in a three-dimensional region, I. M. Vinogradov.

1240. American Mathematical Society. **Twelve Papers on Logic and Algebra.** A.M.S. Translations—Series II, Vol. 59. Providence, R.I.: American Mathematical Society, 1966. 284p. $14.90.
Contents: Approximation of certain transcendental numbers: I, The approximation of logarithms of algebraic numbers. II, The approximation of certain numbers associated with the Weierstrass P-function, N. I. Fel'dman. Simultaneous approximation of the periods of an elliptic function by algebraic numbers, N. I. Fel'dman. Undecidability of the general problems of completeness, decidability and equivalence for propositional calculi, A. V. Kuznecov. Ramified coverings of algebraic curves, Ju. I. Manin. The Tate height of points on an abelian variety: its variants and applications, Ju. I. Manin. Extensions with given points of ramification, I. R. Safarevic. On the general theory of radicals in cateogires, E. G. Sul'geifer. The lattice of ideals of an object of a category, I and II, E. G. Sul'geifer. Regular imbeddings of categories, E. G. Sul'geifer. On a class of models closed with respect to direct products, A. D. Taimanov. Finite automata and the logic of one-place predicates, B. A. Trahtenbrot. On the number of ideal classes and the group of divisor classes, A. I. Vinogradov.

1241. American Mathematical Society. **Seven Papers on Analysis.** A.M.S. Translations—Series II, Vol. 60. Providence, R.I.: American Mathematical Society, 1967. 284p. $14.90.

Contents: Some questions in the variational theory of nonlinear equations in locally convex spaces, Ja. L. Engel'son. Orlicz spaces and nonlinear integral equations, M. A. Krasnosel'skii and Ja. B. Rutickii. A differential operator with spectral singularities, I and II, V. E. Ljance. The expansion of arbitrary functions in terms of eigenfunctions of the operator— $\Delta u + cu$, A. J. Povzner. An analogue of a theorem of Laurent Schwartz, G. E. Shilov. On a theorem of I. M. Gel'fand, I. Singer. The hypergeometric function and the representations of the groups of real second-order matrices, N. Ja. Vilenkin.

1242. American Mathematical Society. **Fourteen Papers on Functional Analysis and Differential Equations.** A.M.S. Translations—Series II, Vol. 61. Providence, R.I.: American Mathematical Society, 1967. 311p. $16.30.
Contents: On a theorem of Liouville concerning integrable problems of dynamics, V. I. Arnol'd. Some applications of spectral sets, I: Harmonic-spectral measure, Ciprian Foias. On dynamical systems in function spaces, S. Fomin. Normally solvable operators and ideals associated with them, I. C. Gohberg, A. S. Markus, and I. A. Fel'dman. Estimates for integral operators in translation-invariant norms, K. K. Golovkin and V. A. Solonnikov. On Hölder continuity of solutions and their derivatives of linear and quasilinear elliptic and parabolic equations, O. A. Ladyzenskaja and N. N. Ural'ceva. The solution of a problem of rotational flow of a compressible fluid in a magnetic field parallel to the flow velocity, G. I. Nazarov. On the exact solutions of certain problems of magneto-hydrodynamics, G. I. Nazarov. Investigation of a transcendental case in the theory of stability of motion, V. A. Pliss. On the spectrum of the anharmonic oscillator, L. A. Sahnovic. Analytic properties of the discrete spectrum of the Schrödinger equation, L. A. Sahnovic. On the spectrum of the radial Schrödinger equation in the neighborhood of zeros, L. A. Sahnovic. On the extensions of symmetric operators depending on a parameter, A. V. Straus.

1243. American Mathematical Society. **Nine Papers on Logic and Group Theory.** A.M.S. Translations—Series II, Vol. 64. Providence, R.I.: American Mathematical Society, 1967. 256p. $13.50.
Contents: Representations of finite groups over an arbitrary field and over rings of integers, S. D. Berman. Algorithmic operators in constructive metric spaces, G. S. Ceitin. On the semigroup of genera in the theory of integer representations, D. K. Faddeev. Finite-dimensional irreducible representation of the unitary and the full linear groups and related special functions, I. M. Gel'fand and M. I. Graev. On the recognition of replaceability in recursive languages, A. V. Gladkii. On the definition of a group by generators and defining relations, A. I. Kostrikin. The structure of unitary representations of a locally compact group in a space Π_1, M. A. Naimark. Metric properties of endomorphisms of compact commutative groups, V. A. Rohlin. Sylow p-subgroups of the general linear group, R. T. Vol'vacev.

1244. American Mathematical Society. **Nine Papers on Partial Differential Equations and Functional Analysis.** A.M.S. Translations—Series II, Vol. 65. Providence, R.I.: American Mathematical Society, 1967. 296p. $15.50.

Contents: Criteria for unicellularity of dissipative Volterra operators with nuclear imaginary components, M. S. Brodskii and G. E. Kisilevskii. Some new investigations in the theory of nonselfadjoint operators, M. S. Brodskii, I. C. Gohberg, M. G. Krein, and V. I. Macaev. Properties of the S-matrix of the one-dimensional Schrödinger equation, L. D. Faddeev. Diffraction by dihedral and polyhedral angles, A. F. Filippov. On a reduction method for discrete analogues of equations of Wiener-Hopf type, I. C. Gohberg and V. G. Ceban. On the solution of nonstationary operator equations, O. A. Ladyzenskaja. Linear equations of second order with nonnegative characteristic form, O. A. Oleinik. A priori estimates for second-order parabolic equations, V. A. Solonnikov. Solvability of boundary problems for quasi-linear parabolic equations of higher order, M. I. Visik.

1245. American Mathematical Society. **Thirteen Papers on Group Theory, Algebraic Geometry and Algebraic Topology**. A.M.S. Translations—Series II, Vol. 66. Providence, R.I.: American Mathematical Society, 1968. 272p. $14.30. Contents: On the representation of numbers by certain quadratic forms and the theory of elliptic curves, A. N. Andrianov. Rational points of a class of algebraic curves, V. A. Dem'janenko. Topological 2-groups with an even number of generators and one complete defining relation, S. P. Demuskin. Appearance of singularities on surfaces of negative curvature, N. V. Efimov. Arithmetic properties of the solutions of a transcendental equation, N. I. Fel'dman. Unitary representations of real slimple Lie groups, M. I. Graev. Metric properties of endomorphisms of compact groups, S. A. Juzvinskii. On homogeneous algebras, A. I. Kostrikin. On rational points of an elliptic curve, A. I. Lapin. Rational Pontrjagin classes: Homoeomorphism and homotopy type of closed manifolds, I, S. P. Novikov. Locally projectively nilpotent subgroups and nil elements in topological groups, V. P. Platonov. A numerical experiment on the calculation of the Hasse invariant for certain curves, T. A. Tuskina. Periodic nilpotent linear groups over the field of rational numbers, R. T. Vol'vacev.

1246. American Mathematical Society. **Nine Papers on Functional Analysis and Partial Differential Equations**. A.M.S. Translations—Series II, Vol. 67. Providence, R.I.: American Mathematical Society, 1968. 288p. $15.10. Contents: Weak convergence in nonlinear elliptic and parabolic equations, Ju. A. Dubinskii. Nonlinear parabolic equations on the plane, Ju. A. Dubinskii. On the impossibility of certain inequalities between function norms, K. K. Golovkin. Relation between the local and global properties of solutions of hypoelliptic equations with constant coefficients, V. V. Grusin. On admissible extensions of the concept of a solution of linear and quasilinear elliptic equations of second order, O. A. Ladyzenskaja and N. N. Ural'ceva. Fourier series of continuous functions relative to bounded orthonormal systems, A. M. Olevskii. Monotone operator functions on a set consisting of an interval and a point, Ju. L. Smul'jan. Convolution equations in a bounded region in spaces with weighted norms, M. I. Visik and G. I. Eskin. Solvability of boundary problems for general elliptic systems, L. R. Volevic.

1247. American Mathematical Society. **Ten Papers on Differential Equations and Functional Analysis.** A.M.S. Translations—Series II, Vol. 68. Providence, R.I.: American Mathematical Society, 1968. 264p. $13.90.
Contents: Majorization of solutions of second-order linear equations, A. D. Aleksandrov. Uniqueness conditions and estimates for the solution of the Dirichlet problem, A. D. Aleksandrov. Majorants of solutions and uniqueness conditions for elliptic equations, A. D. Aleksandrov. The impossibility of general estimates for solutions and of uniqueness conditions for linear equations with norms weaker than in L_n, A. D. Aleksandrov. On Sturm-Liouville differential operators with discrete spectrum, M. G. Gasymov and B. M. Levitan. A priori estimates and certain properties of the solutions of elliptic and parabolic equations, S. N. Kruzkov. On the determination of a Sturm-Liouville equation by two spectra, B. M. Levitan. Degenerate algebras of operators in the Pontrjagin space Π_k, M. A. Naimark. The Ritz formula and quantum defects of the spectrum of the radial Schrödinger equation, L. A. Sahnovic. Dynamical systems with countably-multiple Lebesgue spectrum, II, J. G. Sinai.

1248. American Mathematical Society. **Thirteen Papers on Algebra, Topology, Complex Variables, and Linear Programming.** A.M.S. Translations—Series II, Vol. 71. Providence, R.I.: American Mathematical Soceity, 1968. 236p. $12.50.
Contents: Coefficient dispersion of one-sheeted functions, I. E. Bazilevic. Generalization of a theorem of Hurewicz on the dimension of preimages, V. G. Boltjanskii and P. S. Soltan. Equivalence of systems of integer matrices, D. K. Faddeev. The number of steps in the solution of linear programming problems by the simplex method, E. I. Filippovic and O. M. Kozlov. The behavior of a meromorphic function in the neighborhood of an essentially singular point, V. I. Gavrilov. On some imbedding theorems, Ja. L. Geronimus. Open mapping theorem for uniform spaces, V. L. Levin. Manifolds with free Abelian fundamental groups and their applications (Pontrjagin classes, smoothnesses, multidimensional knots), S. P. Novikov. One-to-one continuous mappings of topological spaces, V. V. Proizvolov. The Pontrjagin-Hirzebruch class of codimensionality 2, V. A. Rohlin. On integral representations belonging to one genus, A. V. Roiter. Problems of linear conjugacy of holomorphic functions of several complex variables, V. S. Vladimirov. Duality on elliptic curves over local fields, II, O. N. Vedenskii.

1249. American Mathematical Society. **Fifteen Papers on Analysis.** A.M.S. Translations—Series II, Vol. 72. Providence, R.I.: American Mathematical Society, 1968. 276p. $14.50.
Contents: On the dispersion of the coefficients of mean p-valent functions, I. E. Bazilevic and N. A. Lebedev. Smoothness of harmonic and analytic functions on the boundary points of a domain, E. P. Dolzenko. Brownian motion in certain symmetric spaces and nonnegative eigenfunctions of the Laplace-Beltrami operator, E. B. Dynkin. Classes of functions and associated integral transforms in the complex domain, M. M. Dzrbasjan and S. A. Akopjan. On some special classes of analytic functions p-valent in the disk $|z| < 1$ and in the circular domain $1 < |z| < +\infty$, T. G. Ezrohi. A variational method in the theory of functions of bounded characteristic, S. A. Gel'fer. The three-sphere

theorem for a class of elliptic equations of high order and a refinement of this theorem for a linear elliptic equation of second order, Ju. K. Gerasimov. On stability of solutions of stochastic differential equations, I. I. Gihman and A. Ja. Dorogovcev. On the spectral functions of a differential equation of second order with operator coefficients, M. L. Gorbacuk. Lacunary power sequences in the spaces C and L_p, V. I. Gurarii and V. I. Macaev. A generalization of the Liouville theorem, Ju. F. Korobeinik. Stability of solutions of differential-difference equations with periodic coefficients, Z. I. Rehlickii. Cauchy problems for parabolic equations degenerating at infinity, G. N. Smirnova. Linear differential equations with randomly perturbed parameters, M. G. Sur. On the convergence of Haar series to $+\infty$, A. A. Talaljan and F. G. Arutjunjan.

1250. American Mathematical Society. **Fourteen Papers on Algebra, Topology, Algebraic and Differential Geometry.** A.M.S. Translations–Series II, Vol. 73. Providence, R.I.: American Mathematical Society, 1968. 260p. $13.70.
Contents: A homotopy criterion for a pointwise mapping, V. P. Kompaniec. Spaces of transfinite dimensionality, B. T. Levsenko. Order and exponent of an elliptic curve, P. A. Medvedev. Symmetry principle for multiply connected domains and some of its applications, I. P. Mitjuk. An example of a nontrivial topologization of the field of rational numbers, A. F. Mutylin. Structure of unitary representations of locally bicompact groups and symmetric representations of algebras in Pontrjagin Π_k space, M. A. Naimark. On Lie groups transitive on compact manifolds, A. L. Oniscik. On the spectral decomposition of topological spaces, B. A. Pasynkov. Characterization of types of conformally flat four-dimensional Riemannian spaces, P. I. Petrov. Duality of functors, R. S. Pokazeeva and A. S. Svarc. On the regularity of the proximity product of regular spaces, V. Z. Poljakov. The asymptotic behavior of the number of closed geodesics on a compact manifold of negative curvature, Ja. G. Sinai. On the classification of n-dimensional vector-bundles over an algebraic curve of arbitrary genus, A. N. Tjurin. On two-dimensional algebraic tori, V. E. Voskresenskii.

1251. American Mathematical Society. **Ten Papers on Analysis.** A.M.S. Translations–Series II, Vol. 74. Providence, R.I.: American Mathematical Society, 1968. 260p. $13.70.
Contents: On the structure of solutions of a system of linear differential equations with almost periodic coefficients, B. F. Bylov. Rational approximations and boundary properties of analytic functions, E. P. Dolzenko. Boundary properties of integrals of Cauchy type and harmonic conjugate functions in domains with rectifiable boundary, V. P. Havin. On differentiable functions of n variables, L. D. Ivanov. Imbedding theorem for a class of functions defined on the entire space or on a half space, I and II, L. D. Kudrjavcev. Theory and some applications of the local Laplace transform, Ju. I. Ljubic and V. A. Tkacenko. On fundamental solutions of elliptic systems, M. I. Matiicuk and S. D. Eidel'man. Approximation by polynomials and analytic functions in the areal mean, S. O. Sinanjan. On Jackson's theorem in L spaces, M. F. Timan.

1252. American Mathematical Society. **Seven Papers on Equations Related to Mechanics and Heat.** A.M.S. Translations–Series II, Vol. 75. Providence, R.I.: American Mathematical Society, 1968. 256p. $13.50.

Contents: Dynamic systems of the first degree of nonroughness in the plane, A. A. Andronov and E. A. Leontovic. The heat equation and the Weierstrass transform on certain symmetric Riemannian spaces, L. D. Eskin. Inverse problem for quantum mechanics equations with energy dependent potentials, V. I. Mal'cenko. Determination of high-energy asymptotics of amplitudes of scattering of quantum mechanical particles by energy dependent potentials, V. I. Mal'cenko. Asymptotic methods of nonlinear mechanics applied to nonlinear differential equations with retarded argument, Ju. A. Mitropol'skii and V. I. Fodcuk. Convolution equations in a half space, R. L. Sahbagjan. Estimates of the solutions of a non-stationary linearized system of Navier-Stokes equations, V. A. Solonnikov.

1253. American Mathematical Society. **Thirteen Papers on Algebra and Analysis.** A.M.S. Translations—Series II, Vol. 76. Providence, R.I.: American Mathematical Society, 1968. 263p. $13.90.
Contents: Limits for the characteristic roots of a matrix (I), Geng Ji (Keng Chi) [G. Gun]. Approximation on the entire real axis to two functions conjugate in the sense of M. Riesz by means of integral operators of a special type, B. L. Golinskii. On distances between finite-dimensional analogues of L_p spaces, V. I. Gurarii, M. I. Kadec, and V. I. Macaev. On Hänkel and Toeplitz matrices and signatures of Toeplitz forms, I. S. Iohvidov. Continuation of Toeplitz forms preserving the number of their positive squares, I. S. Iohvidov. Irreducible representations of the discrete group SL (2,P) which are unitary in an indefinite metric, R. S. Ismagilov. Decomposition of a direct product of irreducible representations of a semisimple Lie algebra into a direct sum of irreducible representations, A. U. Klimyk. Clebsch-Gordan coefficients for the unitary, orthogonal and symplectic groups and Clebsch-Gordan series for the unitary group, A. U. Klimyk. On the integral manifold of a nonlinear system in a Hilbert space, Ju. A. Mitropol'skii and O. B. Lykova. On an orthonormal system and its applications, A. M. Olevskii. Transformative semigroups of transformations, B. M. Sain. Baire functions in the spaces of countable composition, V. E. Sneider. Plurisubharmonic functions in tubular radial domains, V. S. Vladimirov.

1254. American Mathematical Society. **Thirteen Papers on Functional Analysis and Differential Equations.** A.M.S. Translations—Series II, Vol. 79. Providence, R.I.: American Mathematical Society, 1968. 269p. $14.30.
Contents: On an a priori estimate in the theory of hydrodynamical stability, V. I. Arnol'd. Continuation of functions beyond the boundary of a domain with preservation of differential-difference properties in L_p, O. V. Besov. Bases of spaces of continuous functions on compacta, and some geometric problems, V. I. Gurarii. Rate of growth and boundedness of solutions of second-order differential equations with periodic coefficients, T. M. Karaseva. On equivalent norms in the space of square summable entire functions of exponential type, V. Ja. Lin. Generalization of a theorem of Bogoljubov to the case of Hilbert space, O. B. Lykova. On the construction of solutions of linear differential equations with quasiperiodic coefficients by the method of accelerated convergence, Ju. A. Mitropol'skii and A. M. Samoilenko. Weak structural stability of homogeneous systems, V. V. Nemyckii and Ju. V. Malysev. On the asymptotics

of eigenvalues and singular numbers of linear smoothing operators, V. I. Paraska. Periodic solutions of the first boundary problem for parabolic equations, I. I. Smulev. On the estimation of a mixed derivative in $L_p(G)$, Ju. K. Solncev. On linear similarity of certain nonselfadjoint operators to selfadjoint operators and on the asymptotic behavior for $t \to \infty$ of the solution of a nonstationary Schrödinger equation, I. V. Stankevic. On the norms of linear polynomial operators, S. A. Teljakovskii.

1255. American Mathematical Society. **Sixteen Papers on Number Theory and Algebra.** A.M.S. Translations—Series II, Vol. 82. Providence, R.I.: American Mathematical Society, 1969. 264p. $13.90.
Contents: Semigroups given by identities with distinguished elements, A. P. Birjukov. Asymptotic behavior and ergodic properties of solutions of the generalized Hardy-Littlewood equation, B. M. Bredihin and Ju. V. Linnik. Finite groups whose proper subgroups all admit nilpotent partitions, V. M. Busarkin and A. I. Starostin. On systems of congruences, A. A. Karacuba. A functional equation for Dirichlet L-series and the problem of divisors in arithmetic progressions, A. F. Lavrik. On the representation of numbers by positive binary diagonal quadratic forms, G. A. Lomadze. Identical relations on varieties of quasi-groups, A. I. Mal'cev. Some finiteness conditions in the theory of semigroups, L. N. Sevrin. Upper bounds and numerical calculation of the number of ideal classes of real quadratic fields, I. S. Slavutskii. Sets of nonnegative integers not containing an arithmetic progression of length p, Ju. T. Tkacenko. Groups with a normalizer condition for closed subgroups, V. I. Usakov. On a theorem of infinite dimensionality of an associative algebra, E. B. Vinberg. On extension to the left halfplane of the scalar product of Hecke L-series with magnitude characters, A. I. Vinogradov. On the density conjecture for Dirichlet L-series, A. I. Vinogradov. An estimate for a certain sum extended over the primes of an arithmetical progression, I. M. Vinogradov. Invariant measures on Boolean algebras, D. A. Vladimirov.

1256. American Mathematical Society. **Eighteen Papers on Logic and Theory of Functions.** A.M.S. Translations—Series II, Vol. 83. Providence, R.I.: American Mathematical Society, 1969. 284p. $13.30.
Contents: A uniqueness theorem for harmonic functions in a halfspace, I. S. Arson and M. A. Pak. On the reduction to single-valued form of certain functions analytic in domains of finite connectivity, L. E. Dunducenko. On approximation of functions continuous on Jordan arcs, L. I. Kolesnik. Some problems in the structure of measurable functions, S. G. Kozlovcev. On differential properties of measurable functions, S. G. Kozlovcev. Estimates from below for entire functions of finite order, I. F. Krasickov. Some problems of the theory of classes of models, A. I. Mal'cev. On the decrease of functions analytic in a halfplane, Ju. I. Masljukov. On reduction of the decision problem of recursively enumerable sets to the separability problem, A. A. Mucnik. Boundary properties of functions defined on a region with angular points, I, II, and III, S. M. Nikol'skii. Extension of functions of several variables preserving differential properties, S. M. Nikol'skii. On best approximation of functions of class Z_{2k} by certain linear polynomial operators, I. M. Petrov. Extremal properties of certain classes

215

of univalent functions, V. A. Pohilevic. On simultaneous orthogonality of functions on a domain and on its boundary, E. A. Sinev. On some extremal properties of r regular functions bounded in the disk $|z| < 1$, V. A. Turkovskii. Some remarks on the Baire property, I, Yang Zong-pan.

1257. American Mathematical Society. **Twelve Papers on Algebra, Algebraic Geometry, and Topology.** A.M.S. Translations–Series II, Vol. 84. Providence, R.I.: American Mathematical Society, 1969. 276p. $14.50.
Contents: Finiteness conditions in the general theory of groups, S. N. Cernikov. Some problems of Burnside type, E. S. Golod. On polynomials with small prime factors, II, V. I. Hmyrova. On birational forms of rational surfaces, V. A. Iskovskih. Topological imbeddings in a manifold and pseudoisotopy, L. Keldys. Rational surfaces over perfect fields, I, Ju. I. Manin. Invariant measure of a compact ring groups, V. G. Paljutkin. A generalization of the Jacobian variety, A. N. Parsin. On the dimension of increments of bicompact extensions of proximity spaces and topological spaces, I and II, Ju. M. Smirnov. On multiplicative and spectral properties of spatial matrices with nonnegative elements, N. P. Sokolov. Measurable realization of continuous automorphism groups of a unitary ring, A. M. Versik.

1258. American Mathematical Society. **Twelve Papers on Functional Analysis and Geometry.** A.M.S. Translations–Series II, Vol. 85. Providence, R.I.: American Mathematical Society, 1969. 258p. $13.60.
Contents: Infinitesimal deformations of convex surfaces with bush constraints, V. T. Fomenko. On an integral connected with symmetric Riemann spaces of nonpositive curvature, S. G. Gindikin and F. I. Karpelevic. Inequalities for the norms of derivatives in weighted L_p spaces, V. G. Glusko and S. G. Krein. On the theory of linear operators in spaces with two norms, I. C. Gohberg and M. K. Zambickii. Unitary representations of the Lorentz group in a space with indefinite metric, R. S. Ismagilov. Plus-operators in a space with indefinite metric, M. G. Krein and Ju. L. Smul'jan. T-polar representation of plus-operators, M. G. Krein and Ju. L. Smul'jan. Some criteria for the completeness of a system of root vectors of a linear operator in a Banach space, A. S. Markus. Geometry of nested families with empty intersections, and Structure of the unit sphere of a nonreflexive space, D. P. Mil'man and V. D. Mil'man. Geometry of the Möbius strip and differential equations, V. R. Petuhov. Almost-reducible and symmetric almost-reducible spaces with affine connection, N. M. Pisareva. Affinely connected space admitting a transitive group of motions with a completely reducible stationary linear subgroup, N. M. Pisareva.

1259. American Mathematical Society. **Fifteen Papers on Real and Complex Functions, Series, Differential and Integral Equations.** A.M.S. Translations–Series II, Vol. 86. Providence, R.I.: American Mathematical Society, 1970. 282p. $14.40.
Contents: Extremal properties of entire functions of exponential type, N. I. Ahiezer. On the decrease of harmonic functions of three variables, I. S. Arson. On series relating to the systems $\{o(nx)\}$, V. F. Gaposkin. Uniqueness theorem for the Caplygin equation problem, I and II, Dong Guang-chang (Tung Kuang-ch'ang). Imbedding theorems and their application to the solution of elliptic

equations, L. D. Kudrjavcev and S. M. Nikol'skii. Generalization of the Carlson inequality, V. I. Levin and E. K. Godunova. A Tauberian theorem on large indices for the Borel method, V. I. Mel'nik. On a question involving equivalence of conditions for convergence of Fourier series, M. K. Potapov. On the regular approximation of solutions of complex elliptic equations of arbitrary order, E. M. Saak. Differentiability relative to congruent sets, G. H. Sindalovskii. Some estimates for trigonometric series with quasiconvex coefficients, S. A. Teljakovskii. Constructive characterizations of some function classes, R. M. Trigub. On the rapidity of convergence of certain approximate methods of Galerkin type in an eigenvalue problem, G. M. Vainikko. Representability of functions by superposition of functions of a smaller number of variables, A. G. Vituskin.

1260. American Mathematical Society. **Sixteen Papers on Differential and Difference Equations, Functional Analysis, Games and Control.** A.M.S. Translations–Series II, Vol. 87. Providence, R.I.: American Mathematical Society, 1970. 303p. $15.40.
Contents: Extrapolative problems in automatic control and the method of potential functions, M. A. Aizerman, E. M. Braverman, and L. I. Rozonoer. On positive linear operators and hypercomplex systems, I. A. Bahtin. The fundamental solution of Petrovskii-correct problems, V. M. Borok and Ja. I. Zitomirskii. Cauchy's problem for equations of odd order that are correct in the sense of Petrovskii, V. M. Borok and Ja. I. Zitomirskii. Qualitative study of the trajectories of nonlinear systems of difference equations, Ja. V. Bykov and V. G. Linenko. Remark on a finite-automaton game with an opponent using a mixed strategy, M. L. Cetlin. On the asymptotic behavior of solutions of parabolic equations as $t \to \infty$, L. D. Eskin. On certain classes of games and automata games, I. M. Gel'fand, I. I. Pjateckii-Sapiro, and M. L. Cetlin. On fundamental solutions of systems invariant under rotation, G. N. Gestrin. Certain characteristic properties of differential operators of infinite order, Ju. F. Korobeinik. The stabilization of solutions of differential equations in a Banach space and some of its applications, S. A. Maksimov. On polynomials orthogonal with a weight, S. Z. Rafal'son. Imbedding theorems, S. L. Sobolev and S. M. Nikol'skii. On the strong solvability of the Tricomi problem, N. G. Sorokina. Imbedding theorems for weight classes, S. V. Uspenskii. On convergence of multiple expansions in eigenvectors and associated vectors of an operator sheaf, V. N. Vizitei and A. S. Markus.

1261. American Mathematical Society. **Eighteen Papers on Analysis and Quantum Mechanics.** A.M.S. Translations–Series II, Vol. 91. Providence, R.I.: American Mathematical Society, 1970. 310p. $15.80.
Contents: On the existence of a solution of a class of systems of nonlinear integral equations, L. I. Barklon. A direct method for the solution of integral equations, B. G. Gabdulhaev. Selfadjointness of field operators and the moment problem, V. P. Gacok. On the approximation of functions in arbitrary norms, K. K. Golovkin. Imbedding theorems for fractional spaces, K. K. Golovkin. Properties of functions related to their rate of approximability by rational fractions, A. A. Goncar. On saturation classes of linear operators in $L_p(-\infty, \infty)$,

R. G. Mamedov. On the variation of a function and the Fourier coefficients in Haar and Schauder systems, V. A. Matveev. Investigation of a nonlinear singular equation with Hilbert kernel, H. S. Muhtarov. On the divergence of interpolation processes at a fixed point, A. A. Privalov. Some properties of 2-increasing functions, Ju. L. Smul'jan. Indecomposable n-increasing functions, Ju. L. Smul'jan. Entire functions bounded in a locally compact nonarchimedeanly normed field, M. S. Stavskii. On the construction of an S-matrix in accordance with the theory of perturbations, B. M. Stepanov. On norms of trigonometric polynomials and approximation of differentiable functions by linear averages of their Fourier series, II, S. A. Teljakovskii. A method for the approximation of unbounded solutions of systems of quasilinear Volterra integral equations, V. R. Vinokurov. On various mathematical problems in the theory of atomic spectra, G. M. Zislin and A. G. Sigalov. On the spectrum of the energy operator for atoms with fixed nuclei on subspaces corresponding to irreducible representations of a group of permutations, G. M. Zislin and A. G. Sigalov.

1262. American Mathematical Society. **Sixteen Papers on Logic and Algebra.** A.M.S. Translations–Series II, Vol. 94. Providence, R.I.: American Mathematical Society, 1970. 276p. $14.00.
Contents: Completeness of holomorphs of abelian groups with automorphism 2, I. H. Bekker. An associative calculus with an unsolvable problem of equivalence, G. S. Ceitin. Partial universal algebras with prescribed lattices of subalgebras and correspondences, A. A. Iskander. Generalized quasi-universal classes of models, S. R. Kogalovskii. On the semantics of the theory of types, S. R. Kogalovskii. Some remarks on ultraproducts, S. R. Kogalovskii. An abstract characterization of the classes of semigroups of endomorphisms of systems of general type, E. S. Ljapin. On a class of mixed abelian groups with primary periodic part, V. I. Myskin. Some questions in the theory of algorithms, E. S. Orlovskii. Lattice isomorphisms of semigroups decomposable into free products, L. N. Sevrin and V. A. Baranskii. Strong bands of semigroups, L. N. Sevrin. On lattice isomorphisms of commutative holoid semigroups, L. N. Sevrin. Conditions of decomposability of relations into direct products, M. A. Spivak. On potential invertibility of elements of semigroups, E. G. Sutov. On Horn formulas, A. D. Taimanov. A new algorithm for derivability in the constructive propositional calculus, N. N. Vorob'ev.

1263. American Mathematical Society. **Eleven Papers in Analysis.** A.M.S. Translations–Series II, Vol. 95. Providence, R.I.: American Mathematical Society, 1970. 252p. $12.80.
Contents: On unitary coupling of semi-unitary operators, V. M. Adamjan and D. Z. Arov. Approximate calculations of determinants, D. F. Davidenko. The description of all solutions of the truncated power moment problem and some problems of operator theory, M. G. Krein. On the boundedness of solutions of a homogeneous linear equation of second order in Hilbert and Banach spaces, K. S. Mamii. Summation methods for generalized Fourier series, V. A. Marcenko. A theory of integral funnels for dynamical systems without uniqueness, M. I. Minkevic. On sets of uniqueness, S. B. Steckin and P. A. Ul'janov. Parabolic convolution equations in a bounded region, M. I. Visik and G. I. Eskin. On the

theory of fields of direction cones, E. E. Viktorovskii. On the spectrum of a linear differential operator of second order in a Hilbert space of vector functionals with values in an abstract Hilbert space, I and II, F. Z. Ziatdinov.

1264. American Mathematical Society. **Ten Papers on Algebra and Functional Analysis.** A.M.S. Translations—Series II, Vol. 96. Providence, R.I.: American Mathematical Society, 1970. 260p. $13.20.
Contents: Projection of lattices, N. D. Filippov. Almost Cebysev systems of continuous functions, A. L. Garkavi. On Cebysev and almost Cebysev spaces, A. L. Garkavi. On multiplicative representations of J-nonexpansive operator functions, I and II, Ju. P. Ginzburg. Partial idempotent operations associated with ordered sets, V. V. Rosen. 0-rings and LA-rings, B. M. Sain. Basic problems in the theory of projections of similattices, L. N. Sevrin. Ordered algebraic systems (1963-1965), A. A. Vinogradov. Ordered algebraic systems (1966), A. A. Vinogradov.

1265. American Mathematical Society. **Eleven Papers on Algebra, Analysis and Topology.** A.M.S. Translations—Series II, Vol. 97. Providence, R.I.: American Mathematical Society, 1970. 258p. $13.70.
Contents: On dynamical systems without uniqueness as semigroups on non-singlevalued mappings of a topological space, I. U. Bronstein. The concept of motion in a generalized dynamical system, B. M. Budak. On the removal of singularities of analytic functions, E. P. Dolzenko. On a generalization of the theorems of Wiener-Levy type of M. G. Krein, I. C. Gohberg. Fundamental aspects of the representation theory of Hermitian operators with deficiency index (m, m), M. G. Krein. On neutral polyverbal operations, O. N. Macedonskaja. On E. L. Post's "Tag" problem, S. Ju. Maslov. A classification of fixed points, A. N. Sarkovskii. Behavior of a mapping in the neighborhood of an attracting set, A. N. Sarkovskii. J-majorizing and modular operators in J-spaces, Ju. L. Smul'jan. Entire functions bounded in a locally compact nonarchimedeanly normed field, M. S. Stavskii.

1266. Aoki, Masanao. **Introduction to Optimization Techniques: Fundamentals and Applications of Non-linear Programming.** New York: Macmillan, 1971. 335p.
Contents: Minimization of constrained and unconstrained functions. Duality and optimization. Optimization methods. Applications to engineering problems.

1267. Ash, Robert. **Information Theory.** New York: John Wiley, 1965. 339p. $13.50.
Contents: Analysis of channel models. Proof of coding theorems. Coding systems. Statistical properties of information sources.

1268. Ashton, W. D. **The Theory of Road Traffic Flow.** New York: John Wiley, 1966. 178p. $5.00.
Contents: Traffic measurements. Traffic models. Traffic signals. Queueing. Accident statistics.

1269. Babich, V. M., editor. **Mathematical Problems in Wave Propagation Theory.**
Trans. from the Russian. New York: Plenum, 1970. 150p. $15.00.
Contents: Sharply directed propagation of Love-type surface waves, G. P.
Astrakhantsev. Eigenfunctions concentrated in a neighborhood of closed geodesic, V. M. Babich. The calculation of interference waves for diffraction by a
cylinder and a sphere, A. I. Lanin. The expansion of an arbitrary function in
terms of an integral of associated Legendre functions of first kind with complex
index, B. G. Nikolaev. Application of an integral transform with generalized
Legendre kernel to the solution of integral equations with symmetric kernels,
B. G. Nikolaev. Solution of three-dimensional problems for the hyperboloid
of revolution and the lens in electrical prospecting, B. G. Nikolaev. Application
of the Laplace method to the construction of solutions of the Helmholtz equation, B. G. Nikolaev. The problem of constructing solutions of the Neumann
problem for the stationary diffraction of waves from a half space separated by
an inclined boundary into two angular regions with different wave propagation
speeds, B. G. Nikolaev. Eigenfunctions of the Laplace operator on the surface of
a triaxial ellipsoid and in the region exterior to it, T. F. Pankratova. Calculation
of the wave fields for multiple waves near the points of origin, N. S. Smirnova.

1270. Barker, Stephen. **Philosophy of Mathematics.** Englewood Cliffs, N.J.:
Prentice-Hall, 1964. 128p.
Contents: Euclidean geometry. Non-Euclidean geometry. Number and literal-
istic philosophies of number. Transition to a nonliteralistic view of number.

1271. Barron, D. W. **Recursive Techniques in Programming.** New York:
American Elsevier, 1968. 72p. $5.25.
Contents: The ideas of recursion. Examples and applications. Mechanisms for
recursion. Recursion and iteration. List processing.

1272. Basri, Saul A. **A Deductive Theory of Space and Time.** Amsterdam:
North-Holland, 1966. 163p. $7.00.
Contents: Space-time geometry. Space geodesics. Geodesic geometry. Space
geometry.

1273. Batchelder, Paul M. **An Introduction to Linear Difference Equations.**
New York: Dover, 1967. 209p. $2.00.
Contents: Fundamental ideas. Equations of the first order. The Gamma function.
The hypergeometric equation: general case. The hypergeometric equation:
irregular cases.

1274. Batchelor, G. K. **An Introduction to Fluid Dynamics.** New York: Cambridge University Press, 1967. 634p. $14.50.
Contents: The physical properties of fluids. Kinematics of the flow field.
Equations governing the motion of a fluid. Flow of a uniform incompressible
viscous fluid. Flow at large Reynolds number: effects of viscosity. Irrotational
flow theory and its applications. Flow of effectively inviscid fluid with
vorticity.

1275. Beale, E. M. L., editor. **Applications of Mathematical Programming Techniques**. New York: American Elsevier, 1970. 450p. $19.50.
Contents: Linear programming. Economic applications. Matrix generators and output analysers. Non-convex programming methods. Integer programming applications. Geometric programming. Strategic deployment problems. Nonlinear programming. M.R.41 No. 8027.

1276. Beale, E. M. L. **Mathematical Programming in Practice**. New York: John Wiley, 1968. 195p. $5.50.
Contents: Linear programming. Introduction to the simplex method. Transportation problems. Duality and the dual simplex method. Parametric programming. The inverse matrix method. Formulation of linear programming problems. Output from LP calculations. Post-optimal analysis. Lagrange multipliers and Kuhn-Tucker conditions. Quadratic programming. Separable programming. Integer programming. Decomposition. Stochastic programming. M.R.41 No. 1369.

1277. Belevitch, V. **Classical Network Theory**. San Francisco: Holden-Day, 1968. 440p. $19.50.
Contents: Electric networks. Mathematical formulation of n-port transformers. Kirchhoff's laws. The Hermite normal form. Congruence transformations. Synthesis of passive one-ports. Scattering matrices. Positive matrices. Bounded matrices. Degree of a matrix. Recent contributions to the synthesis of n-port networks.

1278. Bellman, Richard. **Introduction to the Mathematical Theory of Control Processes. Vol. I: Linear Equations and Quadratic Criteria**. New York: Academic Press, 1967. 245p. $11.50.
Contents: What is control theory. Second-order linear differential and difference equations. Stability and control. Continuous variational processes: calculus of variations. Dynamic programming. Review of matrix theory and linear differential equations. Multidimensional control processes via the calculus of variations. Multidimensional control processes via dynamic programming. Functional analysis. Miscellaneous exercisis. Volume 2 in preparation. M.R.36 No. 7426.

1279. Bellman, Richard, R. E. Kalaba, H. H. Kagiwada, and M. C. Prestrud. **Invariant Imbedding and Time-Dependent Transport Processes**. New York: American Elsevier, 1964. 266p. $7.50.
Contents: Numerical inversion of Laplace transforms: a one-dimensional neutron multiplication process. Time-dependent diffuse reflection from a slab.

1280. Beltrami, E. J. **An Algorithmic Approach to Nonlinear Analysis and Optimization**. New York: Academic Press, 1970. 235p. $12.00.
Contents: Iterative methods on normed linear spaces. Constrained optimization on E^n. Computational techniques for constrained optimization on E^n. Constrained optimization in function space. Weak convergence in Hilbert space.

1281. Benes, V. E. **Mathematical Theory of Connecting Networks and Telephone Traffic.** New York: Academic Press, 1965. 319p. $12.00.
Contents: Algebraic and topological properties of connecting networks. Rearrangeable networks. Nonblocking networks. A telephone exchange model. Covariance functions of trunk groups. Traffic measurements. A thermodynamic traffic theory. Markov processes. Traffic in connecting networks.

1282. Blaquiere, A., F. Gerard, and G. Leitmann. **Quantitative and Qualitative Games.** New York: Academic Press, 1969. 172p. $11.50.
Contents: Games and plays. Some geometric aspects of quantitative games. Differential quantitative games. Multiple-stage quantitative games. Some geometric aspects of qualitative games. Differential qualitative games. A connection between qualitative and quantitative games. M.R.41 No. 6583.

1283. Boas, M. L. **Mathematical Methods in the Physical Sciences.** New York: John Wiley, 1966. 778p. $13.95.
Contents: Infinite series. Complex numbers. Determinants and matrices. Partial differentiation and multiple integrals. Vector analysis. Fourier series. Ordinary differential equations. Calculus of variations. Gamma, Beta and error functions. Asymptotic series. Sterling's formula. Elliptic integrals and functions. Coordinate transformations. Tensor analysis. Functions of a complex variable. Series solutions of differential equations. Legendre polynomials. Bessel functions. Sets of orthogonal functions. Integral transforms. Partial differential equations. Probability.

1284. Bolotin, V. V. **The Dynamic Stability of Elastic Systems.** Trans. by V. I. Weingarten, I. B. Greszezuk, K. N. Trirogoff, K. D. Gallegos, and M. Friedman. San Francisco: Holden-Day, 1964. 400p. $14.95.
Contents: Regions of dynamic instability. The influence of damping on the regions of dynamic instability. Non-steady-state vibrations. Differential equations for the dynamic stability of elastic systems. The fundamentals of the nonlinear theory of dynamic stability.

1285. Boot, John C. G. **Quadratic Programming: Algorithms, Anomalies, Applications.** Amsterdam: North-Holland, 1964. 213p. $6.50.
Contents: Farkas' theorem. The Kuhn-Tucker conditions. Constraints. Linear programming. The simplex method. The duality theorem. The cosine algorithm. The Theil-van de Panne method for quadratic programming. The Lagrangian capacity method. The Houthakker method. Parametric solution methods.
RECOMMENDED.

1286. Bracken, Jerome, and G. P. McCormick. **Selected Applications of Non-Linear Programming.** New York: John Wiley, 1968. 110p. $8.95.
Contents: Formulation as non-linear programs of problems like: bid evaluation, chemical equilibrium, launch vehicle design, stochastic programming, and sampling.

1287. Butkovskiy, A. G. **Distributed Control Systems.** Trans. from the Russian by Scripta Technica. Edited by George M. Kranc. New York: American Elsevier, 1970. 462p. $24.50.

Contents: Systems for optimal control of objects with distributed parameters. Generalized theory of optimal control of systems with distributed parameters. Optimization of systems described by recurrence relations. Method of moments in optimal control of systems with distributed parameters. Approximate and numerical methods for solving optimal control problems. Optimal heating of large solids. Optimal control of continuous furnaces.

1288. Caianiello, E. R., editor. **Automata Theory.** New York: Academic Press, 1966. 342p. $14.00.

Contents: Generalizations of regular events, V. Amar and S. G. Putzolu. Speed-up theorems and incompleteness theorems, M. A. Arbib. Une application de la théorie des graphes à une problème de codage, C. Berge. Introduction to the CUCH, C. Böhm and W. Gross. Quelques remarques sur la détermination de l'algorithme optimal pour la recherche du plus court chemin dans un graphe sans circuits, M. Borillo. Algebraic theory of feedback in discrete systems, J. R. Büchi. Une méthode de généralisation d'algorithmes de la théorie des graphes en algèbre linéaire, P. Camion. m-Valued logics and M-ary selection functions, A. Caracciolo di Forino. Generalized Markov algorithms and automata, A. Caracciolo di Forino. Synthesis of reliable automata from unreliable components, J. D. Cowan. Diophantine equations and recursively enumerable sets, M. Davis. Recursive functions—an introduction, M. Davis. The reduced form of a linear automaton, A. Gill. State graphs of autonomous linear automata, A. Gill. Sur certains procédés de définition de langages formels, M. Gross. Brain models and thought processes, E. M. Harth. Universal spaces: a basis for studies of adaptation, J. H. Holland. Graphes de transfert des réseaux neuroniques, H. Korezlioglu. Sur certaines chaines de Markov nonhomogènes, J. Larisse and M. P. Schützenberger. Explicability of sets and transfinite automata, L. Löfgren. Introduction to the problem of the reticular formation, W. S. McCulloch and W. Kilmer. Eléments de la théorie générale des codes, M. Nivat. Organigrammes et machines de Turing, L. Nolin. Lectures on classical probabilistic automata, M. O. Rabin. Sur certaines variétés de monoides finis, M. P. Schützenberger. On a family of sets related to McNaughton's L-language, M. P. Schützenberger. Pseudo-variety separability of subsets of a monoid, with applications to finite automata, L. A. M. Verbeek. M.R.33 No. 3855.

1289. Case, Kenneth M., and P. F. Zweifel. **Linear Transport Theory.** Reading, Mass.: Addison-Wesley, 1967. 342p. $17.50.

Contents: The transport equation. Green's function. Reciprocity theorems. The singular eigenfunction. The one-speed transport problem. Applications to standard geometries.

1290. Chorafas, Dimitris N. **Control Systems Functions and Programming Approaches. Volume A: Theory. Volume B: Applications.** New York, Academic Press, 1966. Vol. A, 395p., $16.00. Vol. B, 276p., $11.00.

Contents: VOLUME A. The dynamics of digital automation. Data collection and teletransmission. Numerical, logical, and stochastic processes. Mathematics for systems control. Programming for real-time duty. VOLUME B. Process-type cases and data control. Applications in the metals industry. Guidance for discrete particles.

1291. Chorafas, Dimitris N. **Systems and Simulation.** New York: Academic Press, 1965. 503p. $16.00.
Contents: The mathematics of the simulator. Evaluating industrial systems. Applications with stochastic processes. Research on traffic and cargo problems. Hydrological applications. Simulation by analog means.

1292. Codd, E. F. **Cellular Automata.** New York: Academic Press, 1968. 122p. $8.00.
Contents: Nature of investigation. Outline. Basic definitions. Introductory remark. Cellular space. Configurations. Computation. Construction. Self-reproduction. Symmetries of cellular spaces. Propagation and universality. Preliminary definitions. Propagation in certain 2-state cellular spaces. Universality.

1293. Cole, Julian. **Perturbation Methods in Applied Mathematics.** Waltham, Mass.: Blaisdell, 1968. 260p.
Contents: General non-rigorous survey of perturbation methods applied to problems in differential equations with applications in fluid dynamics and elasticity. RECOMMENDED.

1294. Converse, A. O. **Optimization.** New York: Holt, Rinehart and Winston, 1970. 295p.
Contents: Lagrange multipliers. Discrete and continuous sequential problems. Linear programming. Quadratic programming. Geometric programming.

1295. Curle, N., and H. J. Davies. **Modern Fluid Dynamics. Vol. I: Incompressible Flow.** New York: Van Nostrand-Reinhold, 1968. 308p. $10.50; $5.95pa.
Contents: One-dimensional flow. General equations of motion. Two-dimensional motion. Irrotational flow in three dimensions. Dynamics of real fluids. The Laminar boundary layer in incompressible flow. Turbulent flow.

1296. Danskin, John M. **The Theory of Max-Min and Its Applications to Weapons Allocation Problems.** Econometrics and Operations Research, Vol. V. New York: Springer-Verlag, 1967. 126p. $8.00.
Contents: Von Neumann's principle. A law of the mean. Lagrange multiplier theorem. Max-min problems. Weapon assignment problems. M.R.37 No. 3843.

1297. De Alfaro, V., and T. Regge. **Potential Scattering.** Amsterdam: North-Holland, 1965. 205p.
Contents: Introduction. Mathematical tools. The partial wave boundary conditions at x = 0. The partial wave boundary conditions at infinity. The Jost function and the S-matrix. Yukawian potentials. Physical interpretations of the poles of $S(\lambda; k)$ for physical angular momenta. Asymptotic behavior of

$S(\lambda;k)$ for large λ and fixed k. The analytic properties of the scattering amplituted in the t-plane for real energy. Ordinary dispersion relations. Mandelstam representation. The inverse problem. Generalization of ordinary potential scattering.

1298. Demyanov, V. F., and A. M. Rubinov. **Approximate Methods in Optimization Problems.** 2nd ed. Trans. from the Russian. New York: American Elsevier, 1970. 256p.
Contents: Successive approximation methods. Conditions for an extremum and some problems of optimal-control theory. M.R.39 No. 6137.

1299. Dresher, Melvin. **Games of Strategy: Theory and Applications.** Englewood Cliffs, N.J.: Prentice-Hall, 1961. 186p.
Contents: Games, strategy and saddle-point. The fundamental theorem. Properties of optimal strategies. Games in extensive form. Methods of solving games. Games with infinite number of strategies. Solution of infinite games. Games with convex payoff functions. Games of timing–duals. Tactical air war game. Infinite games with separable payoff functions.

1300. Duffin, Richard J., E. L. Peterson, and C. Zener. **Geometric Programming: Theory and Applications.** New York: John Wiley, 1967. 278p. $12.50.
Contents: Linear and convex programming. Farka's theorem. Primal and dual linear programs. Geometric programming. Primal problems. Duality theorems. M.R.35 No. 5225.

1301. Dyer, Peter, and S. R. McReynolds. **The Computation and Theory of Optimal Control.** New York: Academic Press, 1970. 247p. $13.50.
Contents: Parameter optimization. Optimal control of discrete systems. Optimization of continuous systems. The gradient method and the first variation. The successive sweep method and the second variation. Systems with discontinuities. The maximum principle and the solution of two-point boundary value problems. M.R.41 No. 8095.

1302. Falb, Peter L., and J. L. de Jong. **Some Successive Approximation Methods in Control and Oscillation Theory.** New York: Academic Press, 1969. 240p. $13.50.
Contents: Operator theoretic iterative methods. Representation of boundary value problems. Application to control problems. Application to oscillation problems. M.R.41 No. 9446.

1303. Fletcher, R., editor. **Optimization.** New York: Academic Press, 1969. 354p. $16.00.
Contents: A review of methods for unconstrained optimization, R. Fletcher. Variance algorithms for minimization, W. C. Davidon. Nonlinear optimization by simplex-like methods, E. M. L. Beale. Generalization of the Wolfe reduced gradient method to the case of nonlinear constraints, J. Abadie and J. Carpentier. Programming mathématique convexe, P. Huard. On the solution of a structured linear programming problem in upper bounded variables, V. de Angelis.

Decomposition of a nonlinear convex separable economic system in primal and dual directions, T. O. M. Kronsjö. Large step gradient methods for decomposable nonlinear programming problems, L. E. Schwartz. The application of nonlinear programming to the automated minimum weight design of rotating discs, B. M. E. de Silva. An integral equation approach to second variation techniques for optimal control problems, G. S. Tracz and B. Bernholtz. Computerized hill climbing game for teaching and research, L. P. Hyvarinen and G. M. Weinberg. Review of constrained optimization, D. Davies and W. H. Swann. Acceleration techniques for nonlinear programming, R. Fletcher and A. P. McCann. A constrained minimization method with quadratic convergence, B. A. Murtagh and R. W. H. Sargent. An algorithm for constrained minimization. W. Murray. Nonlinear least squares fitting using a modified simplex minimization method, W. Spendley. Least distance programming, A. W. Tucker. Sufficient conditions for the convergence of a variable metric algorithm, D. Goldfarb. A method for nonlinear constraints in minimization problems, M. J. D. Powell. A modified Newton method for optimization with equality constraints, Y. Bard and J. L. Greenstadt. Variable metric methods and unconstrained optimization, G. P. McCormick and J. D. Pearson. Generalized Lagrangian functions and mathematical programming, J. D. Roode.

1304. Friedrichs, K. O. **Special Topics in Fluid Dynamics.** New York: Gordon & Breach, 1966. 177p. $4.40.
Contents: Vorticity. Vortex filaments and sheets. Airfoils. Unsteady motion of an airfoil of infinite span with a constant profile. Accelerating and oscillatory wings. Airfoil of finite span. Viscous fluid flows. Boundary layer theory. Instability of discontinuous surface. Stability of shear flow.

1305. Gale, David. **The Theory of Linear Economic Models.** New York: McGraw-Hill, 1960. 330p. $10.50.
Contents: Linear programming. The theory of linear programming. Computation: the simplex method. Integral linear programming. Two-person games. Solutions of matrix games. Linear models of exchange. Linear models of production.

1306. Garvin, Walter W. **Introduction to Linear Programming.** New York: McGraw-Hill, 1960. 281p. $8.95.
Contents: The simplex method. The computational procedure. Sensitivity analysis. A gasoline-blending problem. The transportation problem. Unbalance and transshipment. Assignment problems and the caterer problem. A tanker-routing problem. Upper bounds. Statistical linear programming. The revised simplex method. Resolution of degeneracy. Parametric linear programming. A simple economic model. Duality. The warehouse problem.

1307. Gass, Saul I. **Linear Programming: Methods and Applications.** 3rd ed. New York: McGraw-Hill, 1969. 272p. $9.95.
Contents: The general linear-programming problem. The simplex computational procedure. The duality problems of linear programming. The revised simplex method. Degeneracy procedures. Parametric linear programming. Additional computational techniques. The transportation problem. Linear programming and the theory of games.

1308. Gerstenhaber, Murray, editor. **Some Mathematical Questions in Biology.** Providence, R.I.: American Mathematical Society, 1971. 160p. $5.50.
Contents: A statistical mechanics of nervous activity, Jack D. Cowan. Graphical analysis of ecological systems, Robert MacArthur. Extinction, Richard Levins. The temporal morphology of a biological clock, Arthur T. Winfree.

1309. Giles, R. **Mathematical Foundations of Thermodynamics.** Elmsford, N.Y.: Pergamon, 1964. 248p.
Contents: Fundamental concepts. Formal processes. Components of contents. Irreversibility. Mechanical systems and adiabatic processes. Entropy. Topological considerations. Thermodynamic spaces. Equilibrium states and potential. Perfect equilibrium states. Thermodynamics of a rigidly enclosed system. Systems of variable volume electric and magnetic systems. Galilean thermodynamics.

1310. Ginzburg, Abraham. **Algebraic Theory of Automata.** New York: Academic Press, 1968. 165p. $9.00.
Contents: Algebraic preliminaries. Semiautomata. Recognizers (Rabin-Scott automata). Regular expressions. Coverings of automata. Covering by permutation and reset semiautomata. The theory by Krohn and Rhodes. M.R.39 No. 4009.

1311. Glicksman, A. M. **An Introduction to Linear Programming and the Theory of Games.** New York: John Wiley, 1963. 131p. $6.95; $3.95pa.
Contents: Elementary aspects of linear programming. Convex sets in the Cartesian plane and the fundamental extreme point theorem. The simplex method in linear programming. Elementary aspects of the theory of games. Matrix games and linear programming.

1312. Goldstein, Sydney. **Lectures on Fluid Mechanics.** New York: John Wiley, 1960. 309p.
Contents: Kinematics. Dynamics of the general fluid. Electric and magnetic forces. Inviscid fluids. Newtonian fluids and the Navier-Stokes equations. Exact solutions of the Navier-Stokes equations. Boundary-layer theory for incompressible fluids. Boundary layers in gases. Turbulence. Stability. Dynamics of inviscid gases. Mixtures and high temperature effects in gases. Longitudinal waves in an ionized gas.

1313. Greenberg, Harold. **Integer Programming.** New York: Academic Press, 1971. 196p.
Contents: Linear programming. Integer programming. Enumeration and continuous solution methods. Some results on number theory and dynamic programming.

1314. Greenspan, H. P. **The Theory of Rotating Fluids.** New York: Cambridge University Press, 1968. 340p. $15.00.
Contents: Contained rotating fluid motion: linear theories. Contained rotating fluid motion: non-linear theories. Motion in an unbounded rotating fluid. Depth-averaged equations: models for oceanic circulation. Stability.

1315. Greig, D. M., and T. H. Wise. **Hydrodynamics and Vector Field Theory.**
New York: Van Nostrand-Reinhold, 1963. Vol. I, 213p. $3.95. Vol. II, 166p.,
$4.20.
Contents: VOLUME I. Vector theory. Physical fields. Hydrodynamics. Electric
currents in networks. Electromagnetism. Water waves. Viscous flow. VOLUME
II. Two-dimensional boundary value problems: complex potential. Images. Con-
formal transformation. Harmonics. Three-dimensional boundary value problems.
Spherical harmonics. Images. Current function. Stokes' stream function.

1316. Hacker, T. **Flight Stability and Control.** New York: American Elsevier,
1970. 349p. $22.00.
Contents: Mathematical modelling of disturbed flight. Stability equations.
Stability of partially controlled motions. Steady basic motions. Optimization
problems.

1317. Harris, Z. S. **Mathematical Structures of Languages.** New York: John
Wiley, 1968. 230p. $11.95.
Contents: Properties of languages relevant to a mathematical formulation.
Sentence forms. Sentence transformations. Sentences defined by transformations.
Regularization beyond language. The abstract system. The interpretation.
M.R.39 No. 1245.

1318. Hobson, Arthur. **Concepts in Statistical Mechanics.** New York: Gordon &
Breach, 1971. 184p.
Contents: Statistical mechanics. Probability. Classical statistical mechanics.
Quantum statistical mechanics. Irreversibility.

1319. Hyvarinnen, L. P. **Information Theory for Systems Engineers.** New York:
Springer-Verlag, 1968. 205p. $3.80.
Contents: Introduction to the concepts of information and information theory.
Information content and entropy. Coding for noiseless channels. Joint events
and natural languages. Error checking and error correcting codes. Information
theory for the continuous channel.

1320. Isaacs, R. **Differential Games: A Mathematical Theory with Applications
to Warfare and Pursuit, Control and Optimization.** New York: John Wiley,
1965. 384p. $15.95.
Contents: Definitions, formulation, and assumptions. Discrete differential games.
The basic mathematics and the solution technique in the small. Mainly examples:
transition surfaces; integral constraints. Efferent of dispersal surfaces. Afferent
or universal surfaces. Games of kind. Examples of games of kind. Equivocal
surfaces and the homicidal chauffeur game. The application to warfare. Toward
a theory with incomplete information.

1321. Jacobson, David H., and D. Q. Mayne. **Differential Dynamic Programming.** New York: American Elsevier, 1970. 224p. $12.50.
Contents: Continuous time systems. Bang-bang control problems. Discrete-
time systems. Stochastic systems—discrete disturbances. Stochastic systems—
continuous disturbances. M.R.41 No. 8023.

1322. Jauch, J. M. **Foundations of Quantum Mechanics**. Reading, Mass.:
Addison-Wesley, 1968. 299p. $15.00.
Contents: Basic tools from functional analysis and algebra. Basic concepts of
quantum mechanics. Elementary particles.

1323. Jeffrey, A. **Magnetohydrodynamics**. New York: John Wiley, 1966. 252p.
$3.50.
Contents: The fundamental equations of magnetohydrodynamics. Magnetohydro-
dynamic boundary conditions. Incompressible magnetohydrodynamic flow.
Waves and the theory of characteristics. Magnetohydrodynamic simple waves.
Magnetohydrodynamic shock waves. Steady magnetohydrodynamic flow.

1324. Kalman, Rudolf E., Peter L. Falb, and Michael A. Arbib. **Topics in
Mathematical System Theory**. New York: McGraw-Hill, 1969. 384p. $16.50.
Contents: Elementary control theory from the modern point of view. Optimal
control theory. Control system design. Automata theory. Loop-free decomposition of finite automata. Advanced theory of linear systems. M.R.40 No. 8465.

1325. Karreman, H. F., editor. **Stochastic Optimization and Control**. New
York: John Wiley, 1968. 217p. $9.95.
Contents: Introduction to stochastic optimization and control, Stuart E.
Dreyfus. Stochastic programs in abstract spaces, Richard M. Van Slyke and
Roger J. Wets. The concept of invariant set for stochastic dynamical systems
and applications to stochastic stability, Harold J. Kushner. Some problems of
optimal stochastic control, Wendell H. Fleming. Stochastic system identification techniques, A. V. Balakrishnan. Stochastic optimization problems in space
guidance, John V. Breakwell. On the theory of automatic phase control,
William C. Lindsey and Charles L. Weber. Application of optimal control and
estimation theory to missile guidance, Winston L. Nelson. Allocating unreliable units to random demands, Allen Kinger and Thomas A. Brown. M.R.37
No. 7198.

1326. Kaufmann, A., and R. Cruon. **Dynamic Programming: Sequential
Scientific Management**. New York: Academic Press, 1967. 278p. $12.00.
Contents: Discrete dynamic programs with a certain future and a limited
horizon. Discrete dynamic programs with a certain future and an unlimited
horizon. Discrete dynamic programs with a random future and a limited horizon.
Discrete dynamic programs with a random future and an unlimited horizon
(general case). Discrete D. H. Dynamic programs with finite Markovian chains.
Various generalizations.

1327. Keller, J. B., and S. Antman, editors. **Bifurcation Theory and Non-Linear
Eigenvalue Problems**. New York: W. A. Benjamin, 1969. 434p. $12.50; $4.95pa.
Contents: Column buckling, an elementary example of bifurcation, E. L. Reiss.
Bifurcation theory for ordinary differential equations, J. B. Keller. Existence
of buckled states of circular plates via the Schauder fixed point theorem, J. H.
Wolkowisky. Buckled states of elastic rings, I. Tadjbakhsh. Appendix: the
case $n = 1$, S. Antman. A bifurcation problem in super productivity, F. Odeh.

A bifurcation theory for nonlinear elliptic partial differential equations (continued), M. S. Berger. Some positone problems suggested by nonlinear heat generation, H. B. Keller. Bifurcation phenomena in surface wave theory, J. J. Stoker. Perturbation theory of quasiperiodic solutions of differential equations, J. Moser. Some buckling problems in non-linear elasticity, C. Sensenig. Equilibrium states of nonlinearly elastic rods, S. Antman. Nonuniqueness of rectangular solutions of the Berhard problem, P. H. Rabinowitz. Exchange of stability in Couette flow, H. F. Weinberger. Perturbation solutions of some nonlinear boundary value problems, M. Millman.

1328. Kilmister, C. W. **Lagrangian Dynamics: An Introduction for Students.** New York: Plenum, 1968. 136p. $7.50.
Contents: The concept of a general method in dynamics. Co-ordinates and transformations. Examples of the use of Lagrange's equations in deriving equations of motion. Further general applications of Lagrange's equations. Methods of approximation and other topics.

1329. Kirk, Donald E. **Optimal Control Theory: An Introduction.** Englewood Cliffs, N.J.: Prentice-Hall, 1971. 452p.
Contents: Dynamic programming. Calculus of variations. Pontryagin's principle. Numerical methods for optimal trajectories.

1330. Kirzhnitz, D. A. **Field Theoretical Methods in Many-Body Systems.** Trans. from the Russian and edited by D. M. Brink. New York: Pergamon, 1967. 394p. $17.00.
Contents: Occupation-number representation. Hartree-Fock approximations for many-body systems. The Thomas-Fermi approximations. Field theoretical methods of perturbation theory. The S-matrix. Rules for calculating the elements of S-matrices. Green's function techniques. Dilute and dense many-body systems.

1331. Klerer, Melvin, and J. Reinfelds, editors. **Symposium on Interactive Systems for Experimental Applied Mathematics.** New York: Academic Press, 1968. 472p. $19.50.
Contents: Interactive systems from the users point of view. Components of interactive systems. Other topics of interest to interactive systems. Automation of applied mathematics. Implementation of interactive systems.

1332. Korfhage, R. R. **Logic and Algorithms with Applications to the Computer and Information Sciences.** New York: John Wiley, 1966. 194p. $9.95.
Contents: Sets, relations, and mappings. Boolean algebras. The propositional calculus. A view of binary vectors. Algorithms and computing machines. The first order predicate calculus. Formal languages. A brief history.

1333. Kowalik, J., and M. R. Osborne. **Methods for Unconstrained Optimization Problems.** New York: American Elsevier, 1969. 160p. $9.50.
Contents: Direct search methods. Descent methods. Least squares methods. Constrained problems.

1334. Krekó, Béla. **Linear Programming**. Trans. from the German by J. H. L. Ahrens and C. M. Safe. New York: American Elsevier, 1968. 400p. $14.50. Contents: The technique of linear programming. The transportation problem. The general simplex method. The mathematical foundations of linear programming. Linear programming and the theory of games. Practical applications. M.R.38 No.1906.

1335. Krut'ko, Pyotr D. **Statistical Dynamics of Sampled Data Systems**. New York: American Elsevier, 1969. 528p. $25.00. Contents: Mathematical theory—linear systems. Adjoint systems. Statistical analysis of linear systems. Statistical analysis of nonlinear systems. Optimal stationary systems with infinite observation time. Optimal systems with finite observation time. Optimal non-stationary systems.

1336. Kuczma, Marek. **Functional Equations in a Single Variable**. Warsaw: Panstwowe Wydawnictwo Naukowe, 1968. 383p. $9.00. Contents: Existence and uniqueness of continuous solutions. Differentiable solutions. Monotonic solutions. Convex solutions. The Schröder and Abel functional equations. Analytic functions. Iteration group. Schröder, Abel and Levy iterates. Commuting functions. The functional equations of invariant curves.

1337. Kurth, R. **Axiomatics of Classical Statistical Mechanics**. Elmsford, N.Y.: Pergamon, 1960. 180p. Contents: The phase flows of mechanical systems. The initial distribution of probability in its phase space. Probability distributions which depend on time. Time-independent movability distributions. Statistical thermodynamics.

1338. Kushner, Harold. **Stochastic Stability and Control**. New York: Academic Press, 1967. 161p. $7.50. Contents: Stochastic stability. Finite time stability and first exit times. Optimal stochastic control. The design of controls. M.R.35 No. 7723.

1339. Lancaster, Kelvin. **Mathematical Economics**. New York: Macmillan, 1968. 411p. $10.95. Contents: Linear and nonlinear optimizing techniques. Economic models. Equilibrium theory. Balanced growth. Efficient growth. Stability analysis. Set theory. Matrices. Linear equations. Functions. Continuous functions. Fixed point theorems. Differential equations. Difference equations. The calculus of variations.

1340. Larson, Robert E. **State Increment Dynamic Programming**. New York: American Elsevier, 1968. 269p. $14.50. Contents: The conventional dynamic programming computational procedure. Fundamental concepts of state increment dynamic programming. Computational procedure within a block. Procedures for processing blocks. A computer program for problems with four state variables. Applications. Successive approximation. M.R.38 No. 1904.

1341. Lee, E. B., and L. Markus. **Foundations of Optimal Control Theory**. New York: John Wiley, 1967. 576p. $19.95.
Contents: Methods, theory, and practices in optimal control syntheses. Optimal control of linear systems. Optimal control for linear processes with integral convex cost criteria. The maximal principle and the existence of optimal controllers. Necessary and sufficient conditions for optimal control. Control system properties. Controllability, observability, and stability. Synthesis of optimal controllers for some basic nonlinear control processes.

1342. Lefschetz, Solomon. **Stability of Nonlinear Control Systems**. New York: Academic Press, 1965. 150p. $7.50.
Contents: Introductory treatment of dimensions one and two. Indirect controls. Direct controls: linearization multiple feedback. Systems represented by a set of equations of higher order. Discontinuous characteristics. Some recent results of V. M. Popov. Some further recent contributions.

1343. Llewellyn, R. W. **Linear Programming**. New York: Holt, Rinehart, and Winston, 1964. 371p.
Contents: The transportation problem. Matrices and simultaneous linear equations. The simplex algorithm. Performing the calculations. Algebraic simplifications and allied topics. The revised simplex method. The dual problem. Primal-dual methods. The final tableau. Secondary constraints. Some special applications of the simplex method, game theory. Some special transportation problems.

1344. Loomba, N. Paul. **Linear Programming: An Introductory Analysis**. New York: McGraw-Hill, 1964. 284p. $9.95.
Contents: Linear programming and management. The graphical method. Systematic trial and error method. Matrices and vectors. The vector method. The simplex method. The dual degeneracy. The transportation model. The assignment model.

1345. Lyapunov, A. **Stability of Motion**. Trans. by F. Abramovici and M. Shimshoni. New York: Academic Press, 1966. 203p. $9.75.
Contents: Introduction, V. P. Basov. An investigation of one of the singular cases of the theory of stability of motion, I, A. M. Lyapunov. On the problem of the stability of motion, A. M. Lyapunov. An investigation of one of the singular cases of the theory of stability of motion, II, A. M. Lyapunov. An investigation of a transcendental case of the theory of stability of motion, V. A. Pliss.

1346. Lyapunov, A., editor. **Systems Theory Research**. Volume 18. Trans. from the Russian. New York: Plenum, 1968. 320p. $27.50.
Contents: A collection of papers on: Control system theory. Programming and computer experiments. Problems of mathematical economics. Control processes in living organisms. Questions of mathematical linguistics. An operator notation for an English-to-Russian translation algorithm.

1347. Mann, H. B., editor. **Error Correcting Code**. New York: John Wiley, 1968. 231p. $8.95.

Contents: Error correcting codes: a historical survey, F. J. MacWilliams. Combinatorial structures in planetary reconnaissance, Edward Posner. Linear recurring sequences and error-correcting codes, Neal Zierler. Block coding for the binary symmetric channel with noiseless, delayless feedback, E. R. Berlekamp. Some algebraic and distance properties of convolutional codes, James L. Massey. Topological codes, Charles Saltzer. On a class of cyclic codes, S. Lin. Some results on majority-logic decoding, E. J. Weldon, Jr. Unimodular modules and cyclotomic polynomials, P. Camion. Algebraic coding and the Lee metric, Solomon W. Golomb and Lloyd R. Welch. Sequences with small correlation, R. Turyn. M.R.38 No. 3071.

1348. Marcus, Solomon. **Algebraic Linguistics: Analytic Models.** New York: Academic Press, 1966. 254p. $12.00.
Contents: Language and partitions. Linguistic typology. Parts of speech and syntactic types. Grammatical gender. Configurations. Subordination and projectivity.

1349. Martin, W. T., and I. Segal, editors. **Analysis in Function Space.** Cambridge, Mass.: M.I.T. Press, 1964. 218p. $6.00.
Contents: Random theory in classical phase space and quantum mechanics, N. Wiener and G. della Riccia. On function space integrals, M. D. Donsker. A new method of solution for quantum field theory and associated problems, S. F. Edwards. Classical analysis on a Hilbert space, L. Gross. Tempered distributions in functional space, P. Kristensen. Schrödinger particles interacting with a quantized scalar field, E. Nelson. Developments in renormalization theory, A. Salam. Quantum fields and analysis in the solution manifolds of differential equations, I. Segal. Applications of function space integrals to problems in equilibrium statistical mechanics, A. J. F. Siegert. Applications of functional integration to nonrelativistic physical problems, S. F. Edwards. A C*-algebra approach to field theory, D. Kastler. Remarks on the mathematical structure of quantum field theory, R. Haag. Application of functional integrals to Euclidean quantum field theory, K. Symanzik. A physicist sums up, R. E. Peierls.

1350. McLeod, E. B. **Introduction to Fluid Dynamics.** Elmsford, N.Y.: Pergamon, 1964. 254p.
Contents: Fundamental motions. Conservation of mass. Forces acting on a fluid equilibrium. Dynamic equations of motion. Irrotational motion. Integration of Euler's equation in special cases. Flows representable by harmonic functions. Two-dimensional flows. Rectilinear vortices. General vortex motion. Flows with a free boundary. Compressible fluids.

1351. Mulaik, Stanley A. **The Foundations of Factor Analysis.** New York: McGraw-Hill, 1972. 453p. $14.95.
Contents: Factor analysis. Composite variables. Multiple and partial correlation. Fundamental equations. Rotations. Procrustian transformations. Factorial invariance. Multivariate analysis.

1352. Murphy, Ray E. **Adaptive Processes in Economic Systems.** New York: Academic Press, 1965. 209p. $9.50.
Contents: The mathematical model. The primitive adaptive process. Subjective probability. The role of entropy in economic processes. Adaptive economic processes. An adaptive investment model. Multiactivity capital allocation processes. Economic state space. Interactions between decision makers in state space.

1353. Nelson, R. J. **Introduction to Automata.** New York: John Wiley, 1968. 400p. $12.95.
Contents: Number theoretic functions. Formal systems. Turing machines. Finite automata. Decomposition theory. Regular expressions, grammars, and acceptors. M.R.37 No. 5061.

1354. Nikaido, Hukukane. **Convex Structures and Economic Theory.** New York: Academic Press, 1968. 405p. $19.50.
Contents: Mathematical theorems of convexity. Simple multisector linear systems. Balanced growth in nonlinear systems. Efficient allocation and growth. The working of Walrasian competitive economies. Special features of competitive economies: gross substitutability. The Jacobian matrix and global univalence.

1355. Nikodym, O. M. **The Mathematical Apparatus for Quantum-Theories.** New York: Springer-Verlag, 1966. 952p.
Contents: Boolean lattices. Theory of traces. The tribe of figures on the plane. The trace-theorem. The lattice of subspaces of the Hilbert-Hermite space. Tribes of spaces. Double scale of spaces. Maximal normal operators. Tribe of repartition of functions. Permutable normal operators. Approximation of somata by complexes. Vector fields on the tribe and their summations. Quasi-vectors and their summation. Dirac's delta function. Upper and lower (DARS-) summation of fields of real numbers in a Boolean tribe.

1356. Pervozvanskii, A. A. **Random Process in Non Linear Control Systems.** New York: Academic Press, 1965. 341p.
Contents: Statistical linearization. Some non-stationary problems in non-linear feedback systems. Design and performance of self-adjusting systems. Requires knowledge of probability and control theory.

1357. Phillips, O. M. **The Dynamics of the Upper Ocean.** New York: Cambridge University Press, 1966. 270p. $9.50.
Contents: The equations of motion. The dynamics of surface waves. Ocean surface waves. Internal waves. Oceanic turbulence.

1358. Phillipson, G. A. **Identification of Distribute Systems.** New York: American Elsevier, 1971. 154p.
Contents: Definition of the state identification problem. Solution and characterization of the solution of such problems. FOR RESEARCH WORKERS.

1359. Pielou, E. C. **An Introduction to Mathematical Ecology.** New York: John Wiley, 1969. 286p.
Contents: Population dynamics. Spatial patterns in one-species populations. Spatial relations of two or more species. Many-species populations.

1360. Pollard, Harry. **Mathematical Introduction to Celestial Mechanics.** Englewood Cliffs, N.J.: Prentice-Hall, 1966. 111p.
Contents: The conservation laws. The three-body problem. The central force problem. Equilibrium points and stability. Lunar theory.

1361. Radbill, John R., and Gary A. McCue. **Quasilinearization and Nonlinear Problems in Fluid and Orbital Mechanics.** New York: American Elsevier, 1970. 228p. $14.00.
Contents: Linear ordinary differential equations. Quasi-linearization. Solution of a boundary layer equation. Two-component boundary layers on an ablating wall. Computation of electrostatic probe characteristics. Prediction of the stability of laminar boundary layers. Prediction of the stability of laminar pipe flow. Optimum orbital transfer with "bang-bang" control. A generalized subroutine for solving quasilinearization problems.

1362. Reid, Constance. **Introduction to Higher Mathematics for the General Reader.** New York: T. Y. Crowell, 1960. 184p. $3.50.
Contents: Number theory. Euclidean geometry. Non-Euclidean geometry. Algebra. Group theory. Logic. Calculus. Topology.

1363. Saaty, Thomas L., editor. **Lectures on Modern Mathematics.** New York: John Wiley. Vol. 1: 1963, 175p., $6.50. Vol. 2: 1964, 183p., $6.50. Vol. 3: 1965, 321p., $12.75.
Contents: VOLUME 1. A glimpse into Hilbert space. Some applications of the theory of distributions. Numerical analysis. Algebraic topology. Lie algebras. Representations of finite groups. VOLUME 2. Partial differential equations with applications in geometry, L. Nirenberg. Generators and relations in groups— the Burnside problems, Marshall Hall, Jr. Some aspects of the topology of 3-manifolds related to the Poincare conjecture, R. H. Bing. Partial differential equations: problems and uniformization in Cauchy's problem, Lars Gärding. Quasiconformal mappings and their applications, Lars V. Ahfors. Differential topology, John Milnor. VOLUME 3. Topics in classical analysis, Einar Hille. Geometry, H. S. M. Coxeter. Mathematical logic, Georg Kreisel. Some recent advances and current problems in number theory, Paul Erdös. On stochastic processes, Michel Loeve. Random integrals of differential equations, J. Kampe de Feriet.

1364. Saaty, Thomas L. **Mathematical Models of Arms Control and Disarmament: Application of Mathematical Structures in Politics.** New York: John Wiley, 1968. 190p. $10.95.
Contents: Consistency analysis. Equilibrium. Stability. Deterrence. Arms races. Game theory. Meta-games. Bargaining analysis. Illustrations from Vietnam and World War II. M.R.39 No. 5128.

1365. Saaty, Thomas L. **Modern Nonlinear Equations**. New York: McGraw-Hill, 1967. 473p. $15.50.
Contents: Differential calculus in abstract spaces. Implicit function theorems. Fixed-point theorems. Iterative methods of solution nonlinear equations. Functional equations. Non-linear difference equations. Delay-differential equations. Integral equations. Integro-differential equations. Stochastic differential equations. Probability theory and stochastic processes. RECOMMENDED AS A REFERENCE BOOK. M.R.36 No. 1249.

1366. Saaty, Thomas L. **Optimization in Integers and Related Extremal Problems**. New York: McGraw-Hill, 1970. 288p. $16.50.
Contents: Basic concepts: examples of methods and problems. Methods of geometric optimization. Some elementary applications. Optimization subject to Diophantine constraints. Integer programming.

1367. Saaty, Thomas L., and J. Bram. **Nonlinear Mathematics**. New York: McGraw-Hill, 1965. 381p.
Contents: Linear and nonlinear transformations. Non-linear algebraic and transcendental equations. Non-linear optimization. Non-linear programming. Non-linear ordinary differential equations. Automatic control and the Pontryagin principle. Linear and non-linear prediction theory.

1368. Sario, L., and K. Oikawa. **Capacity Functions**. New York: Springer-Verlag, 1969. 361p. $24.00.
Contents: Normal operators. Principal functions. Capacity functions. Modulus functions. Relations between fundamental functions. Mappings related to principal functions, capacity functions, and modulus functions. Extremal slit regions. Degeneracy. Practical tests.

1369. Shisha, Oved, editor. **Inequalities: Proceedings of a Symposium**. New York: Academic Press, 1967. 360p. $18.00.
Contents: Inequalities and the principle of nonsufficient reason, G. Polyá. Inequalities in the differential geometry of surfaces, E. F. Beckenbaugh. A "workshop" on Minkowski's inequality, E. F. Beckenbach. Uncertainty principles in Fourier analysis, N. G. de Bruijn. Inequalities for the sum of two M-matrices, K. Fan. On some inequalitites and their application to the Cauchy problem, A. Friedman. General subadditive functions, R. P. Gosselin. Chebyshevian spline functions, S. Karlin and Z. Zeigler. Some new inequalities and unsolved problems, J. E. Littlewood. Lengths of tensors, M. Marcus. Monotonicity of ratios of means and other applications of majorizations, A. W. Marshall, I. Olkin, and F. Proschan. Ratios of means and applications, B. Mond and O. Shisha. Algebraic inequalities, T. S. Motzkin. The arithmetic-geometric inequality, T. S. Motzkin. Signs of minors, T. S. Motzkin. A priori bounds in problems of steady subsonic flow, L. E. Payne. On spline functions, I. J. Schoenberg. Bounds on difference of means, O. Shisha and B. Mond. Positive-definite matrices, O. Taussky. Inequalitites of Chebyshev, Zolotareff, Cauer, and W. B. Jordan, J. Todd. On the new maximum-minimum theory of eigenvalues, A. Weinstein. Generalizations of Ostrowski's inequality for matrices with dominant principal diagonal, Y. K. Wong. M.R.36 No. 1251.

1370. Shisha, Oved, editor. **Inequalities: Proceedings of the Second Symposium on Inequalities.** New York: Academic Press, 1970. 439p. $22.00.
Contents: Some integral inequalities, N. Aronszajn. Meromorphic plane maps viewed from space, E. F. Beckenbach. A metric space connected with generalized means, G. T. Cargo and O. Shisha. Asymptotic cones and duality of linear relations, Ky Fan. On the inequality $\|f'\|^2 \leq 4\|f\| \cdot \|f''\|$, Robert R. Kallman and Gian-Carlo Rota. Metric inequalities and symmetric differences, John B. Kelly. Dependence of some classical inequalities on the index set, H. W. McLaughlin and F. T. Metcalf. Inequalities for submatrices, Marvin Marcus. Difference and ratio inequalities in Hilbert space, B. Mond and O. Shisha. The real solution set of a system of algebraic inequalities is the projection of a hypersurface in one more dimension, T. S. Motzkin. Singularities of the n-body problem, II, Harry Pollard and Donald G. Saari. Integral inequalitites with boundary terms, Ray Redheffer. Perturbation of zeros of analytic functions, Paul C. Rosenbloom. Perturbation of zeros of analytic functions, II, Paul C. Rosenbloom. Cardinal interpolation and spline functions, I. J. Schoenberg. On two complementary variational characterizations of eigenvalues, William Stenger. Positive-definite matrices and their role in the study of the characteristic roots of general matrices, Olga Taussky. Inequalities expressing degree of convergence of rational functions, J. L. Walsh. Inequalities for generalized convex functions, Zvi Ziegler. On certain lemmas of Marcinkiewicz and Carleson, A. Zygmund. M.R.41 No. 8188.

1371. Sneddon, I. N., and M. Lowengrub. **Crack Problems in the Classical Theory of Elasticity.** New York: John Wiley, 1969. 221p. $14.95.
Contents: Two-dimensional crack problems. Three-dimensional crack problems.

1372. Stratonovich, R. L. **Conditional Markov Processes and Their Application to the Theory of Optimal Control.** Trans. from the Russian by Scripta Technica. Edited by Richard Bellman. New York: American Elsevier, 1968. 367p. $14.75.
Contents: Some auxiliary questions on the theory of Markov processes. The main results of the theory of conditional Markov processes. Application of the theory of conditional Markov processes to the theory of optimal control. Some general results of the theory of optimal control. M.R.36 No. 4912.

1373. Strauss, Aaron. **An Introduction to Optimal Control Theory.** New York: Springer-Verlag, 1968. 153p. $3.50.
Contents: Control problems. Controllability. Restrict controls. Time optimal control. Existence of solutions. The maximum principle. M.R.38 No. 1375.

1374. Taha, H. A. **Operations Research: An Introduction.** New York: Macmillan, 1971. 703p.
Contents: PART I. Linear programming and related topics. PART II. Probability, inventory, and queueing theory. PART III. Nonlinear programming. Appendix A: Review of vectors and matrices. Appendix B: Review of basic theorems in differential calculus. Appendix C: General program for computing Poisson queueing formulas.

1375. Tait, J. H. **An Introduction to Neutron Transport Theory.** New York: American Elsevier, 1965. 142p. $6.00.
Contents: Neutrophysics. The neutron transport equation. Solution of the group equation. The diffusion approximation. Numerical solutions.

1376. Tiffen, R. **Plane Elastic Deformation.** New York: American Elsevier, 1970. 168p. $6.75.
Contents: Tensor analysis. Analysis of stress and strain. The elastic potential function. Solution of problems involving finite elastic deformation. Finite plane strain. Infinitesimal theory. Plane strain and generalized plane stress. Uniqueness theorems of plane strain and generalized plane stress. Functions of a complex variable. Problems involving half-planes and regions that can be mapped conformally on to half-planes. Problems involving regions with circular boundaries.

1377. Tou, Julius T., editor. **Advances in Information Systems Science.** New York: Plenum, 1971. 354p.
Contents: Pattern recognition. Image processing. Computer graphics. Integer programming. Logical design of optimal digital networks.

1378. Towill, D. R. **Transfer Function Techniques for Control Engineers.** Hartford, Conn.: Davey, 1971. 513p.
Contents: Control engineering techniques. Derivation of transfer functions. The root-locus and pole-zero methods. Applications of transfer function techniques to linear and non-linear control systems.

1379. Truesdell, C. **The Elements of Continuum Mechanics.** New York: Springer-Verlag, 1966. 279p. $5.40.
Contents: Dynamical theory. Cauchy's first law. Viscometric flows of incompressible fluids. Elastic isotropic materials. Fading memory. Thermodynamics. Quasi-elastic materials. RECOMMENDED. M.R.35 No. 5166.

1380. Varadarajan, V. S. **The Geometry of Quantum Theory.** New York: Van Nostrand-Reinhold. Vol. I: 1968, 193p., $8.50. Vol. II: 1970, 255p., $11.95.
Contents: VOLUME I. Projective geometry. Boolean algebras. Complemented modular lattices. The orthogonal and unitary groups. The logic of quantum mechanics. The lattice of subspaces of a Banach space. VOLUME II. Measure theory on G-spaces. Systems of imprimitivity. Multipliers. Kinematics and dynamics. Relativistic free particles.

1381. Willems, Jan C. **The Analysis of Feedback Systems.** Cambridge, Mass.: M.I.T. Press, 1971. 188p.
Contents: Analysis of feedback systems. Non-linear operators. Positive operators. Stability criteria. The circle and the Nyquist criteria.

1382. Wood, Paul E., Jr. **Switching Theory.** New York: McGraw-Hill, 1968. 390p. $13.50.
Contents: Boolean algebra. Combinational networks. Minimum-complexity combinational networks. Several topics on combinational networks. Sequential

networks. Minimum-complexity sequential networks. Asynchronous sequential networks. Clocked sequential systems.

1383. Wymore, A. W. **A Mathematical Theory of Systems Engineering: The Elements.** New York: John Wiley, 1967. 353p. $15.95.
Contents: Systems definitions. Modeling of systems. Comparison of systems. Coupling of systems. Subsystems and components. Discrete systems. M.R.40 No. 5285.

1384. Yakowitz, S. J. **Mathematics of Adaptive Control Processes.** New York: American Elsevier, 1969. 174p. $11.00.
Contents: Control processes and dynamic programming. Adaptive control processes. Adaptive control processes in two-arm bandit theory. Adaptive control processes in pattern recognition theory. M.R.39 No. 5209.

1385. Zacharov, B. **Digital Systems, Logic and Circuits.** New York: American Elsevier, 1968. 168p. $3.95.
Contents: Number systems. Boolean algebra. Simplification of Boolean expression. Physical implementation of logic operations. Magnetic core logic. Switching matrixes.

CHAPTER 34

PERIODICALS

1386. **Acta Informatica.** 1971. 4 issues/yr. $29.70. Editor: W. Niegel, Mathematisches Institut, Technische Universität, München, Germany. Publisher: Springer-Verlag, 175 Fifth Avenue, New York, N.Y. 10010.

1387. **Acta Mathematica.** (Hungary) 1950. Quarterly. $16.00. Editor: Gy. Hajos, Akademiai Kiado, Publishing House of the Hungarian Academy of Sciences, Alkotmany u.21, Budapest V, Hungary. Editorial Office: Acta Mathematica, Budapest 502, Postafiok 24.

1388. **Acta Mathematica.** (Sweden) 1882. Monthly. 2v./yr. Kr. 80/vol. Editor: L. Carleson, Institute Mittag-Leffler, Auravagen 17, S-18262 Djursholm, Sweden.

1389. **Acta Scientiarum Mathematicarum.** 1922. Semi-annual. $16.00. Editor: Bela Sz.-Nagy, Bolyai Institute, University of Szeged, Szeged, Hungary.

1390. **Acta Universitatis Carolinae.** (Series Mathematica et Physica) 1959. Semi-annual. Editor: J. Mohr, Universita Karlova, Ovocny trh 3, Prague 1, Czechoslovakia.

1391. **Advances in Applied Probability.** 1969. Semi-annual. $13.00. Editor: J. Gani, Department of Probability and Statistics, The University, Sheffield, 7RH, England.

1392. **Advances in Mathematics.** 1967. Bimonthly. 2v./yr. $48.00/yr. Editor: Gian-Carlo Rota, Department of Mathematics, Massachusetts Institute of Technology, Cambridge, Mass. 02139. Publisher: Academic Press, 111 Fifth Avenue, New York, N.Y. 10003.

1393. **Algebra and Logic.** (Translation of **Algebra i Logika** of the Academy of Sciences of the USSR) 1968. Bi-monthly. $85.00. Publisher: Plenum, 227 West 17th Street, New York, N.Y. 10011.

1394. **Allgemeines Statistisches Archiv.** 1891. Quarterly. $15.00. Editor: W. Huefner, Vandenhoeck and Ruprecht, Postfach 77, 34 Goettingen, West Germany.

1395. **American Journal of Mathematics.** 1878. Quarterly. $15.00. Editor: Wei-Liang Chow, Department of Mathematics, Johns Hopkins University, Baltimore, Maryland 21218. Publisher: Johns Hopkins University.

1396. **American Mathematical Monthly.** 1894. 10 issues/yr. $15.00. Editor: Harley Flanders, Mathematical Association of America, 1225 Connecticut Avenue N.W., Washington, D.C. 20036.

1397. **The American Statistician.** 1947. 5 issues/yr. $3.00. Editor: Morris Hamburg, 806 Fifteenth Street N.W., Washington, D.C. 20005. Publisher: American Statistical Association.

1398. **Annali di Matematica (Pura ed applicata).** 1850. L.8000. Editor: G. Sansone, Zanichelli Editori, Via Irnerio 34, Bologna, Italy.

1399. **Annals of Mathematical Logic.** 1969. Quarterly. $25.00. North-Holland Publishing Company, P.O. Box 3489, Amsterdam, Holland.

1400. **Annals of Mathematical Statistics.** 1930. Bi-monthly. $30.00. Editor: Ingram Olkin, Department of Statistics, Stanford University, Stanford, Calif. 94305. Publisher: Institute of Mathematical Statistics.

1401. **Annals of Mathematics.** 1884. Bi-monthly. $24.00. Editor: Edward Nelson, Fine Hall, Princeton University, Princeton, N.J. 08540. Publisher: Princeton University.

1402. **Applied Mathematics and Mechanics.** (Translation of **Prikladnaia Matematica i Mekhanika.**) 1933. Bimonthly. $90.00. Translation editor: G. Hermann, Stanford University. Publisher: Pergamon Press, Maxwell House, Fairview Park, Elmsford, N.Y. 10523.

1403. **Archive for Rational Mechanics and Analysis.** 1957. 4v./yr. $32/vol. Editor: C. Truesdell, Springer-Verlag, 175 Fifth Avenue, New York, N.Y. 10010.

1404. **The Arithmetic Teacher.** 1954. Monthly. $9.00. Editor: Carol V. McCamman, National Council of Teachers of Mathematics, 1201 Sixteenth Street N.W., Washington, D.C. 20036.

1405. **Artificial Intelligence.** 1970. Quarterly. $22.00. Editor: Bernard Meltzer. Publisher: American Elsevier, 52 Vanderbilt Avenue, New York, N.Y. 10017.

1406. **Australian Journal of Statistics.** 1959. 3 issues/yr. $3.50. Editor: H. O. Lancaster, Department of Mathematical Statistics, University of Sydney, Sydney N.S.W., Australia.

1407. **Biometrics.** 1945. Quarterly. $15.00. Editor: L. A. Nelson, Institute of Statistics, North Carolina State University, Raleigh N.C.27607. Publisher: Biometric Society.

1408. **Bulletin of the American Mathematical Society.** 1894. Bi-monthly. $12.00 for non-members. Editor: Murray Gerstenhaber, Department of Mathematics, University of Pennsylvania, Philadelphia, Pennsylvania 19104.

1409. **Bulletin of the Australian Mathematical Society.** 1969. Bi-monthly. $25.00. Editor: B. H. Neumann, Department of Mathematics, Institute of Advanced Studies, Australian National University, POB 4, Canberra, ACT2600, Australia.

1410. **Bulletin of the Calcutta Mathematical Society.** 1908. Quarterly. $6.50. Editors: M. C. Chaki and S. C. Bose, 92 Acharya Prafulla Road, Calcutta 9, India.

1411. **Bulletin of the Calcutta Statistical Association.** 1947. Quarterly. $5.50. Editor: H. K. Nandi, Calcutta Statistical Association, Calcutta University, New Science Building, 35 B.C. Road, Calcutta 19, India.

1412. **Bulletin of the Operations Research Society of America.** Editor: Paul Gray, Stanford Research Institute, Menlo Park, Calif. 94025. Publisher: Operations Research Society of America.

1413. **Canadian Journal of Mathematics/Journal Canadien de Mathematiques.** 1949. Bi-monthly. $18.00. Editor: G. de B. Robinson, Department of Mathematics, University of Toronto, Toronto, Ontario, Canada. Publisher: Canadian Mathematical Congress.

1414. **Canadian Mathematical Bulletin/Bulletin Canadien de Mathematiques.** 1958. 5 issues/yr. $10.00. Editor: Ronald Bercov, Department of Mathematics, University of Alberta, Edmonton, Alberta, Canada. Publisher: Canadian Mathematical Congress.

1415. **Communications in Mathematical Physics.** 1948. 4 issues/yr. $29.70. Publisher: Springer-Verlag, 175 Fifth Avenue, New York, N.Y. 10010.

1416. **Communications on Pure and Applied Mathematics.** 1948. Bi-monthly. $30.00. Editor: Fritz John, Courant Institute of Mathematical Sciences, New York University, 251 Mercer Street, New York, N.Y. 10012. Publisher: Courant Institute of Mathematical Sciences.

1417. **Computer Graphics and Image Processing.** 1972. Quarterly. $28.00/yr. Editors: Azriel Rosenfeld et al. Computer Science Center, University of Maryland, College Park, Maryland 20742. Publisher: Academic Press, 111 Fifth Avenue, New York, N.Y. 10003.

1418. **Computing.** 8 issues/yr. $71.40. Editor: Erich Bukovics, I. Institut fur Mathematik der Technischen Hochschule in Vien, Karlsplatz 13, A-1040 Wien, Austria. Publisher: Springer-Verlag, 175 Fifth Avenue, New York, N.Y. 10010.

1419. **Computing Reviews.** 1960. Monthly. $35.00; $12.50 for members. Editor: Michael A. Duggan. Publisher: Association for Computing Machinery, 1133 Avenue of the Americas, New York, N.Y. 10036.

1420. **Computing Surveys.** 1969. Quarterly. $40.00. Editor: W. S. Dorn, Department of Mathematics, University of Denver, Denver, Colo. 80210. Publisher: Association for Computing Machinery.

1421. **Contents of Contemporary Mathematical Journals.** 1969. 26 issues/yr. $20.00 ($12.00 to members of A.M.S.). Publisher: American Mathematical Society.

1422. **Czechoslovak Mathematical Journal (Matematicky Ustav Ceskoslovenke Akademie Ved).** 1872. Quarterly. $32.00. Editor: J. Marik, Mathematical Institute of Czechoslovak Academy of Sciences, Zitna 25, Prague, Czechoslovakia. Distributed by Plenum Publishing Corp., 227 West 17th Street, New York, N.Y. 10011.

1423. **Differential Equations.** (English translation of **Differentsialnye Uravneniya.**) 1965. Monthly. $150.00. Editor: N. P. Erugin, The Faraday Press, Inc., 84 Fifth Avenue, New York, N.Y. 10011.

1424. **Discrete Mathematics.** 1971. 4 issues/yr. $17.00. Editor: P. L. Hammer, Centre de Recherches Mathematiques, Universite de Montreal, P.O. Box 6128, Montreal, Canada. Publisher: North-Holland Publishing Company, 305-311 Keizersgracht, P.O. Box 3489, Amsterdam, Holland.

1425. **Duke Mathematical Journal.** 1935. Quarterly. $12.00. Editor: Leonard Carlitz, Department of Mathematics, Duke University, Durham, North Carolina 27706. Publisher: Duke University Press.

1426. **Fibonacci Quarterly.** 1963. Quarterly. $6.00. Editor: Verner E. Hoggart, Jr., Department of Mathematics, San Jose College, San Jose, Calif. Publisher: The Fibonacci Association, c/o Brother Alfred Brousseau, St. Mary's College, California.

1427. **Functional Analysis and Its Applications.** (Trans. of **Funktsional'nyi Analiz i ego Prilozheniya.**) 1967. Quarterly. $95.00. Publisher: Plenum, 227 West Seventeenth Street, New York, N.Y. 10011.

1428. **Fundamenta Mathematicae.** 1920. 4 issues/yr. $12.00. Polish Academy of Sciences, Polish Scientific Publishers PWN, Miodava 10, Warsaw, Poland. Distributor: Ars Polona-Ruch, Krakowskie Przedmiescie 7, Warsawa 1, Poland.

1429. **Glasgow Mathematical Journal.** 1952. Semi-annual. $7.00. Editorial office: c/o Department of Mathematics, University of Glasgow, Glasgow W2, Scotland. Publisher: Longman, 33 Montgomery Street, Edinburgh ET7 5JX, Scotland.

1430. **Illinois Journal of Mathematics.** 1957. Quarterly. $12.00. Editor: Mary-Elizabeth Hamstrom, Department of Mathematics, University of Illinois, Urbana, Illinois 61801. Publisher: University of Illinois.

1431. **Indiana University Mathematics Journal.** 1950. Monthly. $30.00. Editor: Paul R. Halmos, Department of Mathematics, Indiana University, Bloomington, Indiana 47401. Publisher: Indiana University.

1432. **Industrial Mathematics.** 1950. Semi-annual. $6.00 for non-members. Editor: Harry H. Denman, 100 Farnsworth, Detroit, Mich. 48202. Publisher: Industrial Mathematics Society.

1433. **Information and Control.** 1957. 10 issues/yr. 2v./yr. $44.00. Editor: Murray Eden, Research Laboratory for Electronics, M.I.T., Cambridge, Mass. 02139. Publisher: Academic Press, 111 Fifth Avenue, New York, N.Y. 10003.

1434. **Information Sciences.** 1969. Quarterly. $28.00. Editor: John M. Richardson, North American Rockwell Corporation Science Center, Aerospace and Systems Group 1049, Camino Dos Rios, Thousand Oaks, Calif. 91360. Publisher: American Elsevier, 52 Vanderbilt Avenue, New York, N.Y. 10017.

1435. **International Abstracts in Operations Research.** 1960. Bimonthly. $12.50. Editor: H. E. Bradley, The Upjohn Company, Kalamazoo, Mich. 49001. Publisher: International Federation of Operational Research Societies.

1436. **International Journal for Numerical Methods in Engineering.** 1969. Quarterly. $36.00. Editors: O. C. Zienkiewicz and Richard H. Gallagher. Publisher: John Wiley, 605 Third Avenue, New York, N.Y. 10016.

1437. **International Journal of Computer Mathematics.** 1964. 4 issues/vol. $41.00/vol. Editor: Randall Rustin, Courant Institute of Mathematical Sciences, 251 Mercer Street, New York, N.Y. 10012. Publisher: Gordon & Breach, 440 Park Avenue South, New York, N.Y. 10016.

1438. **Inventiones Mathematicae.** 1965. 2v./yr. $29.70/vol. Publisher: Springer-Verlag, 175 Fifth Avenue, New York, N.Y. 10010.

1439. **Israel Journal of Mathematics.** 1963. $18.00. Editors: c/o Department of Mathematics, The Hebrew University of Jerusalem, Jerusalem, Israel. Publisher: Weizmann Science Press, P.O. Box 801, Jerusalem, Israel.

1440. **Journal d'Analyse Mathematique.** 1951. 2v./yr. $25.00. Editor: S. Agmon, Mathematics Institute, Hebrew University, Jerusalem, Israel. Publisher: V. Amira, 3 rue Mapou, Jerusalem, Israel. Distributed in the U.S.A. by Stechert-Hafner, Inc., 31 East Tenth Street, New York, N.Y. 10003.

1441. **Journal of Algebra.** 1964. Monthly. 3v./yr. $78.00. Editor: Graham Higman, Mathematical Institute, 24-29 St. Giles, Oxford OX1 3LB, England. Publisher: Academic Press, 111 Fifth Avenue, New York, N.Y. 10003.

1442. **Journal of Applied Probability.** 1964. 3 issues. $17.00/yr. Editor: J. Gani, Department of Probability and Statistics, The University, Sheffield S3 7RH, England.

1443. **Journal of Approximation Theory.** 1968. Quarterly. $22.00. Editor: Oved Shisha, ARL (LB), Building 450, Wright-Patterson AFB, Ohio 45433. Publisher: Academic Press, 111 Fifth Avenue, New York, N.Y. 10003.

1444. **Journal of Combinatorial Theory.** 1966. Bimonthly. 2v./yr. $32.00. Editor: D. H. Younger, Faculty of Mathematics, University of Waterloo, Waterloo, Ontario, Canada. Publisher: Academic Press, 111 Fifth Avenue, New York, N.Y. 10003.

1445. **Journal of Computer and System Sciences.** 1967. Bimonthly. $30.00. Editor: E. K. Blum, Department of Mathematics, University of Southern California, University Park, Los Angeles, Calif. 90007. Publisher: Academic Press, 111 Fifth Avenue, New York, N.Y. 10003.

1446. **Journal of Differential Equations.** 1965. Bimonthly. $52.00. Editor: Joseph P. LaSalle, Division of Applied Mathematics, Brown University, Providence, R.I., 02912. Publisher: Academic Press, 111 Fifth Avenue, New York, N.Y. 10003.

1447. **Journal of Differential Geometry.** 1967. Quarterly. $12.00. Editor: C. C. Hsiung, Department of Mathematics, Lehigh University, Bethlehem, Penn. 18015. Publisher: Lehigh University. Distributed by the American Mathematical Society.

1448. **Journal of Functional Analysis.** 1967. Bimonthly. $44.00. Editor: Irving Segal, Department of Mathematics, M.I.T., Cambridge, Mass. 02139. Publisher: Academic Press, 111 Fifth Avenue, New York, N.Y. 10003.

1449. **Journal of Mathematical Analysis and Applications.** 1960. Monthly. $128.00. Editor: Richard Bellman, Department of Mathematics, Engineering and Medicine, University of Southern California, University Park, Los Angeles, Calif. 90007. Publisher: Academic Press, 111 Fifth Avenue, New York, N.Y. 10003.

1450. **Journal of Mathematical Physics.** 1960. Monthly. $45.00; $15.00 for members. Editor: M. Hamermesh, Department of Physics, University of Minnesota, Minneapolis, Minn. 55455. Publisher: American Institute of Physics.

1451. **Journal of Mathematical Psychology.** 1964. Quarterly. $35.00. Editor: J. E. Keith Smith, Department of Psychology, University of Michigan, Ann Arbor, Mich. 48104. Publisher: Academic Press, 111 Fifth Avenue, New York, N.Y. 10003.

1452. **Journal of Mathematics and Physics.** 1921. Quarterly. $8.00. Editor: Eric Reissner, Massachusetts Institute of Technology, Cambridge, Mass. 02139.

1453. **The Journal of Natural Sciences and Mathematics.** 1960. Semi-annual. Rs. 15.00. Editor: L. M. Chawla, Department of Mathematics, Government College, Lahore, Pakistan. Publisher: Government College, Lahore, Pakistan.

1454. **Journal of Number Theory.** 1969. Quarterly. $22.00. Editor: Archie W. Addison, Department of Mathematics, Ohio State University, Columbus, Ohio 43210. Publisher: Academic Press, 111 Fifth Avenue, New York, N.Y. 10003.

1455. **Journal of Optimization and Applications.** 1967. Monthly. $44.00/yr. Editor: Angelo Miele, 201 Space Science Building, Rice University, Houston, Texas 77001. Publisher: Plenum, 227 West Seventeenth Street, New York, N.Y. 10011.

1456. **Journal of Symbolic Logic.** 1936. $30.00. Editor: Burton Dreben, Department of Philosophy, Harvard University, Cambridge, Mass. 02138. Publisher: Association for Symbolic Logic. Distributed by the American Mathematical Society.

1457. **Journal of the American Statistical Association.** 1888. Quarterly. $15.00. Editor: Robert Ferber, 806 Fifteenth Street N.W., Washington, D.C. 20005. Publisher: American Statistical Association.

1458. **Journal of the Association for Computing Machinery.** 1953. Quarterly. $25.00. Editor: G. Salton, Department of Computer Science, Cornell University, Ithaca, New York 14850.

1459. **Journal of the Australian Mathematical Society.** 1959. 8 issues/yr. $50.00. Editor: B. Mond, Department of Mathematics, La Trobe University, Bundoora, Victoria 3083, Australia.

1460. **Journal of the Institute of Mathematics and Its Applications.** 1965. Bimonthly. $44.00/yr. Publisher: Academic Press, 111 Fifth Avenue, New York, N.Y. 10003.

1461. **Journal of the Royal Statistical Society, Series A: General.** 1838. 4 issues/yr. $15.00. Publisher: Royal Statistical Society, 21 Bentinck Street, London W1M 6AR, England.

1462. **Journal of the Royal Statistical Society, Series B: Methodological.** 1934. 3 issues/yr. $15.00. Royal Statistical Society, 21 Bentinck Street, London W1M 6AR, England.

1463. **Journal of the Royal Statistical Society, Series C: Applied Statistics.** 1952. 3 issues/yr. $5.00. Editors: M. H. Quenoville and G. B. Wetherill, Royal Statistical Society, 21 Bentinck Street, London W1M 6AR, England.

1464. **Linear Algebra and Its Applications.** 1968. Quarterly. $26.00. Editor: Alan Hoffman, IBM Research Center, Yorktown Heights, N.Y. 10598. Publisher: American Elsevier, 52 Vanderbilt Avenue, New York, N.Y. 10017.

1465. **Mathematica Scandinavica.** 1953. Quarterly. Kr. 80.00. Editor: Svend Bundgaard, Universitets-Parken, Aarhus, Denmark.

1466. **Mathematical Biosciences.** 1967. Quarterly. $22.00. Editor: Richard Bellman, Department of Mathematics, Engineering, and Medicine, University of Southern California, University Park, Los Angeles, Calif. 90007. Publisher: American Elsevier, 52 Vanderbilt Avenue, New York, N.Y. 10017.

1467. **Mathematical Notes of the Academy of Sciences of the U.S.S.R.** (Translation of **Matematicheski Zametki.**) 1967. Monthly. 2v./yr. $75/vol. Publisher: Plenum, 227 West Seventeenth Street, New York, N.Y. 10011.

1468. **Mathematical Operations Research and Statistics.** 1970. Quarterly. $2.00. Editor: Olaf Bunke, Akademie Verlag, Leipziger Str. 3-4, 108 Berlin, East Germany.

1469. **Mathematical Reviews.** 1940. Monthly. $320.00; $40.00 for members of A.M.S. Editorial office: Mathematical Reviews, 416 Fourth Street, Ann Arbor, Mich. 48103. Publisher: American Mathematical Society.

1470. **Mathematical Spectrum.** 1968. Semi-annual. $1.20. Editor: J. Gani, Department of Probability and Statistics, The University, Sheffield S3 7RH, England. Publisher: Oxford University Press, 200 Madison Avenue, New York, N.Y. 10016.

1471. **Mathematical Systems Theory.** 1967. Quarterly. $30.00. Editor: A. J. Lohwater, Department of Mathematics, Case Western Reserve University, Cleveland, Ohio 44106. Publisher: Springer-Verlag, 175 Fifth Avenue, New York, N.Y. 10010.

1472. **Mathematics Bulletin of the University of Moscow.** (Translation of **Vestnik Moskovskogo Universiteta Matematica.**) Bimonthly. Publisher: Faraday Press, 84 Fifth Avenue, New York, N.Y. 10011.

1473. **Mathematics Magazine.** 1926. Bimonthly. $4.00/yr. Editor: G. N. Wollan, Mathematics Sciences Division, Purdue University, Lafayette, Indiana 47907. Publisher: Mathematical Association of America.

1474. **Mathematics of Computation.** 1943. Quarterly. $20.00; $10.00 for members. Editor: Eugene Isaacson, Courant Institute, 251 Mercer Street, New York, N.Y. 10012. Publisher: American Mathematical Society.

1475. **Mathematics of the U.S.S.R.–Izvestija.** (Trans. of **Izvestija Matematiceskaja Serija.**) 1967. Bimonthly. $200.00; $100.00 for members. Editor: Joseph A. Zilber, American Mathematical Society, P.O. Box 6248, Providence, R.I. 02904.

1476. **Mathematics of the U.S.S.R.–Sbornik.** (Translation of **Matematiceskii Sbornik.**) 1967. Monthly. $290.00; $145.00 for members. Editor: S. H. Gould, American Mathematical Society, P.O. Box 6248, Providence, R.I. 02904.

1477. **The Mathematics Student Journal.** 1954. 4 issues/yr. $2.00. Editor: Thomas J. Hill, Department of Mathematics, University of Oklahoma, Norman, Okla. 73069. Publisher: National Council of Teachers of Mathematics.

1478. **The Mathematics Teacher.** 1908. Monthly. $9.00. Editor: Carol V. McCamman, National Council of Teachers of Mathematics, 1201 Sixteenth Street N.W., Washington, D.C. 20036.

1479. **Mathematische Annalen.** 1868. Issued at frequent intervals according to material received. $29.70/vol. of 4 issues. Publisher: Springer-Verlag, 175 Fifth Avenue, New York, N.Y. 10010.

1480. **Mathematische Zeitschrift.** 1918. Quarterly. $30.00/vol. Editor: H. W. Wielandt, University of Tübingen, Mathematics Institute, (74) Tübingen, Germany. Publisher: Springer-Verlag, 175 Fifth Avenue, New York, N.Y. 10010.

1481. **Michigan Mathematical Journal.** 1952. Semi-annual. $12.00; $4.00 for members. Editor: George Piranian, Department of Mathematics, University of Michigan, Ann Arbor, Mich. 48104. Publisher: University of Michigan Press.

1482. **New Publications.** Beginning 1972, **New Publications** will be incorporated into **Contents of Contemporary Mathematical Journals.**

1483. **Notices of the American Mathematical Society.** 1953. 8 issues/yr. $10.00 for non-members. Editor: Gordon Walker. Publisher: American Mathematical Society.

1484. **Notre Dame Journal of Formal Logic.** 1960. Quarterly. $8.00. Editor: Boleslaw Sobocinski, Department of Mathematics, University of Notre Dame, Notre Dame, Indiana. Publisher: Notre Dame University Press.

1485. **Numerische Mathematik.** 1959. Issued at frequent intervals according to the material received. $29.70/vol. of 4 issues. Publisher: Springer-Verlag, 175 Fifth Avenue, New York, N.Y. 10010.

1486. **Ontario Mathematics Gazette.** 1962. 3 issues/yr. $2.00. Editor: M. K. Davison, Ontario Mathematics Commission, 1260 Bay Street, Toronto 5, Ontario, Canada.

1487. **Operations Research.** 1952. Bimonthly. $10.00. Editor: H. J. Miser, P.O. Box 525, Windsor, Conn. 06095. Publisher: Operations Research Society of America.

1488. **Osaka Journal of Mathematics.** 1949. 3 issues/yr. Editors: H. Kudo et al., c/o Department of Mathematics, Osaka University, Toyonaska, Osaka, Japan. Publisher: Department of Mathematics, Osaka University and Osaka City University.

1489. **Pacific Journal of Mathematics.** 1951. Monthly. $32.00; $16.00 for members. Editor: Richard Arens, Department of Mathematics, University of California, Los Angeles, Calif. 90024. Publisher: Pacific Journal of Mathematics, 103 Highland Boulevard, Berkeley, Calif. 94708.

1490. **Periodica Mathematica Hungarica.** 1971. Quarterly. $18.00. Editor: P. Erdös, Budapest V, Realtanoda u. 13-15, Hungary. Distributed by: Collet, Dennington Estate, Wellingborough, Northants, NN8 2QT, United Kingdom.

1491. **Philosophical Transactions of the Royal Society, Series A: Mathematical and Physical Sciences.** 1665. Published irregularly. $40.00. Publisher: Royal Society, 6 Carlton House Terrace, London, SW1Y 5AG, England.

1492. **Proceedings of the American Mathematical Society.** 1950. Monthly. $72.00. Editor: W. H. Fuchs, Department of Mathematics, White Hall, Cornell University, Ithaca, N.Y. 14850.

1493. **Proceedings of the Edinburgh Mathematical Society.** 1883. Semi-annual. $6.00. Oliver & Boyd, Tweeddale Court, Edinburgh 1, Scotland, U.K.

1494. **Proceedings of the National Academy of Sciences (U.S.A.).** 1915. Monthly. $35.00. Editor: J. A. Last, 2101 Constitution Ave., Washington, D.C. 20418.

1495. **Proceedings of the Royal Society, Series A: Mathematical and Physical Sciences.** 1832. $16.00/vol. Publisher: Royal Society, 6 Carlton House Terrace, London, SW1Y 5AG, England.

1496. **Quarterly Journal of Mathematics.** 1930. Quarterly. $17.00/yr. Editor: E. C. Thompson, Clarendon Press, Oxford, England.

1497. **Quarterly of Applied Mathematics.** 1943. Quarterly. $15.00. Editor: Walter Freiberger, Department of Applied Mathematics, Brown University, Providence, R.I. 02902. Publisher: Brown University.

1498. **Research Institute for Mathematical Sciences Publications.** 1965. 3 issues/yr. $18.50. Publisher: Research Institute of Mathematical Sciences, Kyoto University, Kyoto 606, Japan.

1499. **Rocky Mountain Journal of Mathematics.** 1971. Quarterly. $10.00. Editor: William R. Scott, Department of Mathematics, University of Utah, Salt Lake City, Utah 84112. Publisher: Rocky Mountain Mathematics Consortium.

1500. **Scripta Mathematica.** 1933. Quarterly. $4.00. Editor: Abe Gelbart, Department of Mathematics, Yeshiva University, Amsterdam Avenue and 186th Street, New York, N.Y. 10033. Publisher: Yeshiva University.

1501. **Semigroup Forum.** 1970. Quarterly. $41.50. Editors: A. H. Clifford, K. H. Hoffman, and P. S. Mostert. Publisher: Springer-Verlag, 175 Fifth Avenue, New York, N.Y. 10010.

1502. SIAM Journal on Applied Mathematics. 1953. 8 issues/yr. $36.00. Editor: Frederick J. Beutler, Department of Computer, Information, and Control Engineering, University of Michigan, Ann Arbor, Mich. 48104. Publisher: Society for Industrial and Applied Mathematics.

1503. SIAM Journal on Control. 1963. Quarterly. $27.00. Editor: Lucien W. Neustat, Department of Electrical Engineering, University of Southern California, University Park, Los Angeles, Calif. 90007. Publisher: Society for Industrial and Applied Mathematics.

1504. SIAM Journal on Mathematical Analysis. 1970. Quarterly. $27.00. Editor: F. W. J. Olver, Institute for Fluid Dynamics and Applied Mathematics, University of Maryland, College Park, Maryland 20742. Publisher: Society for Industrial and Applied Mathematics.

1505. SIAM Journal on Numerical Analysis. 1964. Quarterly. $27.00. Editor: Werner C. Rheinboldt, Computer Science Center, University of Maryland, College Park, Maryland 20742. Publisher: Society for Industrial and Applied Mathematics.

1506. SIAM Review. 1959. Quarterly. $16.00. Editor: William E. Boyce, Department of Mathematics, Rensselaer Polytechnic Institute, Troy, N.Y. 12181. Publisher: Society for Industrial and Applied Mathematics.

1507. Siberian Journal: Mathematical Journal of the Academy of Sciences of the U.S.S.R., Novosibirsk. (Translation of **Sibirskii Matematicheskii Zhurnal.**) 1969. Bimonthly. $150.00. Publisher: Plenum, 227 West Seventeenth Street, New York, N.Y. 10011.

1508. Soviet Mathematics—Doklady. (Translation of the mathematics section of **Doklady Akademii Nauk SSSR.**) 1960. Bimonthly. $100.00; $50.00 for members. Editor: S. H. Gould, American Mathematical Society, P.O. Box 6248, Providence, R.I. 02904. Publisher: American Mathematical Society.

1509. Studia Mathematica. 1929. 8 issues/yr. $12.00. Publisher: Polska Akademia Nauk, Panstwowe Wydawnictwo Naukowe, U. Miodowa 10, Warsaw, Poland.

1510. Studies in Applied Mathematics. 1921. Quarterly. $8.00. Editor: D. J. Benney, Department of Mathematics, M.I.T., Cambridge, Mass. 02139. Publisher: Massachusetts Institute of Technology.

1511. Technometrics. 1959. Quarterly. $8.00. Editor: Harry Smith, Jr., School of Management, Rensselaer Polytechnic Institute, Troy, N.Y. 12181. Publisher: American Statistical Association.

1512. Theoretical and Mathematical Physics. (Translation of **Teoretcheskaya i Matematicheskaya Fizika.**) 1969. Monthly. $125.00. Publisher: Plenum, 227 West Seventeenth Street, New York, N.Y. 10011.

1513. **Theory of Probability and Its Applications.** (Translation of **Teorija Verojatnostei i ee Primenenija.**) 1956. Quarterly. $37.50. Editor: N. Artin Brunswick, New York University, Courant Institute, 251 Mercer Street, New York, N.Y. 10012. Publisher: Society for Industrial and Applied Mathematics.

1514. **Topology.** 1962. Quarterly. $30.00. Publisher: Pergamon Press, Maxwell House, Fairview Park, Elmsford, N.Y. 10523.

1515. **Transactions of the American Mathematical Society.** 1900. Monthly. $180.00. Editor: Dana Scott, Department of Philosophy, Princeton University, Princeton, N.J. 08540.

1516. **Transactions of the Society of Actuaries.** Editor: Floyd T. Beasley, Society of Actuaries, 208 South La Salle Street, Chicago, Ill. 60604.

1517. **Ukrainian Mathematical Journal.** (Translations of **Ukrainskii Matematicheski Zhurnal.**) 1967. Bimonthly. $135.00. Publisher: Plenum Press, 227 West Seventeenth Street, New York, N.Y. 10011.

1518. **U.S.S.R. Computational Mathematics and Mathematical Physics.** (Translation of **Zhurnal Vychislite'noi Matematiki i Matematicheskoi Fiziki.**) 1962. Bimonthly. $150.00. Translation editor: R. C. Glass. Publisher: Pergamon Press, Maxwell House, Fairview Park, Elmsford, N.Y. 10523.

CHAPTER 35

DIRECTORIES AND GUIDES

1519. American Mathematical Society. **Abbreviations of Names of Journals.**
Providence, R.I.: American Mathematical Society, 1969. 18p. $1.00.

1520. American Mathematical Society. **Assistantships and Fellowships in
Mathematics in 1972-1973.** Providence, R.I.: American Mathematical Society,
1971. 150p.
Contents: This is a Special Issue of the Notices, Vol. 18, No. 8, December 1971.
List of assistantships and fellowships at universities in the United States and
Canada, and tax status of such grants.

1521. American Mathematical Society. **Combined Membership List—1971-1972.**
Providence, R.I.: American Mathematical Society, 1971. 260p. $10.00; $5.00
for members.
Contents: List of members of the American Mathematical Society, Mathematical
Association of America, and Society for Industrial and Applied Mathematics,
1971-1972.

1522. American Mathematical Society. **Grants and Proposals.** Providence, R.I.:
American Mathematical Society. 20p. $1.00.
Contents: List of books, pamphlets, government agencies, and foundations
which one might refer to to obtain information for research funds. It also
includes information on how to write proposals for grants.

1523. American Mathematical Society. **Mathematical Sciences Administrative
Directory—1972.** Providence, R.I.: American Mathematical Society, 1971.
140p. $5.00; $1.00 for members.
Contents: List of professional organizations, government agencies, editors of
mathematical journals, chairmen of departments of mathematics, and industrial
mathematics research groups.

1524. American Mathematical Society. **Mathematical Sciences Employment
Register Lists.** Sponsored by the American Mathematical Society, Mathematical
Association of America, and Society for Industrial and Applied Mathematics.
Published three times a year: January, May, and August.
a) **Summary of Available Applicants Including the Summary of Academic,
Industrial, and Government Openings.** $30.00/yr.; $15.00/single issue.
b) **Summary of Academic, Industrial, and Government Openings.** $12.00/yr.;
$5.00/single issue.
c) **List of Retired Mathematicians Available for Employment.** No charge.
d) **Finding Employment in the Mathematical Sciences.** No charge.

1525. American Mathematical Society. **A Manual for Authors of Mathematical Papers.** Rev. ed. Providence, R.I.: American Mathematical Society, 1970. 21p. $1.00; free to members.

1526. Gould, S. H. **A Manual for Translators of Mathematical Russian.** Providence, R.I.: American Mathematical Society, 1966. 48p. $3.30; $2.48 for members.

1527. Pemberton, J. E. **How to Find Out in Mathematics.** Elmsford, N.Y.: Pergamon, 1964. 168p.
Contents: Careers for mathematicians. The organization of mathematical information. Mathematical dictionaries, encyclopedias and theses. Mathematical periodicals and abstracts. Mathematical societies. Mathematical education. Computers and mathematical tables. Mathematical history and biography. Mathematical books—bibliographies, evaluation and acquisition. Probability and statistics. Operational research. Appendices—sources of Russian mathematical information, mathematics and the government, actuarial science.

1528. Reckenbeil, R. **The Mathematical Journal from a Technical Point of View.** Providence, R.I.: American Mathematical Society. No charge.
Contents: A manual on printing techniques, typesetting, composition procedures, and preparing material for the printer.

1529. Swanson, E. **Mathematics into Type.** Providence, R.I.: American Mathematical Society, 1971. 98p. $4.00; $3.00 for members.
Contents: A manual on the preparation of manuscripts for publication.
ARBA 72.

CHAPTER 36

PROFESSIONAL ORGANIZATIONS
AND GOVERNMENT AGENCIES

Professional Organizations

American Mathematical Society, P.O. Box 6248, Providence, R.I. 02904.

American Statistical Association, 806 15th Street N.W., Washington, D.C. 20005.

Association for Computing Machinery, 1133 Avenue of the Americas, New York, N.Y. 10036.

Association for Symbolic Logic, P.O. Box 6248, Providence, R.I. 02904.

Biometric Society, P.O. Box 5962, Raleigh, N.C. 27607.

Canadian Mathematical Congress, 985 Sherbrooke Street West, Montreal 110, Quebec, Canada.

Committee on the Undergraduate Program in Mathematics: A Committee of the Mathematical Association of America, P.O. Box 1024, Berkeley, Calif. 94701.

Conference Board of the Mathematical Sciences, 2100 Pennsylvania Avenue, N.W., Suite 834, Washington, D.C. 20037.

The Institute of Management Sciences, P.O. Box 6112, Providence, R.I. 02904.

Institute of Mathematical Statistics, California State College, Hayward, Calif. 94542.

Mathematical Association of America, 1225 Connecticut Avenue N.W., Washington, D.C. 20036.

National Academy of Sciences National Research Council: Division of Mathematical Sciences, 2101 Constitution Avenue N.W., Washington, D.C. 20418.

National Council of Teachers of Mathematics, 1201 Sixteenth Street N.W., Washington, D.C. 20036.

Operations Research Society of America, 428 East Preston Street, Baltimore, Maryland 21202.

Rocky Mountain Mathematics Consortium, Mathematics-Physics Building 203, The University of Montana, Missoula, Montana 59801.

School Mathematics Study Group, Cedar Hall, Stanford University, Stanford, Calif. 94305.

Society for Industrial and Applied Mathematics, 33 South Seventeenth Street, Philadelphia, Penn. 19103.

Society of Actuaries, 208 South La Salle Street, Chicago, Ill. 60604.

Government Agencies

Air Force Office of Scientific Research, Air Force Systems Command, 1400 Wilson Boulevard, Arlington, Virginia 22209.

Army Research Office–Durham: Mathematics Division, Applied Mathematics Program, Box CM, Duke Station, Durham, N.C. 27706.

Atomic Energy Commission, Division of Research, Mathematics and Computer Branch, Washington, D.C. 20545.

Department of Health, Education, and Welfare, Office of Education: Bureau of Educational Personnel Development; Bureau of Elementary and Secondary Education; Bureau of Higher Education; National Center for Educational Research and Development. Washington, D.C. 20202.

National Aeronautics and Space Administration, Washington, D.C. 20546.

National Bureau of Standards, Applied Mathematics Division: Mathematical Analysis Section; Operations Research Section; Statistical Engineering Laboratory. Washington, D.C. 20234.

National Institute of Mental Health: Theoretical Statistics and Mathematics Section, Barlow Building, 5454 Wisconsin Avenue, Chevy Chase, Maryland 20015.

National Institutes of Health: Division of Biologics Standards; Division of Computer Research and Technology. Bethesda, Maryland 20014.

National Science Foundation: Mathematical Sciences Section, Office of Computing Activities, Washington, D.C. 20550.

Office of Naval Research, Department of the Navy: Fluid Dynamics Program; Information Systems Program; Mathematical and Information Sciences Division; Mathematics Program; Operations Research Program; Statistics and Probability Program; Structural Mechanics Program. Arlington, Virginia 22217.

CHAPTER 37

A SELECTED LIST OF PUBLISHERS

Academic Press, 111 Fifth Avenue, New York, N.Y. 10003.

Addison-Wesley Publishing Company, Reading, Mass., 01867.

Allyn and Bacon, 470 Atlantic Avenue, Boston, Mass. 02210.

American Elsevier Publishing Company, 52 Vanderbilt Avenue, New York, N.Y. 10017.

Appleton-Century Crofts, 440 Park Avenue South, New York, N.Y. 10016.

W. A. Benjamin, Publishers, Reading, Mass. 01867.

Blaisdell Publishing Company, 275 Wyman Street, Waltham, Mass. 02154.

Cambridge University Press, 32 East 57th Street, New York, N.Y. 10022.

Chelsea Publishing Company, 159 East Tremont Avenue, Bronx, N.Y. 10453.

Courant Institute of Mathematical Sciences, 251 Mercer Street, New York, N.Y. 10012.

Dickenson Publishing Company, 16561 Ventura Boulevard, Suite 215G, Encino, Calif. 91316.

Dover Publications, 180 Varick Street, New York, N.Y. 10014.

Gordon & Breach Science Publishers, 440 Park Avenue South, New York, N.Y. 10016.

Hafner Publishing Company, 866 Third Avenue, New York, N.Y. 10022.

Harper & Row, Publishers, 49 East 33rd Street, New York, N.Y. 10016.

Harvard University Press, Kittridge Hall, 79 Garden Street, Cambridge, Mass. 02138.

D. C. Heath and Company, 125 Spring Street, Lexington, Mass. 02173.

Holden-Day, 500 Sansome Street, San Francisco, Calif. 94111.

Holt, Rinehart and Winston, 383 Madison Avenue, New York, N.Y. 10017.

Intext Educational Publishers, Scranton, Pennsylvania 18515.

Johnson Reprint Corporation, 111 Fifth Avenue, New York, N.Y. 10003.

Kraus Reprint Company, 16 East 46th Street, New York, N.Y. 10017.

The Macmillan Company, 866 Third Avenue, New York, N.Y. 10022.

Markham Publishing Company, 3322 West Peterson Avenue, Chicago, Ill. 60645.

The M.I.T. Press, Cambridge Mass. 02142.

McGraw-Hill Book Company, 330 West 42nd Street, New York, N.Y. 10036.

Charles E. Merrill Publishing Company, 1300 Alum Creek Drive, Columbus, Ohio 43216.

Oxford University Press, 1600 Pollitt Drive, Fair Lawn, N.J., 07410.

Pergamon Press, Maxwell House, Fairview Park, Elmsford, N.Y. 10523.

Plenum Publishing Corporation, 227 West Seventeenth Street, New York, N.Y. 10011.

Prentice-Hall, Englewood Cliffs, N.J. 07632.

Princeton University Press, Princeton, N.J. 08540.

W. B. Saunders Company, West Washington Square, Philadelphia, Pa. 19105.

Springer-Verlag, 175 Fifth Avenue, New York, N.Y. 10010.

Frederick Ungar Publishing Company, 250 Park Avenue South, New York, N.Y. 10003.

Van Nostrand-Reinhold Company, 450 West 33rd Street, New York, N.Y. 10001.

John Wiley & Sons, 605 Third Avenue, New York N.Y. 10016.

AUTHOR INDEX

AUTHOR INDEX

AUTHOR INDEX

262

AUTHOR INDEX

AUTHOR INDEX

AUTHOR INDEX

Johnson, R. L., 206
Johnson, W. G., 15
Jonah, H. F. S., 3
Jones, C. R., 1115
Jones, D. S., 92, 845
Jones, R. L., 997
Jordan, D. W., 92
Jorgenson, P., 960

Kagiwada, H. H., 1279
Kahane, J. P., 1086
Kahn, P. J., 317
Kain, R. Y., 998
Kalaba, R. E., 874, 1279
Kalinin, V. M., 1087
Kalman, R. E., 1324
Kantorovich, L. V., 698
Kaplan, W., 93
Kaplansky, I., 318, 366, 429
Kappos, D., 1088
Kapur, G. K., 999
Karamcheti, K., 542
Karlin, S., 1089, 1090
Karmazina, L. N., 1226, 1227
Karpov, K. A., 1199, 1200, 1201, 1203
Karrass, A., 398
Karreman, H. F., 1325
Kasriel, R. H., 468
Kato, T., 699
Kattsoff, L. O., 16
Katzan, H., Jr., 1000, 1001
Katznelson, Y., 835
Kaufmann, A., 1326
Keller, H. B., 902, 903
Keller, J. B., 1327
Kelley, J. L., 17, 51, 469
Kells, L. M., 772
Kemeny, J. G., 18, 1091
Kempthorne, O., 1153
Kendall, M. G., 1092, 1159, 1160
Kendall, P. C., 70
Kenelly, J. W., 167
Kennison, J. F., 166
Khabaza, I. M., 904, 1181
Khinchin, A. Y., 245, 1067
Khrenov, L. S., 1202
Kiefer, J., 1033
Kilmister, C. W., 1328
Kingman, J. F. C., 1093

Kiokemeister, F. L., 91
Kireyeva, I. Y., 1203
Kirk, D. E., 1329
Kirzhnitz, D. A., 1330
Kiselev, A. I., 773
Kleene, S. C., 168, 169
Klerer, M., 1331
Kluznin, L. A., 1174
Kneebone, G. T., 170, 195
Knopp, A. W., 1091
Knopp, M. I., 246
Kobayashi, S., 543
Kogbetliantz, E. G., 1204
Kolman, B., 113
Konove, C., 5
Kopperman, R. D., 171
Korfhage, R. R., 82, 172, 1332
Koszul, J. L., 430
Kowalik, J., 905, 1333
Krasnosel'skii, M. A., 774, 775
Krasnov, M. L., 773
Kravtsova, R. I., 1185
Kreider, D. L., 776
Krein, M. G., 691
Kreisel, G., 173
Krekó, B., 1334
Kriegh, R. B., 8, 71
Krikorian, A., 1204
Kripke, B., 608
Krishnaiah, P., 1094
Krivine, J. J., 173
Krook, M., 630
Kruse, A. H., 174
Kruse, R. L., 367
Krut'ko, P. D., 1335
Krylov, V. I., 1180
Kuczma, M., 1336
Kuller, R. G., 776
Kunze, R., 315, 716
Kunze, R. A., 626
Künzi, H. P., 906
Kuratowski, K., 470
Kurosh, A. G., 280
Kurth, R., 1337
Kurtz, A., 1175
Kushner, H., 1338
Kuznetsov, P. I., 1183
Kyburg, H. E., Jr., 1096

AUTHOR INDEX

Laatsch, R., 281
Ladyzenskaja, O. A., 777
Laha, R. G., 1041
Lakshmikantham, V., 778
Lambek, J., 368
Lamperti, J., 1097
Lancaster, H. O., 1098, 1161
Lancaster, K., 1339
Lancaster, P., 319
Lanczos, C., 131, 779
Landin, J., 282
Lang, S., 283, 501
Lange, H. F., 1099
Lapa, V. G., 1082
Lapidus, L., 907
Larsen, M. D., 369
Larson, R. E., 1340
LaSalle, J., 780
Lattès, R., 781
Laugwitz, D., 544
Lawson, C., 987
Lax, P. D., 782
Leadbetter, M. R., 1046
Leavitt, T. C. J., 19
Lebedev, A. V., 1162
Lebedev, N. N., 664
Lebesgue, H., 132
Leblanc, H., 175
Lee, E. B., 1341
Leela, S., 778
Lefschetz, S., 780, 783, 1342
Lehmann, E. L., 1080
Leichus, R., 983
Leisenring, A. C., 176
Leitmann, G., 1282
Lemmon, E. J., 177
Leone, F. C., 1085
Leonhardy, A., 20, 52
LeVeque, W. J., 247
Levi, H., 21
Levin, G., 360
Levine, J. L., 554
Levine, N., 1214
Levinson, N., 644
Levison, M., 1002
Lewis, D. J., 93, 284
Lewis, E. M., 23
Liebeck, H., 320
Lightstone, A. H., 178, 179
Lindsay, J. W., 292

Lindstrum, A. O., Jr., 285
Linnik, U. V., 241
Lions, J. L., 781
Little, C. E., 25
Littlewood, J. E., 609
Liu, C. L., 216
Liulevicius, A., 502
Liverman, T. P. G., 846
Llewellyn, R. W., 1343
Lockett, J. A., 874, 953
Loève, M., 1100
Lohwater, A. J., 633, 1176
Long, C. T., 248
Loomba, N. P., 1344
Loomis, L. H., 94
Loos, O., 431
Lorch, E. R., 700
Lorentz, G. G., 908
Lott, R. W., 1003
Louden, R. K., 1004
Lowengrub, M., 1371
Luenberger, D. G., 909
Lukacs, E., 1101, 1102
Lukasiewicz, J., 180
Luke, Y. L., 665
Luk'yanov, A. V., 1205
Lundell, A. T., 471
Lyapunov, 1345, 1346
Lyusternik, L. A., 910, 1206

Maass, H., 249
Macdonald, I. D., 397
Macdonald, I. G., 347
Mackenzie, R. E., 515
Mackie, A. G., 784
MacLane, S., 286, 411
Macon, N., 911
MacRobert, T. M., 666
Maddox, I. J., 701
Maddox, T. K., 22
Maehly, H., 987
Magnus, W., 398, 667, 785
Makarenko, G. I., 773
Maksoudian, Y. L., 1103
Mal'cev, A. I., 321
Malgrange, B., 702
Malyshev, A. V., 250
Mancill, J. D., 53
Manheim, J. H., 133

266

AUTHOR INDEX

Mann, H. B., 251, 1347
Mansfield, M. J., 472
Marchenko, V. A., 724
Marcus, S., 1348
Marden, M., 645
Marder, L., 95
Margaris, A., 181
Marks, J. L., 98
Markus, L., 1341
Markushevich, A. I., 646
Martin, A. D., 322
Martin, F. B., 1153
Martin, J. J., 1104
Martin, W. T., 1150, 1349
Massera, J. L., 786
Massey, F. J., Jr., 1051
Massey, L. D., 1105
Massey, W. S., 473
Maunder, C. R. F., 474, 475
May, J. P., 476
Mayne, D. Q., 1321
Maz'ya, V. G., 659
McCarthy, P. J., 369
McCarty, G., 432
McCormick, G. P., 1286
McCormick, J. M., 1005
McCoy, N. H., 287
McCracken, D. D., 885
McCue, G. A., 1361
McEntire, A. D., 33
McFarland, D., 23
McIntyre, J. E., 862
McKean, H. P., Jr., 1106
McKeeman, W. M., 1006
McLeod, E. B., 1350
McQuisten, R. B., 96
McReynolds, S. R., 1301
Melsa, J. L., 1135
Mendelson, B., 477
Mendelson, E., 182
Menger, K., 519
Meredith, W. B., 1207
Meschkowski, H., 134, 135, 545
Mesztenyi, C., 987
Metel'skii, A. S., 1180
Meyer, H., 97
Meyer, J. P., 288
Middlemiss, R. R., 98
Mihailov, V. A., 1174
Mikhlin, S. G., 787, 788, 789, 863, 912

Mikusinski, J., 622, 847
Miller, K. S., 790, 1107
Miller, R. G., Jr., 1108
Miller, W., Jr., 668
Millington, T. A., 1177
Millington, W., 1177
Milton, R. C., 1109
Mine, H., 1110
Minlos, R. A., 391
Minsky, M., 1007
Mirsky, L., 217
Mirtes, B., 1009
Mitchell, A. R., 913
Mitchell, B., 412
Mitchell, H. H., 234
Mitra, S. K., 332
Mizel, V. J., 322
Mizohata, S., 791
Mock, G. D., 64
Mode, C. J., 1111
Mode, E. B., 1112
Modenov, P. S., 546
Monjallon, A., 24
Monk, J. D., 183
Mood, A. M., 1113
Moon, J. W., 218
Moon, P., 792
Moore, C. G., 25
Moore, J. T., 323, 324
Moore, R. E., 881, 914, 1008
Moore, T. O., 478
Mordell, L. J., 252
Morrey, C. B., Jr., 102
Morrison, D. F., 1114
Morse, A. P., 184
Morse, M., 503
Morton, K. W., 922
Mosher, R. E., 479
Moss, R. M. F., 413
Mostert, P. S., 426
Mostow, G. D., 288
Mostowski, A., 185
Moursund, D. G., 915
Mueller, F. J., 54
Mulaik, S. A., 1351
Mulholland, H., 1115
Munro, W. D., 1021, 1022
Murdoch, D. C., 99
Murnaghan, F. D., 848
Murphy, R. E., 1352

267

AUTHOR INDEX

269

AUTHOR INDEX

Rosskopf, M. F., 554
Roth, K. F., 244
Rotman, B., 195
Rotman, J. J., 403
Rounyak, J., 591
Roy, J., 1041
Royden, H. L., 611
Rozanov, Y. A., 1124, 1132
Rubenstein, M. F., 1014
Rubin, J. E., 196
Rubinov, A. M., 884, 1298
Rudin, W., 612, 613, 652
Rumshiskii, L. Z., 1133
Rund, H., 864
Russell, D. S., 61
Russell, L. J., 1012
Rutherford, D. E., 289
Rutland, L. W., 8, 71
Ryser, H. J., 222
Ryzhik, I. M., 1194

Saaty, T. L., 209, 1134, 1363, 1364, 1365, 1366, 1367
Sachs, J. M., 32
Sagan, H., 865
Sage, A. P., 1135
Sah, C. H., 291
Salvadori, M. G., 1005
Salzer, H. E., 1214
Samarski, A. A., 820
Sammet, J. E., 1015
Sampson, J. H., 288
Samuel, P., 293, 376, 555
Sanchez, D. A., 803
Sanders, P., 33
Sansone, G., 804
Sario, L., 651, 653, 654, 1368
Saulyev, V. K., 923
Sawaragi, Y., 1136
Saxon, J. A., 1016, 1017
Schaaf, W. L., 34
Schachter, H., 105
Schaefer, H. H., 484
Schafer, R. D., 377
Schäffer, J. J., 786
Schechter, M., 712, 731
Scheleifer, A., Jr., 18
Schenkman, E., 404
Schmeidler, W., 713

Schmidt, O. U., 405
Schoenberg, J. J., 924
Schofield, C. W., 1215
Schreier, O., 556
Schubert, H., 485
Schutte, K., 1164
Schwartz, A., 106
Schwartz, J. T., 333, 422, 557, 683, 714, 715
Schwartz, M. H., 655
Scigliano, J. A., 1024
Scott, J. F., 141
Scott, W. R., 406
Scriba, C. J., 142
Seber, G. A. F., 1137
Secrest, D., 930
Seebach, J. A., Jr., 488
Seeley, R. T., 107
Segal, I. E., 626, 716, 1350
Seidenberg, A., 558, 559
Seinfeld, J. H., 907
Seligman, G. B., 378
Senseman, R. W., 1017
Sentance, W. A., 1002
Serre, J. P., 560
Shabat, B. V., 636
Shafarevich, I. R., 231
Shalaevskii, O. V., 1087
Shampine, L., 728
Shanks, M. E., 761
Shapiro, Z. Y., 391
Sheng, C. L., 197
Shephard, G. C., 334
Sheppard, W. F., 1216
Shilov, G. E., 335, 627, 805, 844
Shilov, G. Y., 614
Shisha, O., 1369, 1370
Shkil', N. I., 752
Shockley, J., 259
Shoenfield, J. R., 198
Shorter, E., 1018
Sidak, Z., 1075
Siegel, A., 1150
Siegel, C. L., 260, 656
Sierpinski, W., 261
Simis, A., 379
Simmons, G. F., 486, 806
Simms, D. J., 434
Simone, A. J., 16
Singer, I. M., 506

270

AUTHOR INDEX

Singer, J., 925
Sion, M., 615
Sjoberg, S., 1198
Skelton, J. E., 1019
Slater, L. J., 669, 670
Slesnick, W. E., 76
Slisenko, A. O, 199
Slupecki, J., 200
Smallwood, C. P., 957
Smart, J. R., 98
Smillie, K. W., 1138
Smirnov, A. D., 1217
Smirnov, N. V., 1218, 1219
Smith, G. D., 926
Smith, H., 1053
Smith, M. G., 807, 849
Smolitskiy, K. L., 912
Snapper, E., 561
Sneddon, I. N., 662, 1371
Snell, J. L., 18, 1091
Snyder, D. P., 201
Snyder, M. A., 927
Sobel, M., 1033
Sobolev, S. L., 808
Sokolnikoff, I. S., 562
Solitar, D., 398
Solonnikov, V. A., 777
Solovyev, A. D., 1066
Soni, R. P., 667
Sorani, G., 507
Southworth, R. W., 1020
Spain, B., 108
Spanier, E. H., 487
Sparks, F. W., 57, 58, 103
Spencer, D. E., 792
Sperner, E., 556
Spiegel, M. R., 809
Spitzer, F., 1139
Sprecher, D. A., 616
Springer, C. E., 563
Srinivasan, S. K., 1140
Srinivasiengar, C. N., 143
Stakgold, I., 810
Stallings, J. R., 508
Stanton, R. G., 928
Stark, H. M., 262
Steen, L. A., 488
Steenrod, N. E., 489, 1165
Stein, E. M., 811, 837
Stein, M. L., 1021, 1022

Stein, S. K., 109
Steinhart, R. F., 1023
Stephens, R., 64
Sternberg, S., 94, 564
Stewart, B. M., 263
Stewart, F. M., 336
Stewart, I., 380
Stiefel, E. L., 929
Stoker, J. J., 565
Stoll, R. R., 202, 337
Storer, T., 264
Stratonovich, R. L., 1141, 1372
Strauss, A., 1373
Stromberg, K., 604
Stroud, A. H., 930
Struble, R. A., 812
Stuart, A., 1092
Studden, W. J., 1090
Sullivan, M. W., 144
Sunahara, Y., 1136
Sunder, L., 253
Sunko, T. S., 45
Suppes, P., 203
Suprunenko, D. A., 338
Swan, R. G., 416, 490
Swanson, E. A., 813, 1529
Sweedler, M. E., 381
Szabo, M. E., 145
Szasz, G., 382
Szego, G. P., 732, 814
Sz.-Nagy, B., 711, 717

Taha, H. A., 1374
Tait, J. H., 1375
Talbot, A., 931
Tallack, J. C., 110
Talman, J. D., 671
Tangora, M. C., 479
Tate, J., 228
Taub, A. H., 149
Taylor, E. G. R., 146
Taylor, H. E., 35
Taylor, S. J., 1093
Tedeschi, F. P., 1024
Teplov, F. B., 1205
Terent'ev, N. M., 1192
Thacher, H., Jr., 987
Thomas, C. B., 413
Thomas, G., 111

271

AUTHOR INDEX

Thomas, J. B., 1142
Thomas, T. Y., 566
Thomasian, A. J., 1143
Thompson, G. L., 18
Thorpe, J. A., 506
Tierney, J. A., 112
Tiffen, R., 1376
Todd, H. N., 1130
Tompkins, C. B., 932
Tou, J. T., 1377
Towill, D. R., 1378
Tranter, C. J., 672
Traub, J. F., 933
Trench, W. F., 113
Trèves, F., 718, 815, 816, 817
Tricomi, F. G., 818, 819
Trotter, H. F., 115
Troyer, R. J., 561
Truesdell, C., 147, 1379
Tucker, H. G., 1144, 1145
Tuller, A., 567
Tumarkin, S. A., 1209
Turan, P., 265
Tutte, W. T., 223
Tychonov, A. N., 820
Tyshkevich, R. I., 338
Tzschach, H. G., 906

Urabe, M., 821
Ural'ceva, N. N., 777
Uspenskii, V. A., 236

Vainberg, M. M., 719
Vajda, S., 224, 225
Valentine, F., 568
Vandiver, H. S., 234
Van Heijenoort, J., 204
Vankatarayuda, T., 384
Varadarajan, V. S., 1380
Varga, R. S., 339, 934
Veech, W. A., 657
Vekua, I. N., 822
Venn, J., 148
Verdina, J., 569
Vernon, J., 67
Vesley, R. E., 169
Vilenkin, N. Y., 226, 844
Vinograde, B., 340

Vitalis, J. A., 1195
Vitasek, E., 870
Vogell, B. R., 554
Von Neumann, J., 149
Von Rosenberg, D. U., 935
Von Vega, B., 1223
Vorobyov, N. N., 266
Vulikh, B. A., 720

Wade, T. L., 35
Wadsworth, G. A., 1146
Wahlin, G. E., 234
Wallace, A. H., 491, 509
Walnut, F. K., 1025
Walsh, J. L., 867
Walter, W., 823
Waltman, P. E., 728
Ward, B., 1026
Warmus, M. W., 1224
Warrack, B. D., 482
Wasan, M. T., 1147, 1148
Wasow, W. R., 889, 936
Watson, W. A., 937
Watts, D. G., 1083
Wayland, H., 658
Weeg, G. P., 938
Weil, A., 267
Weinberger, H. F., 799, 824
Weingram, S., 471
Weir, A. J., 530
Weiss, E. A., 268, 417, 1027
Wendroff, B., 939, 940
West, E. N., 1153
Westlake, J. R., 941
Wheeler, B. W., 10
Wheelon, A. D., 1225
Whitehead, G. W., 492
Whitehead, J. H. C., 150
Whiteside, D. T., 151
Whittle, P., 1149
Widder, D., 114
Wielandt, H., 407
Wiener, N., 838, 1150
Wilcox, C. H., 942
Wilcox, H. J., 341
Wilde, C. O., 721
Wilder, R. L., 152, 205
Wilkins, B. R., 943
Wilkinson, J. H., 944, 945

272

AUTHOR INDEX

Willard, S., 493
Willems, J. C., 1381
Willerding, M. F., 36, 62
Williams, F. J., 1170
Williamson, R. E., 115
Willmore, F. E., 6
Wilson, W. L., 932
Winkler, S., 785
Wise, T. H., 1315
Witzgall, C., 987
Wolf, F. L., 1151
Wolf, J. A., 570
Wong, E. T., 337
Wood, P. E., Jr., 1382
Wooton, W., 42
Wren, F. L., 37, 292
Wylie, C. R., Jr., 571, 572
Wylie, S., 462
Wymore, A. W., 1383

Yackel, J., 1072
Yakowitz, S. J., 1384
Yale, P. B., 573
Yano, K., 574

Yanpol'skii, A. R., 910
Yefimov, N. V., 342
Yosida, K., 722
Young, F. H., 38
Young, G. S., 463
Young, L. C., 866
Youse, B., K., 116

Zaccaro, L. N., 15
Zacharov, B., 1385
Zak, J., 408
Zakrevskii, A. D., 976
Zalgaller, V. A., 575
Zariski, O., 293, 576
Zehna, P. W., 206, 1152
Zehnder, C. A., 906
Zelinsky, D., 343
Zemanian, A. H., 850, 851
Zener, C., 1300
Zhurina, M. I., 1226, 1227
Ziebur, A. D., 81
Zimmerman, M., 1178
Zisman, M., 455
Zuckerberg, H. L., 344
Zuckerman, H. S., 255
Zwiefel, P. F., 1289
Zygmund, A., 839
Zyskind, G., 1153

SUBJECT INDEX

SUBJECT INDEX

SUBJECT INDEX

SUBJECT INDEX

278

SUBJECT INDEX

SUBJECT INDEX

SUBJECT INDEX

Universal algebra, *see* Algebra

Valuation theory, 230, 243, 357
Value distribution theory, 654
Variables, random, *see* Random variables
Variance analysis, 1051, 1071, 1085, 1103, 1153
Variations, calculus of, *see* Calculus of variations

Varieties, analytic, 361
Vector analysis, 63-67, 70, 72, 75, 77, 79, 80, 83-85, 87-89, 92, 94-96, 104, 108, 110, 113-115, 122, 542
Vector geometry, *see* Geometry
Vector spaces, *see* Spaces
Vectors, 65, 66, 99, 333

Wave propagation, 1232